PROTEIN–DYE INTERACTIONS:

Developments and Applications

Proceedings of the First International Conference on 'Modern Aspects of Protein–Dye Interaction: Role in Downstream Processing', 24–28 July 1988, Compiègne, France.

PROTEIN–DYE INTERACTIONS:

Developments and Applications

Edited by

M. A. VIJAYALAKSHMI

Université de Technologie de Compiègne, France

and

O. BERTRAND

INSERM U160, Clichy, France

ELSEVIER APPLIED SCIENCE
LONDON and NEW YORK

ELSEVIER SCIENCE PUBLISHERS LTD
Crown House, Linton Road, Barking, Essex IG11 8JU, England

Sole Distributor in the USA and Canada
ELSEVIER SCIENCE PUBLISHING CO., INC.
655 Avenue of the Americas, New York, NY 10010, USA

WITH 54 TABLES AND 126 ILLUSTRATIONS

© 1989 ELSEVIER SCIENCE PUBLISHERS LTD

Softcover reprint of the hardcover 1st edition 1989

British Library Cataloguing in Publication Data

International Conference on 'Modern Aspects
of Protein–Dye Interaction: Role in
Downstream Processing' (*1st: 1988:
Compiègne, France*)
1. Proteins. Chemical analysis. Use of
reactive dyes
I. Title II. Vijayalakshmi, M.A. III.
Bertrand, O.
547.7'5046

ISBN-13: 978-94-010-6989-2   e-ISBN-13: 978-94-009-1107-9
DOI: 10.1007/978-94-009-1107-9

Library of Congress CIP data applied for

Preface

This volume contains the papers and reports presented at the First International Conference on Dye–Protein Interaction, held 24–28 July 1988 at the University of Compiègne, France. This was the first international meeting dealing entirely with dye–protein interaction. The major focus of the conference was on the better understanding of the mechanism of interaction of proteins with different triazine dyes and the synthesis of novel structural dyes having good biomimetic activities. The potentials and limits of their use in biotechnology, mainly for purification, were stressed. Current contributions in developing dye-based affinity methods were highlighted in such areas as affinity partition, affinity precipitation and new support matrices for efficient affinity chromatography, etc.

The interrelation between metal chelates and dyes in terms of their interactions with proteins was underlined. It is our belief that this proceedings volume will be a stimulus for broad and creative applications of dye affinity concepts in many fields of biomedical research and biotechnology.

In addition, a discussion session emphasised the necessity for understanding the toxicological aspects of these dyes, their fragments and their metabolites. This helped to trigger plans for future work, and this topic will be one of the priorities in a future meeting on dye–protein interactions.

The help of the International Scientific Committee, which included Drs C. R. Lowe (UK), G. Kopperschläger (GDR), E. Stellwagen (USA), D. Thomas (France), G. Birkenmeier (GDR), S. Rajgopal-Narayan (USA), J. P. Dandeu (France), D. Muller (France) and E. Dellacherie (France), in organising this meeting is gratefully acknowledged.

We are grateful to the Université de Technologie de Compiègne (UTC), France, for the support and the infrastructural facilities provided for the meeting. Financial support from INSERM (French National Institute for Medical Research), including that for travel grants to the speakers, is gratefully acknowledged. The following organisations, Groupe Français de Bio-Chromatographie (GFBC), Université Paris VII, and the following industries, Pharmacia, IBF, Sanofi, Bertin, Merck, Prolabo and J. T. Baker, provided financial contributions.

Finally, we are indebted to the authors for their important contributions to this volume, to Miss Nathalie Honoré for her secretarial help in preparing the volume, and to Elsevier Science Publishers Ltd for its timely publication.

M. A. Vijayalakshmi
O. Bertrand

Contents

Chapter 4: Dye Ligands in Affinity Partition, Affinity Ultrafiltration and Affinity Precipitation

Chapter 5: Role of Added Ions in Dye–Protein Interactions

ix

Chapter 6: Blood Proteins Purification Using Dye–Ligand Affinity

Chapter 7: Dye–Ligand Affinity Chromatography for the Purification of Plant Proteins

Chapter 8: Dyes in Molecular Biology

List of Contributors

A. Adenier, Groupe de Dynamique des Interactions Macromoléculaires, Institut de Topologie et de Dynamique des Systèmes de l'Université Paris 7, associé au CNRS (UA 34), 1 rue Guy de la Brosse, 75005 Paris, France

E. Algiman, INSERM U160, Hôpital Beaujon, 92118 Clichy Cedex, France

T. Arnaud, IBF-Biotechnics, 92390 Villeneuve la Garenne, France

M. Atreyi, Department of Chemistry, University of Delhi, Delhi 110 007, India

J. Aubard, Groupe de Dynamique des Interactions Macromoléculaires, Institut de Topologie et de Dynamique des Systèmes de l'Université Paris 7, associé au CNRS (UA 34), 1 rue Guy de la Brosse, 75005 Paris, France

B. B. Baskeviciute, ESP 'Fermentas', All-Union Research Institute of Applied Enzymology, Vilnius, Lithuanian SSR, USSR

O. Bertrand, INSERM U160, Hôpital Beaujon, 92118 Clichy Cedex, France

A. D. Bharucha, Department of Biochemistry, Faculty of Medicine, Laval University, Quebec, Canada G1K 7P4

G. Birkenmeier, Institute of Biochemistry, Karl-Marx-University Leipzig, Liebigstrasse 16, 7010 Leipzig, German Democratic Republic

P. Boivin, INSERM U160, Hôpital Beaujon, 92118 Clichy Cedex, France

E. Boschetti, IBF-Biotechnics, 35 Avenue Jean Jaurès, 92390 Villeneuve la Garenne, France

N. Burton, Institute of Biotechnology, University of Cambridge, Downing Street, Cambridge CB2 3EF, UK

P. G. H. Byfield, Endocrinology Research Group, Clinical Research Centre, Harrow HA1 3UJ, UK

F. Cadelis, Laboratoire de Technologie des Séparations, Université de Technologie de Compiègne, BP 649, 60206 Compiègne, France

L. Carrier-Malhotra, Department of Biochemistry, Faculty of Medicine, Laval University, Quebec, Canada G1K 7P4

R. Charbonneau, Department of Biochemistry, Faculty of Medicine, Laval University, Quebec, Canada G1K 7P4

Y. D. Clonis, Institute of Biotechnology, University of Cambridge, Downing Street, Cambridge CB2 3EF, UK

S. Cochet, INSERM U160, Hôpital Beaujon, 92118 Clichy Cedex, France

B. Dastugue, Laboratoire de Biochimie Médicale, Faculté de Médecine, Université de Clermont Ferrand 1, Clermont Ferrand Cedex, France

S. Dilmaghanian, Institute of Biotechnology, University of Cambridge, Downing Street, Cambridge CB2 3EF, UK

G. Dodin, Groupe de Dynamique des Interactions Macromoléculaires, Institut de Topologie et de Dynamique des Systèmes de l'Université Paris 7, associé au CNRS (UA 34), 1 rue Guy de la Brosse, 75005 Paris, France

M. J. Easton, School of Biological Sciences and Environmental Health, Thames Polytechnic, Wellington Street, London SE18 6PF, UK

R. R. Fisher, Department of Chemical Engineering BF-10, University of Washington, Seattle, Washington 98195, USA

S. S. **Flaksaite,** ESP 'Fermentas', All-Union Research Institute of Applied Enzymology, Vilnius, Lithuanian SSR, USSR

A. A. **Glemz(h)a,** ESP 'Fermentas', All-Union Research Institute of Applied Enzymology, Vilnius, Lithuanian SSR, USSR

M. J. **Harvey,** Blood Products Laboratory, Elstree, Hertfordshire, UK

A. G. **Hitchcock,** Blood Products Laboratory, Elstree, Hertfordshire, UK

P. **Hughes,** British Bio-technology Ltd, Brook House, Watlington Road, Cowley, Oxford OX4 5LY, UK

K. **Huse,** Institute of Biochemistry, Karl-Marx-University Leipzig, Liebigstrasse 16, 7010 Leipzig, German Democratic Republic

M. V. **Jagannadham,** Department of Biochemistry, University of Iowa, Iowa City, Iowa 52242, USA

L. **Jervis,** Department of Biology, Paisley College of Technology, High Street, Paisley PA1 2BE, Scotland, UK

G. **Johansson,** Department of Biochemistry, Chemical Center, University of Lund, PO Box 124, S-221 00 Lund, Sweden

V. A. **Kadusevicius,** ESP 'Fermentas', All-Union Research Institute of Applied Enzymology, Vilnius, Lithuanian SSR, USSR

D. J. **Karalyte,** ESP 'Fermentas', All-Union Research Institute of Applied Enzymology, Vilnius, Lithuanian SSR, USSR

J. **Kirchberger,** Institute of Biochemistry, Karl-Marx-University Leipzig, Liebigstrasse 16, 7010 Leipzig, German Democratic Republic

G. **Kopperschläger,** Institute of Biochemistry, Karl-Marx-University Leipzig, Liebigstrasse 16, 7010 Leipzig, German Democratic Republic

T. **Kriegel,** Institute of Biochemistry, Karl-Marx-University Leipzig, Liebigstrasse 16, 7010 Leipzig, German Democratic Republic

Y. **Kroviarski,** INSERM U160, Hôpital Beaujon, 92118 Clichy Cedex, France

K. C. **Kyriacou,** Department of Chemical Engineering BF-10, University of Washington, Seattle, Washington 98195, USA

G. **Lévesque,** Department of Biochemistry, Faculty of Medicine, Laval University, Quebec, Canada G1K 7P4

C. R. **Lowe,** Institute of Biotechnology, University of Cambridge, Downing Street, Cambridge CB2 3EF, UK

R. **Loy,** Schleicher & Schuell Inc., 10 Optical Drive, Keene, New Hampshire, USA

B. **Machiels,** Department of Chemical Engineering BF-10, University of Washington, Seattle, Washington 98195, USA

S. **McLoughlin,** Institute of Biotechnology, University of Cambridge, Downing Street, Cambridge CB2 3EF, UK

R. P. **Marcisauskas,** ESP 'Fermentas', All-Union Research Institute of Applied Enzymology, Vilnius, Lithuanian SSR, USSR

J. A. **Mazza,** VILMAX, Santiago del Estero 366, 1075 Capital Federal Buenos Aires, Argentina

J. E. **More,** Blood Products Laboratory, Elstree, Hertfordshire, UK

Y. **Moroux,** IBF-Biotechnics, 35 Avenue Jean Jaurès, 92390 Villeneuve la Garenne, France

J. E. **Morris,** Department of Chemical Engineering BF-10, University of Washington, Seattle, Washington 98195, USA

A. H. L. **Mulder,** Chemistry Department, Baylor University, Waco, Texas, USA

D. **Muller,** LRM, CSP Avenue J. B. Clément, 93430 Villetaneuse, France

S. R. **Narayan,** Department of Pharmaceutical Chemistry, University of Kansas, Lawrence, Kansas, USA

M. **Naumann,** Institute of Biochemistry, Karl-Marx-University Leipzig, Liebigstrasse 16, 7010 Leipzig, German Democratic Republic

P. Outumuro, VILMAX, Santiago del Estero 366, 1075 Capital Federal Buenos Aires, Argentina

J. Pearson, Institute of Biotechnology, University of Cambridge, Downing Street, Cambridge CB2 3EF, UK

J.-H. J. (I.-G. I.) Pesliakas, ESP 'Fermentas', All-Union Research Institute of Applied Enzymology, Vilnius, Lithuanian SSR, USSR

J. Porath, The Biochemical Separation Center, Biomedical Center, University of Uppsala, Box 577, S-751 23 Uppsala, Sweden

S. Price, Blood Products Laboratory, Elstree, Hertfordshire, UK

M. V. R. Rao, Department of Chemistry, University of Delhi, Delhi 110 007, India

B. Riahi, Laboratoire de Technologie des Séparations, Université de Technologie de Compiègne, BP 649, 60206 Compiègne, France

R. Rines, Chemistry Department, Baylor University, Waco, Texas, USA

E. R. Robertson, Department of Biology, Paisley College of Technology, High Street, Paisley PA1 2BE, Scotland, UK

J. Rott, Blood Products Laboratory, Elstree, Hertfordshire, UK

X. Santarelli, LRM, CSP Avenue J. B. Clément, 93430 Villetaneuse, France

P. V. Scaria, Department of Chemistry, University of Delhi, Delhi 110 007, India

M. A. Schwaller, Groupe de Dynamique des Interactions Macromoléculaires, Institut de Topologie et de Dynamique des Systèmes de l'Université Paris 7, associé au CNRS (UA 34), 1 rue Guy de la Brosse, 75005 Paris, France

R. K. Scopes, Centre for Protein and Enzyme Technology, La Trobe University, Bundoora, Victoria 3083, Australia

W. H. Scouten, Chemistry Department, Baylor University, Waco, Texas, USA

C. V. Stead, Affinity Chromatography Ltd, 187 Victoria Avenue East, Blackley, Manchester M9 2HW, UK

E. Stellwagen, Department of Biochemistry, University of Iowa, Iowa City, Iowa 52242, USA

D. Stewart, Institute of Biotechnology, University of Cambridge, Downing Street, Cambridge CB2 3EF, UK

S. Subramanian, Miles Inc., Elkhart, Indiana, USA

O. F. Sudz(h)iuviene, ESP 'Fermentas', All-Union Research Institute of Applied Enzymology, Vilnius, Lithuanian SSR, USSR

A. Truskolaski, INSERM U24, Hôpital Beaujon, 92118 Clichy Cedex, France

M. R. Ven Murthy, Department of Biochemistry, Faculty of Medicine, Laval University, Quebec, Canada G1K 7P4

J.-L. Viallard, Laboratoire de Biochimie Médicale, Faculté de Médecine, Université de Clermont Ferrand 1, Clermont Ferrand Cedex, France

M. A. Vijayalakshmi, Laboratoire de Technologie des Séparations, Université de Technologie de Compiègne, BP 649, 60206 Compiègne, France

R. J. Yon, School of Biological Sciences and Environmental Health, Thames Polytechnic, Wellington Street, London SE18 6PF, UK

V. V. Zutautas, ESP 'Fermentas', All-Union Research Institute of Applied Enzymology, Vilnius, Lithuanian SSR, USSR

INTRODUCTION

A SHORT HISTORICAL REVIEW ON THE INTERACTION BETWEEN DYE-STUFFS AND BIOLOGICAL MATTER

Jerker Porath, The Biochemical Separation Center, Biomedical Center, University of Uppsala, Box 577, S-751 23 Uppsala, Sweden

In the aftermath of the 1968 student revolution, there was a certain contempt for established knowledge and traditions. The pendulum is now swinging in the opposite direction and even young scientists realize the importance of knowing and appreciating the roots of science, that is, the achievements of their own forerunners. Not only is it intellectually stimulating to obtain a widened perspective regarding the present position of science in society, but the history of science may tell us about discoveries in the distant past which can, sometimes, be transformed into modern forms for solving present day problems. With these thoughts in mind, I would like to give a very brief historical account of the prerequisites for dye-stuff based chromatography. I shall consider three relevant branches: the chemistry of dyes, dye affinity for biological matter and group selective adsorption as applied to biomolecules.

Dye-Stuffs Chemistry

Man's first contact with dyes and pigments is literary true and concerns their affinity for his skin. It is archeologically supported (1). Stone age man, like primitive people up to the present, painted their skin. Painting of corpses was included in ancient funeral rites. The oriental women of the Bronze Age attempted to charm their men by dyeing their hair with henna, the color principle of Lawsonia inermis and that fashion is to some extent still in vogue among females. To enlighten the dull and monotonous everyday trivialities our ancesters dressed themselves in textiles, dyed with extracts of madder (alizarin) and Isatis tinctoria (indigo) (2). According to

Bible the Hebrews under patriarchal times were using kermes - a red antraquinone derivative - and the Phoenicians traded in the imperial purple from Tyros (dibromoindigo).

Man must adjust his life according to the sources available to him in his own environment and for the American indians it became natural to collect certain logs in order to extract brilliantly colored compounds. So, from the indians we inherited the use of brazilin and hematoxylin to dye wool fabrics, later to be abandoned for that purpose but instead introduced as tools in histology. One of the pioneers in the field of plant dyes was the French lipid chemist Michel Eugène Chevreul, the director of the Gobelins, a dye factory founded in 1662 (3). In the early nineteenth century he also studied the chemistry of quercitrin and morin. Chevreul died at the age of 103. His life span overlapped the time that foreshaddows the advent of the era of synthetic dyes.

The first synthetic dyes appeared during the eighteenth century. The invention or discovery of picric acid is fading back into obscurity but it was probably the first one in the series of millions of dyes to come. The isolation of aniline is a milestone in the history of dyes as is the 18 year old William Henry Perkin's synthesis of Aniline purple, 'Perkin's mauve' or mauveine of 1856 (4). As is so often the case, important and original discoveries are not easily traced back to their origin. O. Unverdorben seems to be the first dye investigator to have prepared aniline (5) initially as a dry--distillation product of indigo in 1826 under the name of 'crystalline', but later to be renamed after an-nil, the arabic word for indigo. Aniline was rediscovered at least twice during the following 15 years but the immense importance of the substance as the building block of a new era of organic chemistry first became apparent after Perkin's debut. Isolation of mauveine from a dirty product is an excellent example of a serependipitous discovery: Perkin tried to synthezise quinine from aniline! Like Michael Faraday, once a young

assistant to Humphry Davy, Perkin, as A.W. Hofmann's pupil, eventually surpassed in knowledge and in skill his teacher - as it should
rightly be and, unfortunately, so seldom seems to be the case now-
adays. Perkin senior was also a clever businessman. He took patents,
founded a factory that became known also for its excellent marketing
and information service to the customers - unheard activities at
that time. About seventy five years after Unverdorben's isolation of
aniline from indigo, following Kekulé's, Bel's and van't Hoff's
epoch making contributions to structural organic chemistry, the synthesis of indigo from aniline became feasible. The circle was closed.
Whenever possible, we should follow the latters lead: from simplicity to complexity.

The dye industry is concerned with the development of coloring materials for textiles and printing, but pioneers of histology and
bacteriology also became dye consumers.

Biological Staining

F.V. Raspail, working in Paris in the early part of last century,
may perhaps be considered as the founding father of histology and
histochemistry (6). But he had many forerunners. Furthest away in
time was Anthony Leevenhoek, the 'father of the microscope'. He
tried with modest success to use saffron to enhance the contrasts of
his microbial objects.

After Perkin's synthesis of mauve the time was ripe for synthetic
dyes to enter the scene of biology and medicine. Bencke was the
first in the line to use aniline dyes (7) and E. Klebs the first, in
1868, to detect an enzyme (peroxidase) by staining (8). Mischer's
isolation of nuclear chromatin in 1873 with the aid of methyl green
must be mentioned and, of course, Koch's and Ehrlich's seminal contributions during the 1880's (9). With the aid of methylene blue
Robert Koch was able to discover the tubercle and colera bacteria.

Paul Ehrlich's early work on vital dyes turned him into chemotherapy, a science which he himself founded in the latter part of the last century and which kept him occupied for the rest of his life. Also, in this field, the problems related to the specific affinities of dyes for biological matter are encountered. Starting out from N.O. Witt's theory of relationships between the color of dyes and chemical constitution (10), Ehrlich postulated the presence in the biologically active substances of toxophoric and haptophoric groups with affinity for biological counterparts - the receptors (11).

Ehrlich used trypanosomes, the cause of African sleeping sickness, as targets for his intended 'magic bullets'. Hundreds of dyes were tested and their action was recorded although only a minor portion of the results was published. Starting with the well-known methylene blue and benzopurpurin he later came across substances of considerable potency which became widely known as trypane red, trypane blue and trypaflavine.

Already during the phlogiston period, adsorption of dyes and other substances was explained according to 'physical theory' or 'chemical theory'. Chevreul, who supported the former, coined the expression 'capillary affinity' for the force that binds dyes to tissues and Ostwald considered dyeing to be effected by 'mechanical affinity'. In this context it is of interest to mention the controversy between Svante Arrhenius and Paul Ehrlich. Arrhenius postulated that a chemical equilibrium was attained between a toxin and its antitoxin whereas Ehrlich considered the combination to be complete which in modern terminology must mean the involvement of covalent binding. Modern dyes may be fixed to a fiber by covalent attachment (reactive dyes) or by 'physical' adsorption, which in present terms can be formulated as attachment by non-covalent bonds. Dye based chromatography, in fact, utilizes both kinds of attachment: chemical fixation of the dye to a solid support and adsorption of soluble ligates (analytes).

In a review on the interactions between dye-stuffs and biological matter it is justified to mention the 'photodynamic action' discovered in 1900 by O. Raab (12): Biomolecules may be damaged by light in the presence of dyes. This phenomenon, an apparent risk factor in the adsorption process, can perhaps be used to advantage, for example in thin layer chromatography, provided that it can be effectively controlled.

Adsorption of Biomolecules

One of the roots of enzyme separation technology is to be found in the development of adsorption methods and, as said, adsorption is closely connected to the applied chemistry of dyes and colored matter. Chromatography as introduced by Tswet (13), and its forerunner, capillary adsorption, were known (14) but largely overlooked in the first three decades of our century. Willstätter, one of the leading figures in organic chemistry, made extensive use of batchwise adsorption to purify enzymes. He was in fact so successful that he obtained highly active enzyme preparations which he thought were devoid of proteins (15). He was not favorably inclined towards Tsvet's chromatography, and, his authority and opinion were serious obstacles to the career of the lonely working Tswet - a fact that delayed the full appreciation and acceptance of chromatography, a powerful tool for chemistry and biology.

In a survey, Lars Sundberg and I traced back the first attempts to use specific adsorption techniques for purification of enzymes to Starkenstein (16) and perhaps G. Hedin. These early experiments were made just after the turn of the century. Several biochemists followed in their footsteps and among the pioneers of bioaffinity chromatography, Campbell, Leuscher and Lerman (17) and Arsenis and McCormick (18), must be mentioned. By using dinitrophenyl- and isoalloxazine derivatives of cellulose as adsorbents they initiated the use of colored ligands in chromatography. The birth of modern chromatography is usually connected with Cuatrecasas et al. (19). Two requirements for its success were at hand in 1967: 1) a suitable carrier and 2) an efficient and sufficiently reliable coupling procedure (20).

It is not my task to review early work on the specific topic of the present symposium. Many of the pioneers are present here today and it is up to them and us, their followers, to show the advantages of using dye ligands in preference to alternatives. Let me just make some comments. Ehrlich formulated his affinity theory but its value for predicting action from structure was rather limited. Empirically, through trial and error, he had to screen thousands of related and unrelated substances to find therapeutica as active as desired. We have not advanced much further in predicting adsorption selectivity from the nature of the dye ligands. Dyes are of complex molecular structure and their interaction with biomacromolecules are not easy to understand even with help of modern valence theory.

From the history of chemotherapeutics we may learn to go from simplicity to complexity, a scientific 'loadstar' worth being pointed out repeatedly. Domagk and his associates introduced prontosil, the first sulfonamide drug, in the early 30s. A few years later Trefouel and coworkers in France, Nitti and Bouvet and others showed that antibiotic activity was retained in the simple degradation products of the azo dyes. Personally, I believe it very well worthwhile to study simple π-electron-rich molecules as ligands for group fractionation of biomolecules . Eighteen years ago Nermin Fornstedt and I used the dye-constituent sulfanilic acid coupled to agarose for group fractionation (21) (incidently, epoxy-coupling was used for the first time to couple affinity ligands to polymer matrices). The subsite contribution to the affinities of such ligands is more easily accessible to rational chemical interpretation. However, we have to be prepared to accept arguments in favour of dye-ligand based chromatography. Synthesis of biomimetic dyes (22), introduction of affinity partitioning (23-24) and precipitation (25) with soluble dye-coupled polymers, systematic chromatographic screening procedures (26) and other techniques to come may, hopefully, provide us with extremely valuable tools. By studying complex adsorption behaviour we may also have a fair chance to discover unknown chemical interactions. Diving deep into dye-ligand-based chromatography may eventually yield unexpected profits.

Acknowledgements

I thank Professor Christofer Lowe for his valuable linguistic suggestions and Erna and Victor Hasselblad Foundation and Pharmacia-LKB for financial support.

References

1. K. McLaren, The Colour Science of Dyes and Pigments, Adam Hilger Ltd., Bristol 1983, p. 1.
2. Ibid, pp. 4-11.
3. J.R. Partington, A History of Chemistry, Vol. IV, MacMillan & Co., London 1964, p. 246.
4. British Patent 1984 (1856).
5. J.R. Partington, A History of Chemistry, Vol. IV, MacMillan & Co., London 1964, p. 183.
6. A.G.E. Pearse, Histochemistry, Theoretical and Applied, 2nd edn., J. & A. Churchill Ltd., London 1960, p. 1.
7. A. Bencke, Korrespbl. Ver. Gemeinsch. Arbeiten 59 (1862) 980.
8. E. Klebs, Z. med. Wiss. 6 (1868) 417.
9. P. Ehrlich, "Encyclopädie der Mikroskopischen Technik", Urband and Schwarzenberg, Berlin and Wien 1903.
10. O.N. Witt, Berichte IX (1876) 522 and XXI (1888) 321.
11. P. Ehrlich, Chemotherapeutics, Scicentific Principles, Methods and Results, Lancet II (1913) 445-451.
12. O. Raab, Z. Biol. 39 (1900) 524.

13. L.S. Ettre, 75 Years' of Chromatography, a Historical Dialogue,
    (L.S. Ettre and A. Zlatkis, Eds.), Elsevier Publ. Sci. Co.,
    Amsterdam 1979, p. 483.
14. F. Goppelsroeder, Capillaranalyse, Buchdruckerei Emil Birkhäuser,
    1901.
15. R. Willstätter, "Untersuchungen über Enzyme" (Erster Band),
    Verlag Julius Springer, Berlin 1928, p. 11.
16. E. Starkenstein, Biochem. Z. 24 (1910) 210.
17. D.H. Campbell, E. Leuscher and L.S. Lerman, Proc. Nat. Acad.
    Sci. U.S. 37 (1951) 575.
18. C. Arsenis and D.B. McCornick, J. Biol. Chem. 239 (1964) 3093.
19. P. Cuatrecasas, M. Wilcheck and C.B. Anfinsen, Proc. Nat. Acad.
    Sci. U.S. 61 (1968) 636.
20. J. Porath, R. Axén and S. Ernback, Nature 215 (1967) 1491.
21. J. Porath and N. Fornstedt, J. Chromatogr. 51 (1970) 479-489.
22. C.R. Lowe, These Proceedings.
23. G. Kopperschläger, These Proceedings.
24. G. Johansson, These Proceedings.
25. M.A. Vijayalakshmi, These Proceedings.
26. R. Scopes, These Proceedings.

Chapter 1

Dye Structures and their Relevance to Protein Recognition

BIOMIMETIC DYES IN BIOTECHNOLOGY

C.R.Lowe, N.Burton, S.Dilmaghanian, S.McLoughlin,
J.Pearson, D.Stewart and Y.D.Clonis

Institute of Biotechnology
University of Cambridge
Downing Street
Cambridge
CB2 3 EF
U.K.

I   INTRODUCTION

Downstream processing refers to all the technologies that are responsible
for the production of pure products after fermentation. Therefore, if one
were to produce, for example, a protein, one would start with an
appropriate bio-reactor containing native or engineered cells. The first
step would be to separate the cells from the broth; if the product is an
intra-cellular one, one would subsequently disrupt the cells to release the
intra-cellular products and then embark on a clarification process to obtain
a clear extract containing the protein of interest. The next step is to
apply a whole series of high resolution purification techniques, particularly
chromatographic steps, prior to subsequently ending up with pure protein.
Therefore, downstream processing entails the execution of primary recovery
stages followed by a series of high-resolution steps where we add value to
the final product and then hopefully end up with pure homogeneous protein.

Special interest is focused in this report on the high resolution stages of the process leading to pure product and particularly those steps involving the most refined version of chromatography, affinity chromatography (1). The technique of affinity chromatography exploits small ligands which bind specifically and reversibly to the protein of interest. The appropriate small ligand is covalently attached to a suitable solid support matrix in such a way that we can establish that as a column. A mixture of proteins, is then applied and adsorption is allowed to take place. Only that protein which specifically recognises the small ligand will be attached to the matrix. Subsequently, the remaining non-bound proteins are washed off and by changing the conditions we can then effect desorption of the desired protein. One may distinguish three main components in the system:

(i)   The solid support itself, which is often a carbohydrate water soluble polymer or other materials as discussed below,

(ii)   the ligand, which is designed to interact specifically and reversibly with the protein of interest, and then invariably, especially if the ligand is small, we may insert a spacer molecule between the ligand and the support matrix.

II   The Dye-Ligands

The choice of ligand is very important and prior to relatively recently, the main ligands selected were biological ligands which interacted selectively with the protein of interest. However, biological ligands have certain deficiencies, mainly involved in the cost of them for potential large-scale applications, the difficulties in immobilisation and in the general mobility of biological molecules. Attention has been paid in selecting quasi biological ligands which would mimmick natural biological ligands but have advantages in terms of ease of immobilisation and stability. One of the main groups that we have selected is the textile dyes (2). These organic molecules are potentially applicable to textiles in printing industries. They are available in very large quantities and comprise two principle components: a chromophor which gives rise to the colour, and the reactive component which allows that colouring matter to be covertly attached to an appropriate support, be it a textile, a paper or a carbohydrate material. In terms of the chromophor there are four principle classes: the anthraquinone dyes which have the characteristic anthraquinone chromophor and generally give rise to bright blue shades, and typically display a λ_{max} in the 600/650 nm range. The second class, the azo-dyes, have the characteristic azo-bond in them and are often orange, yellow and red dyes, with λ_{max} 380 - 450nm or thereabouts. The third principle class are the copperphthalocynines with very characteristic copperphthalocynine chromophor plus many other parts of chromophor as well and these are generally turquoise and green dyes of λ_{max} 615-650nm. The final class are the metal complexes, either a one-to-one metal complex or one-to-two complex, that is, one metal to two chromophors. These tend to give most of the other colours in the range. Of course one can get various mixed chromophors from various combinations. The reactive group in all the cases is based on cyanuric chloride which is tri-s-chlorotriazene. By substituting just one of the cyanuric acid chlorines we obtain the class of dichlorotriazene dyes which are very reactive at room temperature, or by further replacing the second chlorine we obtain the monochlorotriazene dyes which are far less reactive and generally require hot conditions for attachment to the matrix material.

There are several advantages of using dye-ligands(2), particularly for large scale applications, they are inexpensive materials compared to biological ligands and essential commodity chemicals and are available in tonnage quantities worldwide. Secondly, there is a wide range of chromophors available. One single company may produce up to 90 different types of chromophors for potential interaction with proteins. More important, they are biologically, chemically, and photochemically stable. This has a distinct advantage, particularly with respect to certain proteins because adsorbents based on dyes are potentially sterilisable in situ with no degradation of the ligand itself. Because they are reactive materials they are very easily immobilised to hydroxyl-polymers, generally by a single step process, and adsorbents display high capacity and of course are easily re-usable.

In addition to that, it has been found over the last few years that they have very broad binding capability in terms of the complimentary proteins, in fact, by now there should be several thousand different types of proteins which would interact with an immobilised textile dye. For example, oxidoreductases, phosphokinases, and nearly all co-enzyme dependent enzymes are being shown to interact with textile dyes, as well as hydrolases, various transferases, a number of proteins which interact with mono and polynucleotides, synthetases, hydroxylases, nearly all of the glycolitic enzymes, phosphatases, a whole variety of blood proteins and other non-enzyme proteins to name a few. The literature in this area reveals that nearly all of these proteins and enzymes have been shown to interact with one textile dye, Cibacron blue F3GA or C.I.Reactive Blue 2. This is an anthraquinone dye so it has the characteristic anthraquinone group linked to a diaminobenzylsulphonate. The next ring of its structure is the chlorotriazine followed by a terminal ring, in the case of Reactive Blue 2, a benzyl sulphonate. Therefore, three sulphonate groups make this compound anionic as well.

II   The Interaction of Dyes with Proteins

It is generally observed that these materials bind to active sites of enzymes and there is a fair amount of evidence in the literature now that this is so. This evidence is derived from a number of studies, using classic enzyme kinetics (3). It is easy to demonstrate that the Reactive Blue 2 is competitive to a variety of natural biological ligands. Likewise, different spectroscopy can be used to demonstrate the specificity of binding of these dyes to active sites of proteins (4), and, likewise circular dichroism and other techniques such as affinity labelling and x-ray diffraction. Thus, a dichlorotrianine, which is an analogue of Reactive Blue 2, inactivates counter-enzymes in an irreversible fushion as a function of time at pH 8.5(3), but not the methoxylated dye analogue, in other words once the reactivity of the dye is removed, there is no irreversible activation. It has been demonstrated with alcohol dehydrogenase that one molecule of dye is incorporated per subunit of 40,000 molecular weight, and from the inactivated enzyme, after removing excess dye and subsequently hydrolysing the protein, it is possible to separate the one peptide, by classical reverse phase HPRC, which contained the dye material attached to the thiol of cystain-174 at the active site (5). X-ray diffraction studies have demonstrated how dyes bind to the enzymes compared to the natural co-enzyme. The natural co-enzyme NAD as bound to a shallow crevice on the surface of the protein with the hydrogens bonded to various groups. The pyranophosphate backbone forms an isoelectrostatic interaction with hydrogen-47. Finally, the adenine is located relatively close to the centre of the protein and in a position where it can interact with the zinc atom which is then close to cystain 174. The dye Reactive Blue 2 is bound in a similar position in terms of the anthraquinone, but the adenine moiety with the sulphonate is bound in such a position which expose them to the solvent: instead of it coming down towards the centre of the protein, in the case of the dye, it more or less lays across the surface with the terminal ring in a slightly different position to that of the natural co-enzyme. So there is very good evidence that these particular dyes do bind with active sites and they do appear to mimick the binding of natural ligands,in this case co-enzyme NAD. Therefore, we can exploit these dyes in cases where we would exploit the natural co-enzyme.

IV   The Application of Dyes in Protein Purification

A reactive textile dye, Procion Yellow MX-AG was immobilised to a polyhydroxyl matrix, agarose. To this yellow adsorbent it was then applied a crude extract of E. coli containing the enzyme IMP dehydrogenase; this protein was present in about 10 per cent of the cell content. All the inert protein passes through the adsorbent, whereas a single protein was eluted afterwards in a 0-1M potassium chloride gradient in the presence of a small concentration of ethelene glycol just to remove or reduce hydrophobic interactions (6). This is virtually homogeneous IMP dehydrogenase in almost 100 per cent recovery of enzyme activity and it shows what can be done and what selectivity these dyes may have. That is just one of many examples that could be quoted for the application of dye-affinity chromatography under classical low pressure conditions.

In addition to attaching these dyes directly to agarose we can also attach them to silica matrices and therefore use them in a high performance mode. In this particular case, one may refer to two ways of immobilising dyes, one of which the dye is attached directly by the chlorotriazene to silanized silica, and a second, where the aminoalkyl-dye analogue is attached to activated silanised silica. Typically, ten microns mactroporous silica is used and this is then set up in a classical high performance column. A variety of examples on enzyme resolution and purification can be found in the literature where dye-ligand HPAC was employed. However, there is only one example of process-scale HPAC using triazine dyes (7). A French axial compression column was packed, in this case, with 20 micron and 280 Å silica particles. To this was attached the blue dye Procion blue MX-R to give a bed of 3.3 1 blue adsorbent. A relatively crude extract of rabbit muscle was then applied in order to isolate L-lactate dehydrogenase. The inner protein passed straight through the column and subsequently almost homogeneous enzyme was recovered in NADH in a single step with very good recovery. Therefore,it is potentially possible to scale to a process mode high performance affinity chromatography using these textile dyes (7).

A novel and very interesting approach is the recent work that aims to generate affinity adsorbents based on fluorocarbon support matrices. The advantages of fluorocarbons are very simple in that they are totally immissible in water, there is actually no solubility whatsoever, they have very low to zero toxicity; they are used as blood substitutes, they are virtually totally chemically stable and inert and in many cases, in both liquid and soluble form, they have very high densities.
The basic problem with fluorocarbons stems from the fact that they are chemically inert and one has to device techniques to chemically modify these materials in order to adsorb or attach dye-ligands to their surface. One way to resolve this problem, is to modify the dye in such a way that it is converted into a fluoro-dye analogue, that is to have at the dye-ligand's end a fluorocarbon tail. The idea being that the fluorocarbon tail will show remarkable affinity for the solid or liquid perfluorocarbon-based material, whereas the dye will reside on the surface and therefore be potentially interactable with the protein of interest. Recent studies have

shown that the stability of these materials is such that they can actually be incubated in concentrated sodium hydroxide at 50°C with no degradation of the matrix or the ligand. Therefore it seems that these new materials may well find very considerable applications for the purification of therapeutic proteins.

Most of the work that has been done has exploited classical textile dyes. However, since the textile dye is not a perfect mimick of the natural material, a fair amount of work has been done over the last few years aiming to develop these textile dyes so that they form much better mimicks of the natural biological molecules so that we can actually use them to a much greater advantage. By using computer-aided graphics it has been possible to synthesise dyes that show much better mimick for the natural co-enzymes. Thus, a purpose-synthesized Reactive Blue 2 analogue was immobilised to agarose and packed to a column (8). A crude extract of horse liver alcohol dehydrogenase (ADH)was continuously applied and fully saturated the column to such an extent that all the enzyme and all the proteins are passed straight through. The adsorbent was washed and subsequently eluted, with a linear NADH gradient and afforded alcohol dehydrogenase of very acceptable purity in a single step (8). So by using these computer-aided design facilities to design these dyes, one can get a far better purification than using the original parent dyes. Recent work has shown that in fact one can make these dyes such good mimicks of the natural co-enzyme, for example, in the case for binding ADH, that you can actually make them catalytically active as well. A nicotinamide analogue of the parent Reactive Blue 2 dye is catalytically active with ADH. The anthraquinone part is placed very much in the same enzyme pocket as the adenine moiety in the shallow hydrophobic crevices of ADH. The sulphonate and the amino group on the anthraquinone are exposed to solvent as one would expect. The bridging ring system starts to dip then towards the centre in the co-enzyme binding pocket of the enzyme and it is believed that there are such electrostatic interactions to bring about steering of the co-enzyme analogue into the site. Both the chlorotriazene and the nicotinamide ring are placed in exactly the right position so it will interact with the substrate.

An interesting and effective technique introduced by Mosbach (Sweden) is the selective precipitation of proteins using bifunctional affinity ligands, what he referred to as affinity precipitation. In this case, a bifunctional ligand which is capable of interacting with the subunits of a multi subunit protein and providing that the ratio is right, it brings about formation of a cross link matrix, in other words the protein is precipitated. This depends on the selectivity of the ligand, therefore, one can achieve very selective precipitation, as has been shown for purifying rabbit muscle lactate dehydrogenase using a highly specific Reactive Blue 2 analogue (9).

All that work so far has been achieved using negatively charged dyes. Interest is now paid to alternatives to those, as certain proteins bind cationic substrates. A series of triazine dyes which have cationic solubilising groups (e.g. arginine) have been studied and found that they bind selectively to certain proteins, particularly proteolytic enzymes such as trypsin-like proteases (10).

An activated crude beef pancreatic extract is applied to such a cationic triazine dye immobilised to agarose (10). Chymotrypsin passes straight through and shows no affinity to this particular ligand, whereas the trypsin is adsorbed and then subsequently eluted with glycine/HCE pH 2.1 and this is, as far as we are aware, the best single-step purification of trypsin that is so far reported in the literature based on these Cationic materials.

It is believed that dye-ligands will continue to remain popular media in protein and enzyme purification technology.

REFERENCES

1. CLONIS, Y.D. and LOWE, C.R. Affinity Chromatography, in: Scientific Foundations of Clinical Biochemistry (William, D.L. *et al*, eds) William Heinemann Medical Books Ltd., London, vol.1, 2nd edition, chapter 25, in press.

2. CLONIS, Y.D., ATKINSON, A, BRUTON, C.J. and LOWE, C.R. (Eds) Reactive Dyes in Protein and Enzyme Technology, Macmillan Press, Basingstoke, UK, 1987.

3. CLONIS, Y.D. and LOWE, C.R. (1981) Biochim. Biophys. Acta, 659, 86 - 98.

4. CLONIS, Y.D., GOLDFINCH, M.J. and LOWE, C.R. (1981) Biochem.J., 197, 203 - 211.

5. SMALL, D.A.P., LOWE, C.R., ATKINSON, T. and BRUTON, C.J. (1982). Eur.J.Biochem, 128, 119 - 123.

6. LOWE, C.R., HANS, M., SPIBEY, N. and DRABBLE, W.T. (1980) Analyt.Biochem., 104, 23 - 28.

7. CLONIS, Y.D., JONES, K. and LOWE, C.R. (1986) J Chromatogr., 363, 31 - 36.

8. LOWE, C.R., BURTON, S.J., PEARSON, J.C., CLONIS, Y.D. and STEAD, C.V. (1986) J.Chromatogr. 376, 121 - 130.

9. PEARSON, J.C., BURTON, S.J. and LOWE, C.R. (1986) Analyt.Biochem., 158, 382 - 389.

10. CLONIS, Y.D., STEAD, C.V. and LOWE, C.R. (1987) Biotechnol.Bioeng. 30, 621 - 627.

STRUCTURE, PREPARATION AND CHEMISTRY OF REACTIVE DYES

C. V. STEAD
Affinity Chromatography Ltd.
187, Victoria Avenue East, Blackley, Manchester   M9 2HW   UK

ABSTRACT

Facets of the basic chemistry of the various types of reactive dyes relevant to their use in dye-affinity chromatography are discussed and the wide variety of chromogens used in reactive dyes are reviewed. Disadvantages for the biochemist resulting from dependence upon dyes marketed purely for the textile industry are mentioned. Computer-aided study of the interaction between proteins and dye chromogens has suggested improved dye structures for chromatographic application. The preparative methods used in the manufacture of some conventional textile dyes are outlined together with the manner in which they may be varied to yield structures specifically targetted for protein interaction.

CHEMISTRY OF CHLOROTRIAZINYL DYES

Reactive dyes [1,2] for use on the cellulosic fibres, cotton and viscose, were introduced in 1956 to enable dyeings of high fastness to washing to be obtained by a simple application method, this effect being achieved by formation of a covalent bond between the dye and the substrate. The dye can be viewed as being composed of two units, one of which is the chromophore which is responsible for the colour, the other unit being the reactive system whose purpose is to form the covalent bond with the fibre.

Two different types of reactive system are used in this technology. One is the halogenoheterocyclic type, which is used in the Procion (ICI) and Cibacron (Ciba-Geigy) ranges; the other relies upon an activated double bond and is found in the Remazol range of Farbewerke Hoechst. The first reactive dyes for cellulose marketed were of the former type, containing a dichloro-s-triazine group and these dyes constitute what is now the Procion MX range (I). The key intermediate in their preparation is cyanuric chloride (2,4,6-trichloro-1,3,5-triazine) whose three labile chlorine atoms can be replaced in a stepwise manner. The Procion MX dyes are readily prepared by condensation of this intermediate with a sulphonated, water soluble chromogen which contains an amino group (represented by D-NH$_2$ in the structures below). They are quite stable at neutral pH values but are susceptible to hydrolysis on both the acid and alkaline side. To guard against any possibility of hydrolysis, a buffer, usually mixed phosphates is incorporated in the dye powder. Additionally, sodium chloride or other diluent is added to regulate the amount of dye present in the commercial product and a de-dusting agent (e.g. dodecylbenzene) is often added to ensure clean handling properties. The Procion MX dyes react readily with a carbohydrate substrate at a temperature of 30-40°C. Replacement of a second chlorine gives a monochloro-s-triazine and here the reaction temperature is about 85°C; addition of a buffer is now unnecessary but the other additives are incorporated in the commercial product. This type of dye (II) is to be found in the Procion H, Procion P and Cibacron ranges. They were introduced in 1957 and were followed in the mid 1960's by a number of bismonochloro-s-triazine dyes (III) and (IV) which are to be found in the Procion H-E range.

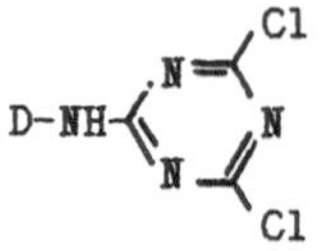

(I)

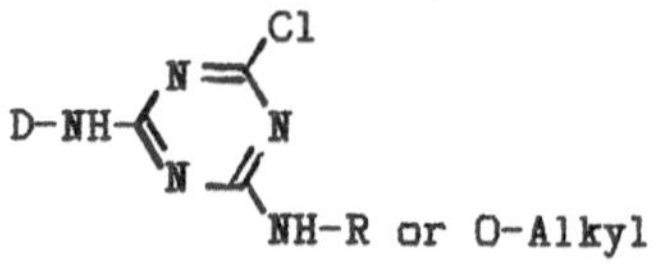

(II; R = H or Aryl)

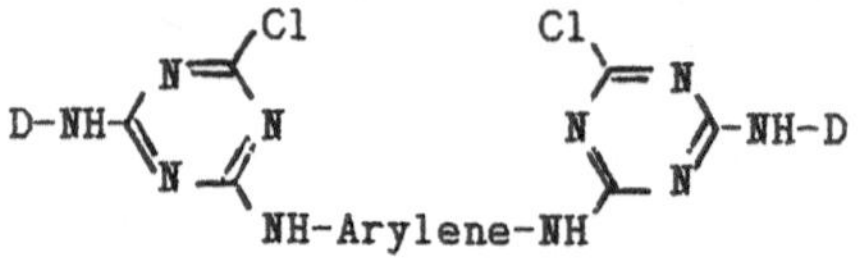

(III)

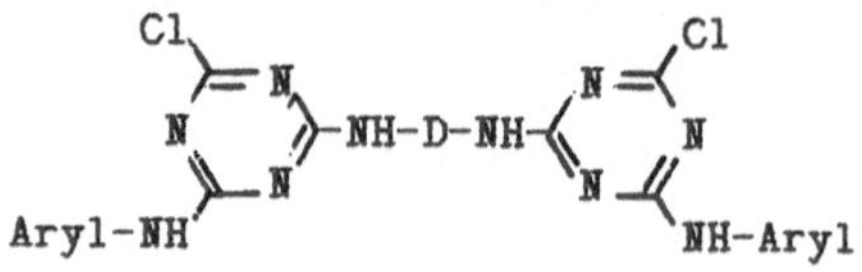

(IV)

The mode of action of all these types involves a nucleophilic displacement of the labile chlorine atom by the ionised carbohydrate substrate.

$$\text{D-NH} - \underset{\underset{X}{\overset{N}{\bigvee}}}{\overset{\overset{Cl}{\overset{N}{\bigwedge}}}{}} \quad \xrightarrow{\text{Carbohydrate-O}^-} \quad \text{Dye-NH} - \underset{\underset{X}{\overset{N}{\bigvee}}}{\overset{\overset{O\text{-Carbohydrate}}{\overset{N}{\bigwedge}}}{}}$$

This process is carried out in water and it is of interest to consider the effects which operate in textile dyeing, where the carbohydrate substrate is cellulose. In the dyeing process, cotton fabric is immersed in an aqueous solution of the dye and sodium chloride (about 6% by weight on the volume of water) is added. This causes the dye to be pushed out of the aqueous solution by a common ion effect and physically adsorbed onto the fibre by formation of hydrogen bonds between the carbohydrate and electron rich nitrogen and oxygen atoms contained within the dye molecule. When satisfactory adsorption has occured, usually after about 45 minutes, an alkali, usually sodium carbonate, is added to raise the pH to about 10.5. Ionisation of both the water and the carbohydrate takes place generating hydroxyl or carbohydrate-O$^-$ ions and either of these can attack the dye. If the former is the attacking species, hydrolysis of the reactive system occurs and the dye is lost to the dyeing process. Attack by the latter causes fixation of the dye to the substrate. These reactions are complete in about an hour and the ratio of the two possible reactions is the all important factor determining the technical utility of a reactive dye. When it is borne in mind that between five and twenty times as much water as cotton is present in the system it would seem that the process should be a failure but two effects swing the balance away from hydrolysis and in favour of fixation. First, the dye is physically adsorbed on the fibre prior to the reaction stage and this results in between 60 and 85% of the dye lying on the fibre and being ideally situated for reaction with the fibre. Secondly, the pH within the fibre is maintained mainly by cellulose-O$^-$ ions which outnumber hydroxyl ions by about 25:1 within the fibre. These two effects swing the balance in favour of fixation and result in 60% or more of the dye becoming attached to the fibre in the majority of cases. Of course, there is always some hydrolysis and the process must be completed by washing away any loose dye as well as the various inorganic materials which have been employed. These considerations apply equally well when a reactive dye is being combined with, say, agarose where an ideal technique with a Procion MX dye would be to stir the carbohydrate in a 5% dye solution, add sodium chloride (6g per 100ml solution) and allow to equilibrate for about an hour before raising the pH to 10.5-11.0 with a suitable alkali. Reaction should then be allowed to proceed for about 2 hours at room temperature and the dyed agarose then copiously washed, preferably with very dilute sodium carbonate solution. In the case of a hot dyeing monochlorotriazine dye of the Procion H or Cibacron type, the fixation stage can be carried out at room temperature if the reaction time is considerably extendeed, say to about four days.

In the case of a dichlorotriazine dye there are two chlorine atoms on the triazine ring but this does not give the dye two chances to achieve fixation. If hydrolysis occurs giving the hydroxytriazine, ionisation of the hydroxyl group (V) under the alkaline conditions being emloyed completely deactivates the second chlorine atom. If, however, fixation has occured, there is the possibility of further reaction of the species (VI; X = Cl) to give either (VI; X = OH) by hydrolysis or to give (VI; X = O-Carbohydrate) thus cross-linking the carbohydrate. Generally in the reaction conditions used to attach the dye to agarose, the second chlorine will remain intact. Hydrolysis of the second chlorine is favoured by more severe treatments whilst cross-linking is favoured by prolonged reaction times using mildly alkaline conditions (pH 8-9).

(V)   (VI)

In addition to this straightforward method of attachment to the substrate, the versatility of reactive dyes allows other techniques of attachment. Thus, reaction of a monochlorotriazine dye with an aliphatic diamine, e.g. 1,6-diaminohexane, introduces a side chain carrying an amino group (VII) which can then be reacted with suitably activated substrate.

(VII)

Such techniques have allowed the evolution of high performance liquid affinity chromatography techniques for protein separation. Derivatisation of microparticulate silica with glycidyl trimethoxysilane, reaction of the epoxysilylated support with 1,6-diaminohexane followed by attachment of a dichlorotriazinyl dye (Procion Blue MX-R) gave a matrix (VIII) which, in a 2m x 15 cm column, allowed rabbit muscle lactate dehydogenase to be purified at the rate calculated to be 1 KG/46 days [3].

OTHER HALOGENOHETEROCYCLIC REACTIVE DYES

In the years since 1956, numerous other halogenoheterocycles have been used to provide reactive dye ranges as the various dyestuff manufacturers have sought to enter the field [4]. These reactive systems all function in a similar manner to the chlorotriazine type. The most important are listed in TABLE 1, wherein an indication of the level of reactivity is given.

TABLE 1

Reactive dyes based on halogenoheterocyclic reactive systems

Reactivity	Range	Manufacturer	Reactive system
High	Procion MX	ICI	dichlorotriazine
Moderate	Levafix E-A	Bayer	fluoro-5-chloropyrimidinyl
	Levafix E	Bayer	dichloroquinoxaline
	Cibacron F	Ciba-Geigy	monofluorotriazine
Low	Procion H	ICI	monochlorotriazine
	Procion P	ICI	monochlorotriazine
	Procion H-E	ICI	monochlorotriazine
	Cibacron	Ciba-Geigy	monochlorotriazine
	Drimarene	Sandoz	trichloropyrimidinyl
	Levafix P	Bayer	5-chloro-6-methyl-2-sulphonylpyrimidinyl

VINYL SULPHONES AND RELATED DYES

The second general class of reactive dyes utilises an activated double bond as reactive system. This class is dominated by the Remazol dyes which contain a 2-sulphatoethylsulphone group which eliminates sodium sulphate during the application process forming forming a vinyl sulphone (IX). Ionised carbohydrate then adds across the activated double bond in this system thus covalently binding the dye to the substrate. The level of reactivity of these dyes falls is intermediate between that of the mono- and di-chlorotriazines.

$$D\text{-}SO_2\text{-}CH_2\text{-}CH_2\text{-}OSO_3H \longrightarrow D\text{-}SO_2\text{-}CH=CH_2 \quad (IX)$$

$$(IX) + Carbohydrate\text{-}O^{\ominus} \longrightarrow D\text{-}SO_2\text{-}CH^{\ominus}\text{-}CH_2\text{-}O\text{-}Carbohydrate$$

$$\downarrow H^{\oplus}$$

$$D\text{-}SO_2\text{-}CH_2\text{-}CH_2\text{-}O\text{-}Carbohydrate$$

A simple variant on this system is found in the Remazol D dyes which have a 2-sulphatoethylsulphonamide group rather than the 2-sulphatoethylsulphone. The reaction mechanism for the Remazol D dyes parallels that for the Remazol dyes.

CHROMOGENS USED IN REACTIVE DYES

In the foregoing discussion of reactive systems, the symbol D has been used to denote the chromogen of the dye. The function of the chromogen is to supply the colour and since the fixation to the substrate is totally dependent upon the reactive system, no limits are imposed on the type of chromogen which may be used. Although in practice the anthraquinone, phthalocyanine, triphendioxazine, formazan, and especially the metallised and unmetallised mono- and dis-azo chromogens dominate the field this still means that a very wide range of dye structures are to be found in the manufacturers ranges, all of them carrying a sufficient number of sulphonic acid groups to give about a 5% solution of the dye in water. Manufacturers do not normally disclose the exact structures of their dyestuffs but the colour of a particular dye often allows it to be assigned to a particular structural class and an idea of its general shape to be gained. The various types of structures used to cover the whole shade range have been listed elsewhere [5].

DISADVANTAGES FOR THE BIOCHEMIST

It must be remembered that reactive dyes are marketed solely for the textile trade and the manufacturer has no real interest in their biochemical application. This introduces certain pitfalls for the biochemist, an obvious one being that he may find a dye in which he is interested has been withdrawn from the manufacturers range. Indeed, this has happened with the withdrawal of both Cibacron Blue F-3GA and Procion Blue H-B by their respective manufacturers. Dyestuff names can also present an area of mystery for the biochemist who may fail to appreciate that the numbers and letters at the end of the name are of crucial importance. Thus there are ten different red Procion dyes on the current ICI range and these are differentiated from one another by the letters at the end of the name. The problem may be fully appreciated when it is considered that Procion Red MX-G and Procion Red MX-7B are both monoazo dyes but they possess different diazo components, different coupling components and different reactive systems.

Certain dyes are manufactured as deliberate mixtures. An example is Procion Blue H-B which is a mixture of two dyes differing in that the non-coloured substituent on the triazine ring carries a sulphonic acid group in either the 3- or 4-position. No final purification stage is used in manufacture since the aim is not to produce a pure chemical compound but rather to produce a powder capable of giving consistent performance in the textile dyeing process. The dye will thus contain small amounts of by-products from the chemical reactions involved in its manufacture. The marketed powder also contains an amount of a diluent, usually sodium chloride, added to ensure consistent dyeing performance. A de-dusting agent, such as dodecylbenzene, may be added to give clean handling properties and, in the case of a Procion MX dye, a buffer will also be incorporated to ensure adequate stability. This makes it difficult for the biochemist to do quantitative work and also introduces the far less likely possibility that one of the minor components might react strongly with the protein he is interested in.

In the preparation and use of a dye-affinity matrix an unusual snare can await the biochemist. When a column is being waashed, either to free it from loose dye or between uses, the tendency may be to wash the column with distilled water. This could have a deleterious effect since copious washing with distilled water will elute the sodium counter-ions away from the sulphonic acid groups on the dye, resulting in a low internal pH within the agarose. This generated acidity may result in fragments of the agarose backbone carrying the dye breaking away or, even, fission of the dye-agarose bond. It is interesting to speculate how many of the dye-leakage problems encountered by biochemists result from this effect.

SYNTHESIS OF DYES FOR BIOCHEMICAL USE

Research work in this area would obviously benefit from the synthesis of dyes specifically designed for application in dye-affinity chromatography rather than relying on dyes produced for the textile trade. The ease with which this can be done depends upon the particular chromophoric type. Thus, the preparation of phthalocyanine, formazan or triphendioxazine dyes employs highly specialised technology and is best left alone by the non-expert. On the other hand, the azo series revolves around application of diazotisation and coupling techniques and this is a fairly straightforward area to work in. Thus, whilst all reactive dyes are solubilised by sulphonic acid groups and are thus anionic in nature, novel dyes can readily be prepared containing positively charged groups. An example is the dye (X) containing an amidinium group. This is simply made by diazotising 4-aminobenzamidine, coupling the diazonium salt with 3-methylaniline and condensing the resulting aminoazo compound with cyanuric chloride [6].

(X)

In dye-affinity chromatography, particular interest has centred around the anthraquinone based dyes due to the similarity in shape and charge distribution between these dyes and NAD⁺. Anthraquinone based dyes represent an intermediate case, being more difficult to prepare than the azo compounds but still reasonable within certain limitations. The starting point for their preparation is invariably the key intermediate 1-amino-4-bromoanthraquinone-2-sulphonic acid (XI) which condenses readily with amines in the presence of a copper catalyst. In the preparation of Procion Blue H-B and Cibacron Blue F3G-A the amine is chosen to be 1,4-diaminobenzene-2-sulphonic acid, the condensation taking place on the unhindered 4-amino group and leaving the 1-amino group free to be condensed with cyanuric chloride. Further condensation with the relevant sulphonated aniline produces the commercial dyes.

(XI)

Variations on the anthraquinone moiety are impractical but variations on the central diamine and the terminal substituent on the triazine ring are readily achieved. Computer aided modelling studies of the commercial dyes and the co-enzyme shows that these dyes are shorter and less flexible than NAD⁺ in the central area and this has prompted the synthesis of dyes more structurally similar to the co-enzyme. Thus, the amine (XII), wherein the more basic aliphatic amino group is acylated to prevent its preferential condensation, can be reacted with 1-amino-4-bromoanthraquinone-2-sulphonic acid yielding (XIII). Hydrolysis of the acetylamino group and condensation of cyanuric chloride with the product and then with a sulphonated aniline yields dyes (XIV) which possess a much greater structural similarity to NAD⁺ [7, 8].

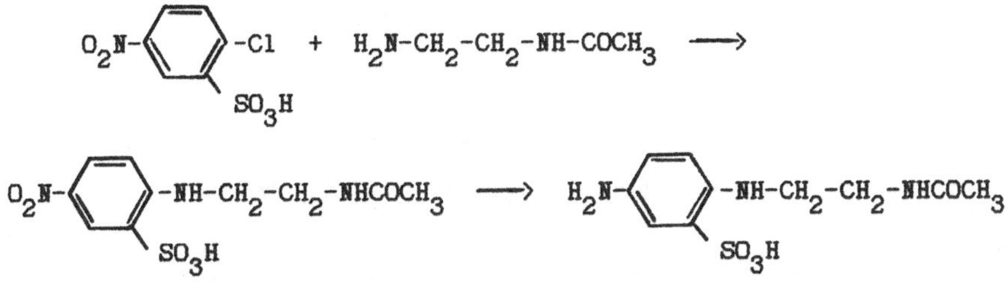

$$O_2N-\langle\bigcirc\rangle-Cl \quad + \quad H_2N-CH_2-CH_2-NH-COCH_3 \quad \longrightarrow$$

with SO_3H on the ring.

$$O_2N-\langle\bigcirc\rangle-NH-CH_2-CH_2-NHCOCH_3 \quad \longrightarrow \quad H_2N-\langle\bigcirc\rangle-NH-CH_2-CH_2-NHCOCH_3$$

with SO_3H substituents.

(XII)

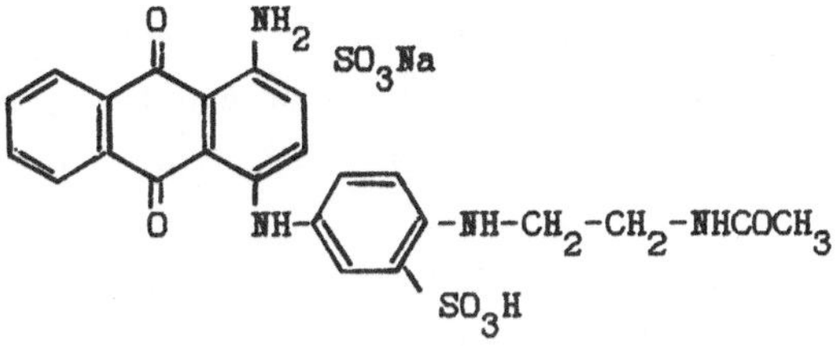

(XIII)

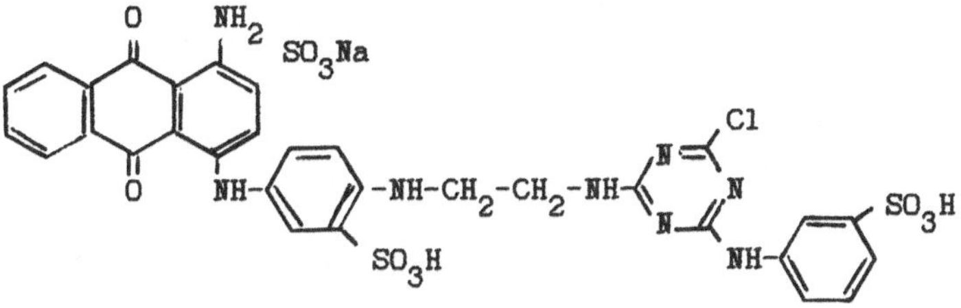

(XIV)

REFERENCES

1. Siegel, E., Schundehutte, K-H. and Hildebrand, D. The Chemistry of Synthetic Dyes, Volume VI, Reactive Dyes, ed., K.Venkataraman, Academic Press, New York and London, 1972.

2. Beech, W.F., Fibre Reactive Dyes, Logos Press, London, 1970.

3. Jones, K., Process Scale HPLC comes of Age, Speciality Chemicals, May, 1986.

4. Stead, C.V., Halogenated Heterocycles in Reactive Dyes, <u>Dyes and Pigments</u>, 1982, **3**, 161-171.

5. Stead, C.V., The Use of Reactive Dyes in Protein Separation Processes, <u>J. Chem. Tech. Biotechnol.</u>, 1987, **37**, 55-71.

6. Lowe, C.R. and Stead, C.V., Reactive Cationic Compounds, <u>U.K. Patent Application</u> 8,517,778.

7. Lowe, C.R., Burton, S.J., Pearson,, J., Clonis, Y.D. and Stead, C.V., Design and Application of Biomimetic Dyes in Biotechnology, <u>J. Chromatography.</u> 1986, **376**, 121-130.

8. Burton, S.J., Lowe, C.R. and Stead, C.V., Anthraquinone Derivatives, <u>U.K. Patent Application</u> 8,517,779.

DYE FRAGMENTS IN PROTEIN-DYE INTERACTIONS

F. CADELIS and M.A. VIJAYALAKSHMI
Université de Technologie de Compiègne B.P. 649 Compiègne, France

SUNANDA R. NARAYAN
Department of Pharmaceutical Chemistry, University of Kansas, U.S.A.

ABSTRACT

The molecular recognition between the triazine dyes and the nucleotide dependant enzyme systems is known. Though, the ionic, hydrophobic and charge-
transfer interactions are known to be the basis of this recognition, a
knowledge of the functional groups, their positions and the minimum structure necessary for the reversible association for an easy exploitation in
Affinity Chromatography of these enzymes is useful. Thus, in this study, we
report our data on the inactivation of two classes of such enzymes, with
dye fragments as well as with some structural isomers. The triazine dye,
Cibacron Blue F3GA (CBF3G-A) and its fragments, 1-amino-4-(4'-aminophenyl-
amino)-anthraquinone 2-3'-disulphonic acid (ASSO) and bromaminic acid have
been studied kinetically as inhibitors of firefly luciferase, an ATP - dependant dimeric enzyme. The importance of stereospecificity has been addressed by studying the interaction of Cibacron Brilliant Blue (CBB-II), the
steric isomer of CBF3G-A with luciferase. The specificity of the anthraquinone ring of CBF3G-A was compared with that of the Xanthene ring of a much
smaller dye molecule, 2',4',5',7'-tetraiodofluorescein (TIF). These studies
suggest that the ASSO portion of the dye seems to be the essential part of
the entire dye molecule which participates in the specific binding to the
ATP dependant enzyme.

Another triazine dye Procion Red HE-3B and two structural variants of
the same were tested with lactate dehydrogenase, a NAD dependant enzyme, in

terms of inhibition kinetics. The values of dissociation constants (K_D), maximal rate of inactivation (k_3) and constants of inhibition (K_I) were calculated. From the data, it could be shown that the variations in the terminal ring play an important role in the specificity of protein dye interactions.

INTRODUCTION

Since the observation that Blue Dextran (a dextran conjugate of the triazine dye, CBF3G-A) binds to various proteins, considerable interest has developed in its chromophore (1). Immobilized dyes have been in the limelight since then playing a significant role in the purification of proteins by conventional affinity chromatography (2), by High Performance Liquid Affinity Chromatography (3, 4), by Affinity Partitioning (5 -7) etc...

Due to the great interest in the application of immobilized CBF3G-A in down stream processing, many investigators have examined their interactions with this dye and its derivatives. The specificity of dye enzyme interactions was apparent even in the very early experiments with phosphofructokinase and pyruvate kinase since the dye-enzyme complexes could be dissociated by low concentrations of substrates or effectors of the respective enzymes.

In 1972, Bohme et al. reported that the ASSO part of the dye molecule is responsible for the specific interaction with phosphofructokinase (8). The similarities in shape, aromaticity and charge distribution between this part of the dye and the ATP molecule and the inhibition of the enzyme by the dye, described as competitive with respect to ATP led to the suggestion that the dye might be conceived as structurally analogous to ATP.

In addition to an apparent specific complex formation, CBF3G-A also forms unspecific complexes with proteins. This second type of dye-enzyme interaction is based on the ability of the dye to act as a weak cation exchanger owing to its sulphonic acid groups. In certain cases, mixed types of dye-enzyme complexes do exist comprising of specific and unspecific interactions.

Thompson et al.(1) originally proposed a hypothesis for the basis of
the interaction between the monochlorotriazinyl sulphonated polyaromatic
dye, CBF3G-A and certain kinases and dehydrogenases. They suggested that
the dye functioned as a nucleotide or coenzyme analogue. It was also proposed that the dye displayed specific affinity for those proteins possessing
a supersecondary structure termed the "dinucleotide fold". Since many proteins known to lack this structure, displayed affinity for immobilized
CBF3G-A, hydrophobic, ionic or charge transfer interactions were attributed
to protein-dye interactions (8, 9). In 1979, Edwards and Woody (10) indicated that the dyes are not highly specific coenzyme analogues since only a
part of CBF3G-A structure is required to compete with the coenzyme and they
do not directly address the extent to which the dyes bound to different enzymes have similar conformations.

We hence carried out some dye-protein interaction studies with firefly luciferase, an ATP dependant dimeric enzyme to find out if the entire
dye molecule is required for the biospecific binding of luciferase. The stereospecific nature of protein-dye interaction was also addressed. In order
to answer these questions, CBF3G-A, CBB-II, the steric isomer of CBF3G-A,
ASSO and bromaminic acid were compared kinetically as inhibitors of luciferase relative to its substrate ATP. Chromatographic studies were conducted
to confirm our kinetic data. We also studied the interaction of a much smaller dye molecule, TIF, with firefly luciferase. TIF was chosen since the
xanthene ring of TIF was reported to bind to the adenine pocket of NAD in
lactate dehydrogenase in a manner analogous to the binding of the anthraquinone ring of CBF3G-A to the adenine pocket of NAD in alcohol dehydrogenase (11). The structure of the dyes used in firefly luciferase studies are
given in Fig. 1.

In 1979, Biellmann et al.reported that modifications of the terminal
ring of CBF3G-A could bring about the most significant variations in the
binding affinity of the dye to proteins (12). In 1986, Lowe et al.studied
the affinity of various analogues of CBF3G-A (the parent dye substituted at
the terminal ring, bridging ring and the anthraquinone moiety) for horse liver alcohol dehydrogenase. They suggested that small substituents bind more

tightly than more bulky species and especially so if substituted in the ortho or meta positions of the terminal arylamine ring with a neutral or anionic group.

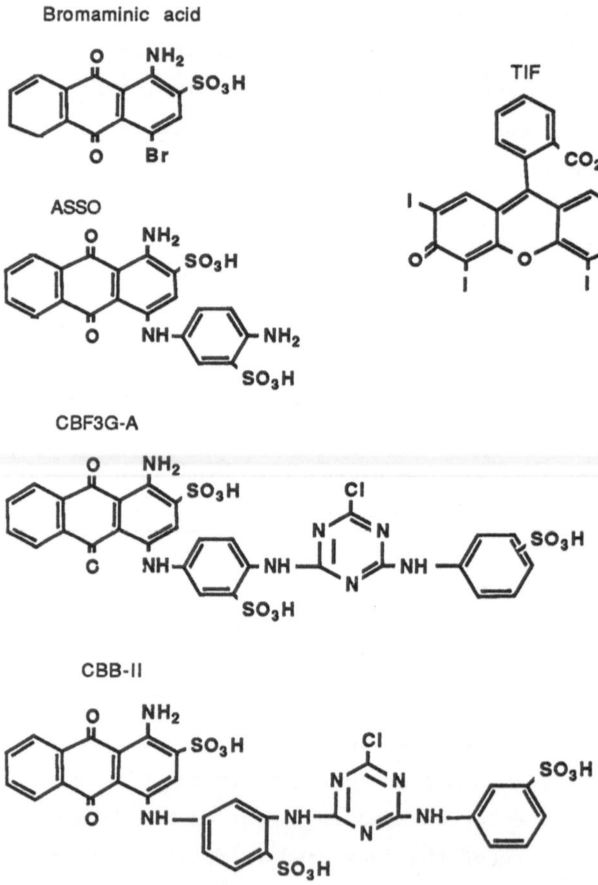

Figure 1.   Structure of the anthraquinone dyes used in our studies.

Hence we undertook to study a red dye, Procion Red HE-3B and two structural variants shown in Fig. 2, in terms of their binding to lactate dehydrogenase, a NAD dependant enzyme in order to follow the increase in their specificity due to the modifications at the terminal rings. In Dye 1, both the terminal sulphonic groups of the parent dye was removed and in Dye 2, there was an additional amino benzene ring on both the terminal ends of the dye.

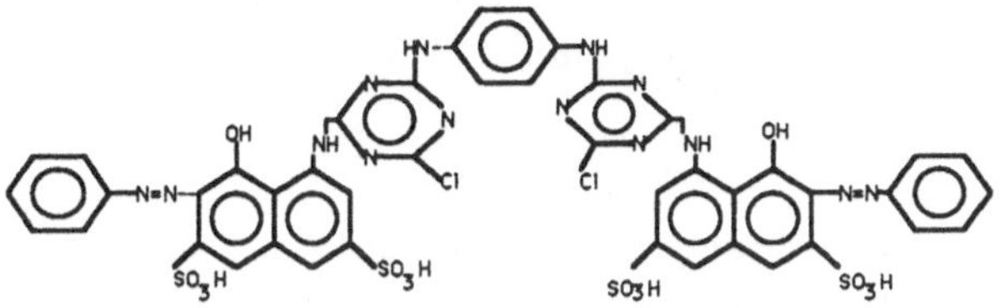

DYE 1

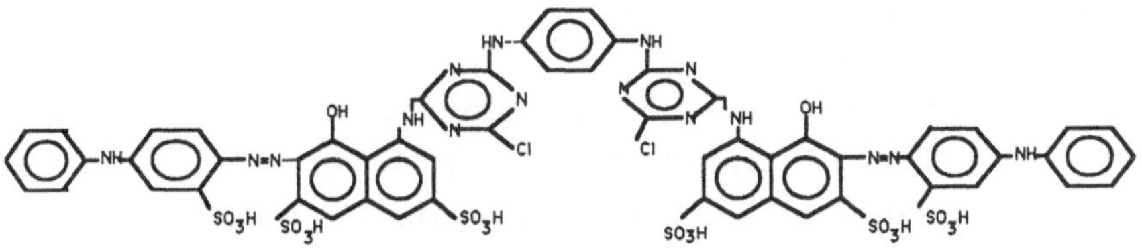

DYE 2

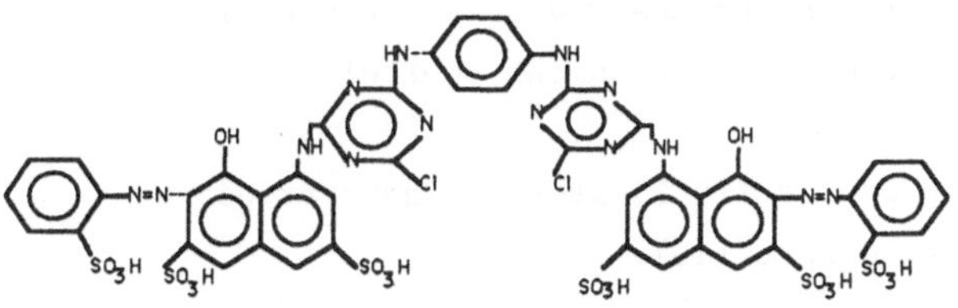

PROCION RED HE-3B

Figure 2.   Structure of Procion Red HE-3B and the variants Dye 1 and Dye 2.

EXPERIMENTAL

Materials

Firefly luciferase (EC 1.14.14.3) was from Boehringer, France, and lactate dehydrogenase (EC 1.1.1.27) was from Sigma. Ciba-Geigy provided CBF3G-A and CBB-II. Procion Red HE-3B was obtained from ICI Organics Divisions, Manchester, U.K. and the two derivatives of Red HE-3B, Dye 1 and Dye 2 were a generous gift from Dr Mazza, Vilmax, Buenos Aires, Argentina. All other chemicals employed were of reagent grade.

Methods

Luciferase activity was measured by a nucleotimeter (13). The unit of activity is the maximum intensity of light recorded in millivolts per picogram of ATP per milligram of protein at 562 nm. The lactate dehydrogenase was assayed at 25°C as described by Bergmeyer (14). One unit of enzyme activity is defined as the amount that catalyses the conversion of 1μmol of substrate to product per minute at 25°C.

Inactivation of lactate dehydrogenase by free triazine dyes were performed in 0.05 M Phosphate buffer, pH 7.5 at 25°C. The reaction vial contained in a total volume of 1 ml; enzyme (1 unit, 25°C) and dye as mentioned in the individual cases. The rate of dye inactivation was followed by periodically removing samples (100 μl) and assaying for enzyme activity. Initial rates of inactivation were deduced from plots of $\log_{10}$ (% of activity remaining) versus time (min) for several dye concentrations and the slopes and intercepts of secondary double reciprocal plots were calculated. A sample without the dye served as the control.

Inactivation of firefly luciferase was carried out in a similar manner as described elsewhere (15).

RESULTS AND DISCUSSION

Firefly luciferase studies: The dissociation constant (K_D), the inhibition constant (K_I) and the maximum rate of inactivation (k_3 min^{-1}) are calculated according to the method already published (15).

Table 1 suggests that the steric isomer CBB-II behaves very differently from the other dyes with a k_3 value of 0.004 min^{-1}, the least among the dyes studied, with similar K_I value as that of CBF3G-A but with a higher K_D value. The most potent inhibitor among them seems to be ASSO and CBF3G-A. Inactivation studies with TIF yields a K_D value of 16.66 μM, a K_I of 400 μM and a k_3 value of 0.1 (min^{-1}). The xanthene ring thus seems to be less specific for luciferase.

TABLE 1
k_3, K_D and K_I of dyes for firefly luciferase

DYE	k_3 (min^{-1})	K_D (μM)	K_I (μM)
Bromaminic acid	0.074	6.66	6.09
ASSO	0.083	2.56	16.06
CBF3G-A	0.006	2.13	111
CBB-II	0.004	4.16	108
TIF	0.014	16.66	400

From these studies and those reported by Deluca (16), it seems that dyes bind to luciferase in a hydrophobic/ionic manner with stereospecificity playing a significant role when the dye structure is large enough to spread across the active site of the enzyme molecule. Chromatography on CBF3G-A and CBB-II revealed that CBF3G-A was a better ligand for the purification of firefly luciferase. It gave about 10 fold purification and about 200% recovery in enzyme activity. CBF3G-A, the aromatic chromophore of blue dextran is found to be a better inhibitor of luciferase compared to its steric isomer, CBB-II with K_I values in the same order but with appreciable difference in the ability of the dyes to inhibit firefly luciferase. Our initial observations (13) that stereospecificity plays an important role in protein-dye interaction is supported by the fact that we did not observe any interaction of luciferase with Procion Blue HB, the m,p isomer of CBF3G -A. So according to Kopperschlager, "the better the arrangement of ionic groups of the dye fits the arrangement of the corresponding functional group of the enzyme, the stronger is the complex binding." For luciferase, as for yeast hexokinase and rabbit muscle lactate dehydrogenase (8, 17), the ASSO portion of the dye, CBF3G-A seems to be the most active inhibitor.

From our studies we tend to conclude that the tail end of CBF3G-A is non-specific and binds only when the catalytic site of the enzyme is large enough and properly oriented to accomodate it, irrespective of the substrate requirements or an overall protein folding pattern. As far as luciferase is concerned, ASSO seems to be the essential part of the entire dye molecule which participates in the binding and for the rest of the dye binding to occur, a precise orientation of the dye chromophore is required. Our studies also support the concept proposed by Edwards and Woody (10) that "the several functional groups and conformational freedom attributable to the dye chromophores allow them to insert aromatic rings into hydrophobic pockets of enzyme protein. While the other portions of the dye molecule and the protein assume conformational freedom, some specificity is achieved". It is also worthwhile to note that the affinity of these dyes to luciferase is stronger in solution compared to when it is immobilized on a support.

Lactate dehydrogenase studies : Procion Red HE-3B and the two structural derivatives of the same dye, Dye 1 and Dye 2 inactivitate lactate dehydrogenase. The time course for the inactivation of rabbit muscle LDH by the three dyes are shown in Fig. 3. The Dye 1 in which the terminal sulphonic groups were removed did not show much inhibition even in the presence of a higher dye concentration of 50 μM. Inhibition was more pronounced in the case of Dye 2 and Procion Red HE-3B, even at concentrations as low as 3.0 μM when compared with Dye 1. The maximal rates of inactivation (k_3), dissociation constants (K_D) and inhibition constants (K_I) of the dyes for LDH are shown in Table 2. These three dyes reversibly inactivate LDH with similar rates of maximal inactivation (k_3) but with markedly different affinities. The parent dye, Red HE-3B and Dye 2 had similar K_D values which was almost 12 times much smaller than the K_D values for Dye 1. This clearly shows that the terminal sulphonic groups plays an important role in the

specificity of these dyes for LDH. Thus, the Dye 1 in which the terminal
sulphonic groups were removed showed decreased affinity for LDH compared to
the parent dye Red HE-3B and Dye 2.

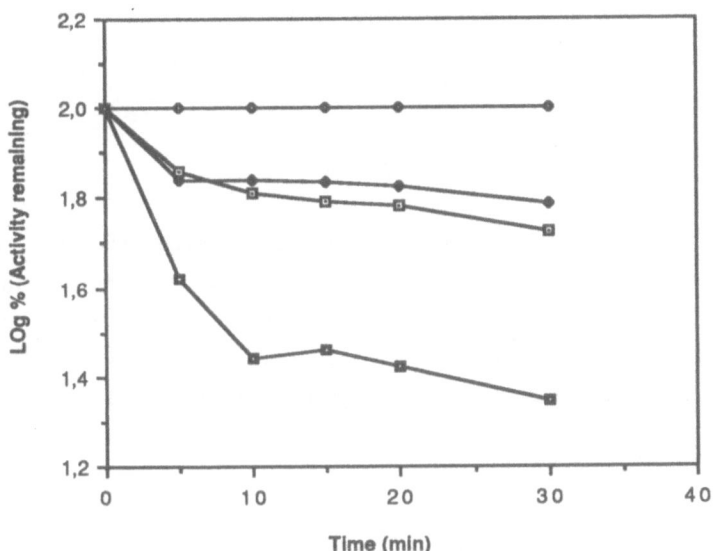

Figure 3.   Time course for the inactivation of rabbit muscle lactate dehy-
             drogenase by Procion Red HE-3B, Dye 1 and Dye 2 at pH 7.5 and
             25°C. ● , Control; ● , Red HE-3B (3.12 µM); ▣ , Dye 1 (50 µM);
             ▣ , Dye 2 (3.12 µM).

TABLE 2
k_3, K_D and K_I of dyes for lactate dehydrogenase

DYE	k_3 (min-1)	K_D (µM)	K_I (µM)
Dye 1	0.111	143.43	0.621
Dye 2	0.333	11.12	0.017
RED HE-3B	0.166	12.93	–

Moreover, the pattern of inhibition obtained for these dyes shows that Dye 1 and Dye 2 are competitive vis-a-vis the coenzyme NAD. The Procion Red HE-3B showed a complicated mode of inhibition pattern. Thus it is evident that the modification of the terminal ring, including an increase in the aromatic groups can make the parent non-specific dye, bind to the NAD dependant enzyme (LDH) in a specific competitive manner.

CONCLUSIONS

From these data and others already available, it is clear that the specificity of the dyes towards the nucleotide dependant enzymes depends on the functional groups and their spatial arrangements. Thus the charge and the bulkiness of the terminal groups are important in designing very useful dyes for the purification and other biotechnological applications of these dyes.

ACKNOWLEDGEMENTS

The kind gift of Dye 1 and Dye 2 and collaboration of Drs Mazza and Outumuro of the Vilmax Company, Argentina are thankfully acknowledged. We thank also Dr J.F. Hervagault of the Laboratoire de Technologie Enzymatique and Dr G. Kirchberger from KMU, Leipzig GDR, for useful discussions and comments on the manuscript.

REFERENCES

1. Thompson S.T.; Cass K.H. and Stellwagen E. Blue Dextran-Sepharose : An affinity column for the dinucleotide fold in proteins. Proc. Natl. Acad. Sci., 1975, 72, 669.

2. Lowe C.R. and Pearson J.C. Affinity chromatography on Immobilized Dyes. Methods in Enzymol., 1984, Vol. 104, Ed. W.B. Jakaby Acad. Press, N.Y. pp. 97-113.

3. Small D.A.P.; Atkinson A. and Lowe C.R. Preparative High-Performance Liquid Affinity Chromatography. J. Chromatogr., 1983, 266, 151.

4. Lowe C.R.; Burton S.J.; Pearson J.C. and Clonis Y.D. Design and application of bio-mimetic dyes in biotechnology. J. Chromatogr., 1986, 376, pp. 121-130.

5. Kopperschlager G. and Birkenmeier G. Affinity partitioning: A new approach for studying dye-protein interactions. J. Chromatogr., 1986, 376, pp. 141-148.

6. Hayet M. and Vijayalakshmi M.A. Affinity precipitation of proteins using bis-dyes. J. Chromatogr., 1986, 376, pp. 157-161.

7. Johansson G. Affinity partitioning. Methods in Enzymol., 1984, Vol. 104, Ed. W.B. Jakoby Acad. Press, N.Y. pp. 356-364.

8. Bohme H.J.; Kopperschlager G.; Schultz J. and Hofmann E. Affinity chromatography of Phosphofructokinase using Cibacron Blue F3G-A. J. Chromatogr., 1972, 69, pp. 209-214.

9. Wilson J.E. Applications of blue dextran and Cibacron Blue F3G-A in purification and structural studies of nucleotide-requiring enzymes. Biochim. Biophys. Res. Comm., 1976, 72, pp. 816-823.

10. Edwards R.A. and Woody R.W. Spectroscopic studies of Cibacron Blue and Congo Red bound to dehydrogenases and kinases. Evaluation of dyes as probes of the dinucleotide fold. Biochim., 1979, 18, pp. 5197-5204.

11. Tucker R.F.; Babul J. and Stellwagen E. Protein dye affinity chromatography using immobilized tetraiodofluorescein. J. Biol. Chem., 1981, 256, 10993.

12. Biellmann J.F.; Samama C.I.; Branden C.I. and Eklund E. X-Ray studies of the binding of Cibacron Blue F3GA to liver alcohol dehydrogenase. Eur. J. Biochem., 1979, 102, 107.

13. Rajgopal S. and Vijayalakshmi M.A. Interaction of firefly luciferase with triazine dyes. J. Chromatogr., 1983, 280, pp. 77-84.

14. Bergmeyer H.U. L-Lactate dehydrogenase, UV-assay with pyruvate and NADH. Methods of Enzymatic Analysis, 1974, Vol. 2 edited by Hans Ulrich Bergmeyer, Academic Press, N.Y., p. 574.

15. Rajgopal S. and Vijayalakshmi M.A. Metal-ion mediated interaction of luciferase with tetraiodofluorescein. J. Chromatogr., 1984, 315, pp. 175-184.

16. Deluca M. Biochemistry, 1969, 8, pp. 160-166.

17. Stellwagen E. Use of blue dextran as a probe for the NAD domains in proteins. Acts Chem. Res., 1977, 10, pp. 92-98.

Chapter 2

Physico-chemical Aspects of Dye–Protein Interactions

REACTIVE DYES AS A SPECTRAL PROBE FOR PROTEIN FOLDING

M. V. JAGANNADHAM and EARLE STELLWAGEN
Department of Biochemistry
University of Iowa
Iowa City, IA 52242 USA

ABSTRACT

A single blue dye was covalently attached to the active site of ribonuclease by reaction of the protein with Procion Blue MX-R, C.I. 61205
reactive blue 4, for 30 minutes at pH 8.5 and 35 degrees. This attachment resulted in the loss of over 95% of the catalytic activity of the
enzyme without disruption of its folded structure as judged by size
exclusion chromatography and by circular dichroic measurements. Addition of increasing concentrations of the denaturant guanidinium chloride
to the protein-dye conjugate first generated visible difference spectra
characteristic of the free dye followed by spectra at higher denaturant
concentrations characteristic for displacement of the conjugated dye
from the active site by the competitive inhibitor 2'-CMP. Formation of
the latter difference spectra coincided with the denaturation of the
protein-dye conjugate as judged by far ultraviolet circular dichroic
measurements. The kinetics of refolding the denatured protein-dye conjugate were observed at 650 nm in 2 M guanidinium chloride maintained at
pH 6 and 10 degrees. The decrease in absorbance fit well with a single
first order reaction having a half-time of 1080 seconds. This value is
within a factor of two of the half-time reported for the refolding of
unconjugated ribonuclease observed using either tyrosine absorbance or
fluorescence measurements.

INTRODUCTION

The spontaneous folding of a denatured protein into its native or biofunctional structure likely proceeds stepwise through a series of partially folded intermediate structures. Definition of these intermediate
structures requires observation of the folding process using a series of
spectral probes which sense different aspects of folding such as the
appearance of secondary structure, the formation of compact tertiary
structure and the formation of biofunctional site(s). A potential general spectral probe which senses formation of a biofunctional site is a

textile dye such as reactive blue 2 whose occupancy of the site produces a change in visible spectrum [1]. However, the requisite second order reaction required for formation of the noncovalent protein:dye conjugate may obscure the kinetics of formation of the biofunctional site. By contrast, covalent attachment of a textile dye to a site prior to denaturation should generate spectral changes whose kinetics directly reflect the formation of the site. This manuscript describes results of such measurements using ribonuclease as a model protein and Procion Blue MX-R, Color Index 61205 reactive blue 4, as the dye.

MATERIALS AND METHODS

Materials

Bovine pancreatic ribonuclease, type XII-A, and rabbit muscle lactate dehydrogenase, type II, were purchased from Sigma. Procion Blue MX-R was a gift from Dr. Vivian Stead formerly with Imperial Chemical Industries. Purified guanidinum chloride was purchased from Heico.

Methods

Protein-dye conjugates were obtained by reaction of the protein and dye in 100 mM Tris-HCl buffer, pH 8.5, maintained at 35 degrees as described by Clonis and Lowe [2]. The catalytic activity or ribonuclease was measured spectrophotometrically using cyclic CMP as the substrate [3]. The reaction with dye was quenched by addition of an amount of DTT equimolar with total dye and the blue components in the reaction mixture were fractioned into two populations by exclusion chromatography using a 1.3 x 12 cm column of Sephadex G-25 and water as the mobile phase. The blue population having the larger hydrodynamic volume defines the protein-dye conjugate and was pooled and lyophilized. The protein content in the conjugate was determined using BCA Protein Assay Reagent from Pierce and the unconjugated protein as a standard while the dye content of the conjugate was determined spectrophotometrically using an extinction of $14.9 \text{ mM}^{-1} \text{ cm}^{-1}$ at 625 nm [4]. Absorbance and circular dichroism measurements were obtained using Aviv model 14DS and 60DS spectrometers, respectively, located in the University of Iowa Protein Structure Facility. High performance size exclusion chromatographic measurements were obtained using a 300 mm TSK-125 column, an Isco pump and two Isco variable wavelength flow detectors. The concentration of guanidinium chloride solutions was measured using an Abbe Mark II digital refractometer.

RESULTS

The inactivation of ribonuclease following continued exposure of the
protein to Procion Blue MX-R at pH 8.5 and 35 degrees is illustrated in
Figure 1.

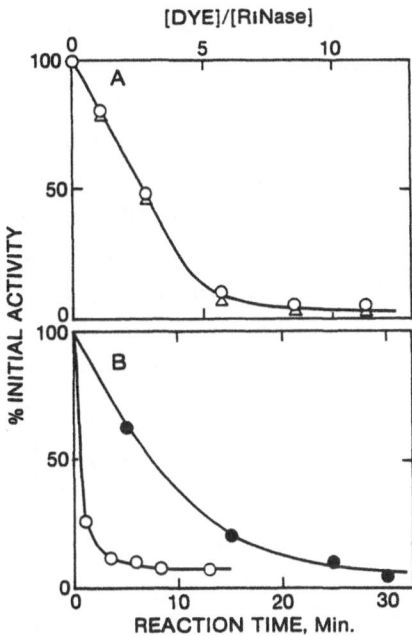

Figure 1. Inactivation of ribonuclease by reaction with Procion Blue
MX-R. The experimental values illustrated in panel A were obtained
after reaction for 30 minutes, circles, and 60 minutes, triangles, using
a solution containing 2 mg/ml protein. The experimental values illustrated in panel B were obtained using a solution containing either 2
mg/ml protein and a total dye/protein molar ratio of 6.75, open circles,
or 12.3 mg/ml protein and a molar ratio of 8.8, filled circles.

It should be noted that over 95% of the initial catalytic activity of
the protein is irreversibly lost by reaction with the dye under these
conditions. Unfortunately, the ability of a competitive inhibitors to
protect the protein from irreversible inactivation by dye cannot be
demonstrated owing to the insignificant binding of the competitive inhibitors at pH 8.5 [5]. Analysis of the inactivated protein following its
fractionation from the reacted dye indicated an average conjugation of
0.85 dyes per protein. This retained dye could not be removed by
subsequent exclusion chromatography of the denatured protein in the

presence of excess guanidinium chloride.

The structure of protein was not significantly perturbed by conjugation of a single dye. The protein-dye conjugate eluted from an HPLC
size exclusion column as a single component having a fractional retention of 0.58 in 0.5 M NaCl containing 50 mM cacodylate buffer, pH 6, a
fractional retention identical with that of the unmodified protein. The
shape of the elution profile of the protein-dye conjugate was identical
when observed at 280 nm and at 600 nm. The molar ellipticity of the
conjugated protein observed in 50 mM cacodylate buffer, pH 6, was -6,800
deg cm^2 dmol^{-1} at 220 nm which compares favorable with the value of
6,960 deg cm^2 dmol^{-1} measured for the unconjugated protein. The
dependence of the molar ellipticity of the conjugated and unconjugated
proteins on the concentration of the denaturant guanidinium chloride are
compared in Figure 2. It can be seen that both proteins exhibit the

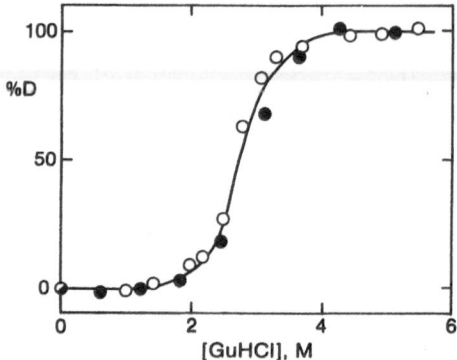

Figure 2. Equilibrium denaturation transitions. The ordinate indicates
the fractional change in molar ellipticity measured at 220 nm generated
by increasing concentrations of the denaturant guanidinium chloride,
abbreviated GuHCl. The open circles indicate measurements obtained
using solutions of the unconjugated protein at a concentration of 0.8
mg/ml and a cell having an optical path of 1 mm. The filled circles
indicate measurements obtained using solutions of the conjugated protein
at a concentration of 0.36 mg/ml and a cell having an optical path of 2
mm. The unconjugated protein had an average molar ellipticity of -1,670
deg cm^2 dmol^{-1} in excess denaturant while the conjugated protein a
value of -1,830 deg cm^2 dmol^{-1}. All solutions contained 50 mM cacodylate buffer, pH 6.

same cooperative structural transition between 2 to 4 M denaturant.

Thus, inactivation of the ribonuclease by conjugation of a single dye does not appear to perturb either the secondary or the tertiary structure of the protein.

The presence of the denaturant caused a modest enhancement in the visible absorbance spectrum of the protein-dye conjugate. This enhancement is best seen in the difference spectra illustrated in Figure 3.

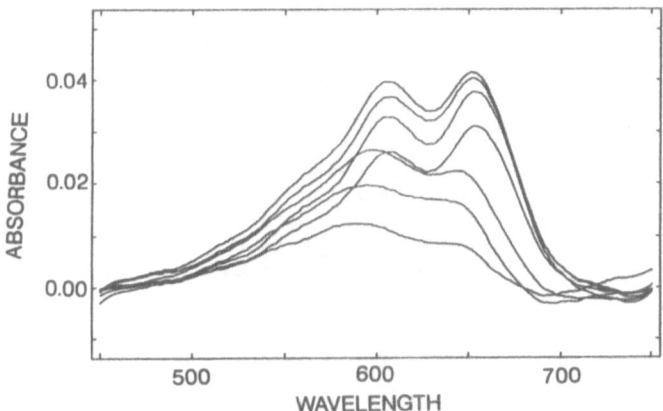

Figure 3. Visible difference spectra of the protein-dye conjugated in guanidinium chloride. All solutions contained 23 μM of the protein-dye conjugate and 50 mM cacodylate buffer, pH 6.0. The illustrated difference spectra were obtained in 0.5, 1.8, 2.4, 3.1, 3.7, 4.2 and 5.1 M denaturant reading upwards at 650 nm, using 0.0 M denaturant as the reference. All measurements were obtained at 23 degrees in cuvets having optical paths of 10 mm.

These difference spectra illustrate two distinct profiles: one having a broad maximum at about 590 nm with a shoulder at 645 nm observed at denaturant concentrations below 2 M, and a second having well defined maxima at 605 and 650 nm observed in denaturant concentrations above 3 M. The protein-dye difference spectra observed in low concentrations of denaturant resemble the difference spectra observed following addition of denaturant to the free or unconjugated dye, Figure 4, having a broad maximum at about 600 nm and a shoulder at about 640 nm. The increase in the difference absorbance at 600 nm is linearly dependent upon denaturant concentration, suggesting that these difference spectra likely result from perturbation of the dye absorbance by changes in the properties of the bulk solvent. The difference spectra observed for the pro-

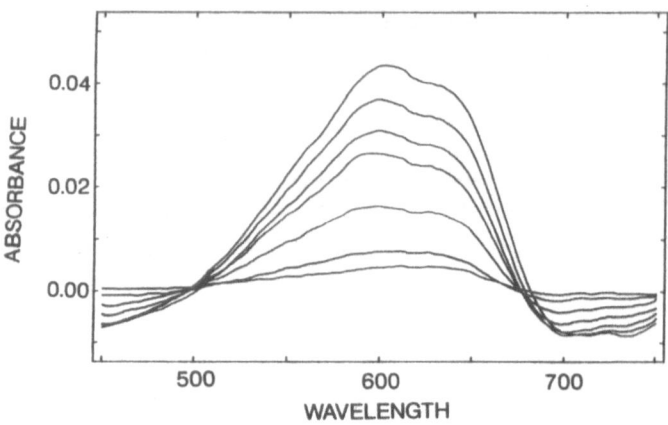

Figure 4. Visible difference absorbance spectra of the unconjugated dye
in guanidinium chloride. The illustrated difference spectra were
obtained in 1.2, 1.7, 2.6, 3.7, 4.4, 5.3 and 6.5 M denaturant, reading
upwards at 600 nm, using 0.6 M denaturant as the reference solution.
All measurements were obtained at 23 degrees using solutions containing
28 μM dye and cells having optical paths of 10 mm.

tein-dye conjugate in more concentrated denaturant resembles the difference spectra observed following addition of the competitive inhibitor
2'-CMP to the protein-dye conjugate in the absence of denaturant, Figure
5, having distinct maxima at 602 and 654 nm. This difference spectrum

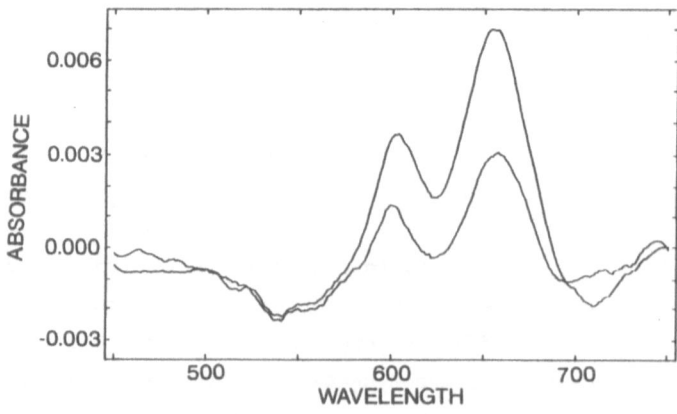

Figure 5. Visible difference spectra of the protein-dye conjugate in
competitive inhibitor. The illustrated difference spectra were observed
in 11 and 22 mM 2'-CMP reading upward at 600 nm. All measurements were
obtained using a 23 μM solution of the protein-dye conjugate in 50 mM
cacodylate buffer, pH 5.5, and 23 degrees using a cell with an optical
path of 10 mm.

likely results from the displacement of the conjugated dye from the catalytic site by the competitive inhibitor which has a Ki value of 2 mM [6]. Unfortunately, the limited solubility of the competitive inhibitor does not facilitate demonstration of the maximal difference in the visible spectrum of the protein-dye conjugate associated with displacement of the conjugated dye from the catalytic site.

Measurements of the kinetics of the change in the visible absorbance of the free and conjugated dye at 650 nm resulting from changes in the concentration of guanidinium chloride were initiated by manual mixing at pH 6 and 10 degrees. A kinetic profile for the refolding of the denatured protein-dye conjugate in 2 M guanidinium chloride is shown in Figure 6. The decrease in absorbance fit well with a single first order

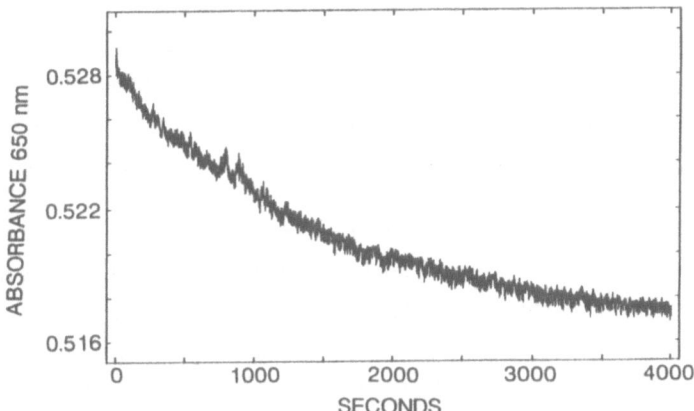

Figure 6. A kinetic profile for the ·refolding of the denatured protein-dye conjugate. A solution of protein-dye conjugate was denatured in excess guanidinium chloride and then diluted in a 10 mm optical cell with buffer to give a solution containing 42 μM protein-dye conjugate, 2.0 M denaturant and 50 mM cacodylate buffer, pH 6.0. This solution was maintained at 10 degrees and its absorbance continuously monitored at 650 nm. A limiting absorbance of 0.5166 was observed after 5000 seconds.

reaction having a half-time of 1080 seconds. This half-time is within a factor of the value observed for the dominant slow-folding form of unconjugated ribonuclease observed using either tyrosine absorbance or fluorescence measurements [7]. The same change in absorbance has a half-time of 375 when refolding is observed in 1.5 M denaturant (not illustrated), a value comparable to the half-time of 280 seconds observed for the unconjugated protein.

DISCUSSION

In order to validate our experimentation with ribonuclease, we first reacted Procion Blue MX-R with rabbit muscles lactate dehydrogenase as described by Clonis an Lowe [2]. We were able to effect the irreversible inactivation of the catalytic activity of this enzyme with excess dye, the protection against inactivation afforded by NADH and the purification of the enzyme-dye conjugate described by these investigators. We find that the catalytic pocket for ribonuclease also can be selectively occupied by single covalently bound dye which inactivates the enzyme but does not perturb the structure of the native protein. While the presence of a competitive inhibitor for ribonuclease does not prevent inactivation during conjugation owing to the pH limitations of conjugation and complexation, addition of the competitive inhibitor to the protein-dye conjugate at a pH optimal for inhibitor binding generates a spectral change appropriate to displacement of the conjugated dye from the inhibitor binding site.

The denaturation difference spectra of the protein-dye conjugate are complex owing to the effects of both the denaturant and the protein structure on the absorbance of the conjugated dye. Since the denaturant contribution should be completed upon mixing, any changes observed following completion of the mixing should reflect changes in the protein structure. The correspondence of the refolding of the conjugated and unconjugated protein using conditions in which the kinetics of refolding are essentially monophasic validates this expectation.

These results indicate that conjugation of reactive dyes such as Procion Blue MX-R can provide spectral probes which report the status of the local structure about the biofunctional sites of proteins. In contrast to dissociable competitive inhibitors, conjugated dyes can report changes in the local structural environment about biofunctional sites in the presence of rigorous conditions such as in high concentrations of denaturant or at elevated temperatures. The visible spectral changes distinguish the conjugated from the ultraviolet spectral changes expressed by the intrinsic protein chromophores such as the peptide bond and the aromatic side chains. The seeming ubiquity of the affinity of biofunctional sites for reactive textile dyes suggests that these dyes can be used to selectively label a wide range of proteins to observe both the equilibrium and kinetic parameters of structural changes. Such sites include both catalytic and allosteric sites on enzymes and binding sites on recognition and transport proteins. A range of dichlorotriazine dyes are available to maximize the change in absorbance accompanying the structural change of a protein-dye conjugate. Rapid kinetic changes can be observed using stopped-flow mixing.

ACKNOWLEDGEMENT

This investigation was supported by U.S. Public Health Service research grant GM22109 from the Institute of General Medical Sciences.

REFERENCES

1. Thompson, S.T. and Stellwagen, E., Binding of cibachron blue F3GA to proteins containing the dinucleotide fold. _Proc. Natl. Acad. Sci. U.S.A._, 1976, 73, 361-365.

2. Clonis, Y.D. and Lowe, C.R., Triazine dyes, a new class of affinity labels for nucleotide-dependent enzymes. _Biochem. J._, 1980, 191, 247-251.

3. Crook, E.M., Mathias, A.P. and Rabin, B.R., Spectrophotometric assay of bovine pancreatic ribonuclease by the use of cytidine 2':3'-phosphate. _Biochem. J._, 1960, 74, 234-238.

4. Rajgopal, S. and Vijalakshmi, M.A., Studies on the interaction of firefly luciferase with triazine dyes. _J. Chromatogr._ 1986, 355, 201-210.

5. Hummel, J.P., Van Ploeg, D.A. and Nelson, C.A., The interaction between ribonuclease and mononucleotides as measured spectrophotometricaly. _J. Biol. Chem._, 1961, 236, 3168-3172.

6. Richards, F.M. and Wyckoff, H.W., Bovine pancreatic ribonuclease. In The Enzymes, ed., P.D. Boyer, Academic Press, New York, 1971, pp. 647-806.

7. Schmid, F.X., Mechanism of folding of ribonuclease A. Slow refoldling is a sequential reaction via structural intermediates. _Biochemistry_, 1983, 22, 4690-4696.

IONIC AND APOLAR INTERACTIONS OF CIBACRON BLUE F3GA WITH MODEL
COMPOUNDS AND PROTEINS   APPLICATION TO PROTEIN PURIFICATION

S. SUBRAMANIAN
Miles Inc., Elkhart, Indiana, USA

ABSTRACT

Studies of the interactions of Cibacron Blue F3GA in aqueous solutions with
organic solvents, salts, oligo-, and poly-peptides by visible absorption
difference spectroscopy indicate that the dye-sodium chloride (or any other
salt) interaction results in a difference spectrum that has a characteristic
positive peak at 690 nm and negative double minima at 630 and 585 nm. Such
a "salt-like" or "ionic" spectrum is also obtained with polycations such as
spermine, spermidine, oligolysines, polylysine, polyarginine, and protamine.
In a striking contrast, the difference spectrum of the dye in binary aqueous solvents containing dioxan or t-butyl alcohol at moderately high concentrations, measured against water, displays a positive peak and shoulder at
655 and 610 nm, respectively with a small negative contribution at wavelen-
gths less than 550 nm. We assign this type of spectrum to a "nonpolar" interaction of the dye with the organic cosolvent molecules. Such spectral
characteristics have been widely noticed in several studies involving Cibacron Blue and proteins, thereby providing a basis for a reasonable interpretation for the interaction of the dye with the proteins. An understanding
of the forces involved in dye-protein interactions is essential for protein
purification studies using the technique of dye-ligand chromatography. Complete separation of pepsin and chymosin present in calf rennet has been a-
chieved using Blue agarose chromatography. The separation appears to result
from a combination of hydrophobic and electrostatic interactions of chymosin with the dye ligand. Differential surface hydrophobicity of the two enzymes may also play a key role in the separation.

INTRODUCTION

The serendipitous finding that a number of enzymes co-eluted with Blue Dextran when subjected to gel filtration on a Sephadex column spawned a whole
new technique of protein purification wherein the dye, Cibacron Blue F3GA,
was used as a "universal pseudo-affinity ligand" attached to different insoluble matrices (1, 2). Cibacron Blue is a sulfonated polyaromatic triazine dye which has polar, ionic, and hydrophobic segments and as such is
capable of interacting with polar, ionic, and apolar groups in other mole-

cules whether they are small molecules or pendants of macromolecules.

Cibacron Blue affinity matrices were found useful in the purification of several dehydrogenases, kinases, and other nucleotide-binding enzymes. In fact, Cibacron Blue with its aromatic rings and sulfonate groups is a structural mimic of NAD. This led to the hypothesis that the binding of Cibacron Blue ligand was specific to what was called the "NAD binding domain" (3). However, several other proteins which do not bind nucleotides do bind the dye. It is thus apparent that Cibacron Blue combines its structural and stereochemical features in such a way as to be able to offer a pseudospecificity to a wide variety of proteins. It now appears that the blue dye can bind with an apparent specificity to any protein that possesses a cluster of aromatic/apolar groups and/or geometrically spaced positively charged groups for proper interaction with the aromatic rings and/or sulfonate groups of the dye molecule.

It is essential to understand the physico-chemical basis of the interaction of the dye with its environment in order to make effective use of the dye ligand in protein purification studies. Spectroscopy has been the widely used tool for this purpose. Although circular dichroism (induced) and nuclear magnetic resonance spectroscopy have been used in isolated cases, visible spectroscopy has been the most thoroughly studied with respect to the interaction of the dye with its solvent environments and proteins.

In this paper we describe the spectral features of the dye in the diverse solvent environments of salts, organic solvent components, polypeptides, and polyamines. In addition, selected dye-protein interactions will be described both in solution and protein binding to immobilized dye. The differential interaction of the two acid proteases, chymosin and pepsin, with the dye ligand is manifested in complete separation of the two proteins present in calf rennet.

MATERIALS AND METHODS

Cibacron Blue F3GA was obtained from Sigma. Its concentration was determined using a molar extinction coefficient of 13,600 at 610 nm(4). Cibacron Blue agarose was purchased from Pierce Chemical Co., Rockford, IL. Dioxan was gold labeled product from Aldrich. J.T. Baker and Company supplied t-butyl alcohol (analytical grade) while urea (ultra pure) was from Schwarz/ Mann. All other salts were analytical grade from several suppliers. Spermine, spermidine, putrescine and several other diamines were obtained from Sigma. All oligolysines were from Vega Biochemicals. Polylysine, polyarginine, and protamine chloride were all Sigma products.

Chicken liver dihydrofolate reductase was prepared and assayed as described by Kaufman and Kemerer (5). Chymosin and porcine pepsin were products of Sigma. Calf rennet extract was supplied by Marschall Products Division of Miles Inc.

Difference spectra and difference spectral titrations were performed at 25 $\pm$ 0.1 C using a Cary 219 spectrophotometer. Dye-ligand chromatography of calf rennet was performed using either 1.5 cm X 10 cm or 6 cm X 14 cm column at a flow rate of 30 ml/hr. The crude rennet and the separated rennet fractions of pepsin and chymosin were analysed for the relative chymosin and pepsin contents by the procedure recommended by the International Dairy Federation (6). The milk-clotting activities of rennet, chymosin, and pepsin were determined using standard procedure. Proteolytic activities were determined at 37 C using denatured hemoglobin as substrate. The effectiveness of separation of chymosin and pepsin was evaluated (7) using an arbitrary clotting to proteolysis ratio (C/P).

RESULTS

The visible absorption spectrum of the dye in water has a broad maximum in the region 610-630 nm. The spectra in binary aqueous solutions show varying magnitudes of hyper- and hypo-chromicity with or without shifts in the wavelength of maximum absorption. It is instructive to examine the difference spectra of the dye in such binary aqueous solutions (measured against water) The difference spectra of the dye in 2.8M NaCl, 7M urea, 50%(V/V) aqueous dioxan, and 91%(V/V)t-butyl alcohol in water are shown in Figure 1.

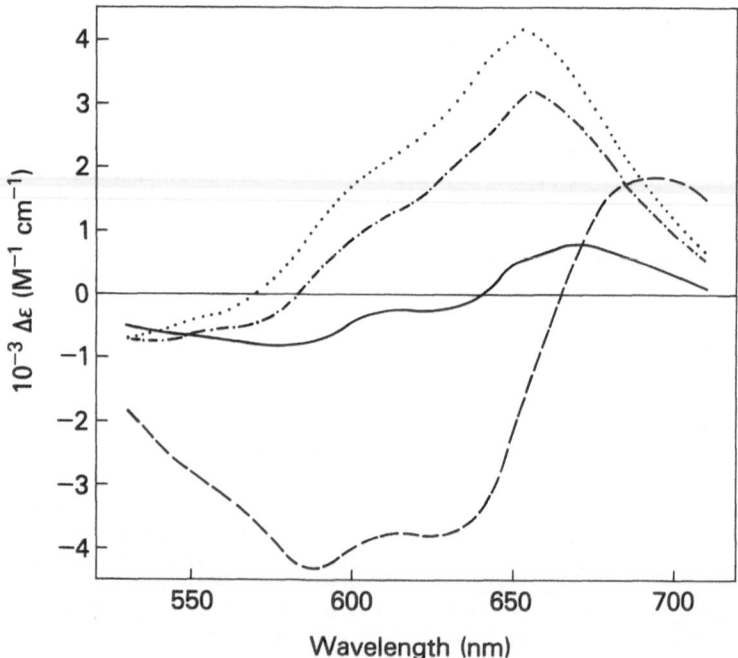

Figure 1.   Difference spectra of Cibacron Blue in aqueous binary solvent mixtures versus water. ---2.8M NaCl vs water; dye concentration, 30uM. _____ 7M urea vs water, dye concentration, 8.2uM. _ • _ • _ • 50% (V/V) aqueous dioxan, dye concentration, 15uM. ..... , 91% t-butyl alcohol-9% water (V/V) versus water; dye concentration, 10.9 uM. Note that the spectra have been normalized with respect to the dye concentrations.

The difference spectrum in NaCl solution versus water has a positive peak at 690 nm and double minima at 630 and 585 nm. For several salts with chloride as the common anion the A(690-585 nm) increases in the order Li<Na<Ammonium<K=Rb=Cs<Mg at an ionic strength of 0.1. The difference spectrum in 7M urea, however, is rather inconspicuous displaying small perturbations. The spectra in aqueous dioxan and aqueous t-butyl alcohol are very similar to each other, with a positive peak at 655 nm and shoulder at 610 nm.

Since a buffer solution is commonly employed in biochemical studies, the spectra of the dye in 1.5M NaCl solution and 50% aqueous dioxan were measured against the same concentration of the dye in 0.1M potassium phosphate at pH 7.4. These are shown in Figure 2. Once again the spectra are similar to those in Figure 1 except for the fact that buffer salt present in the reference cell diminishes the spectral magnitude in the case of sodium chloride and enhances it for aqueous dioxan.

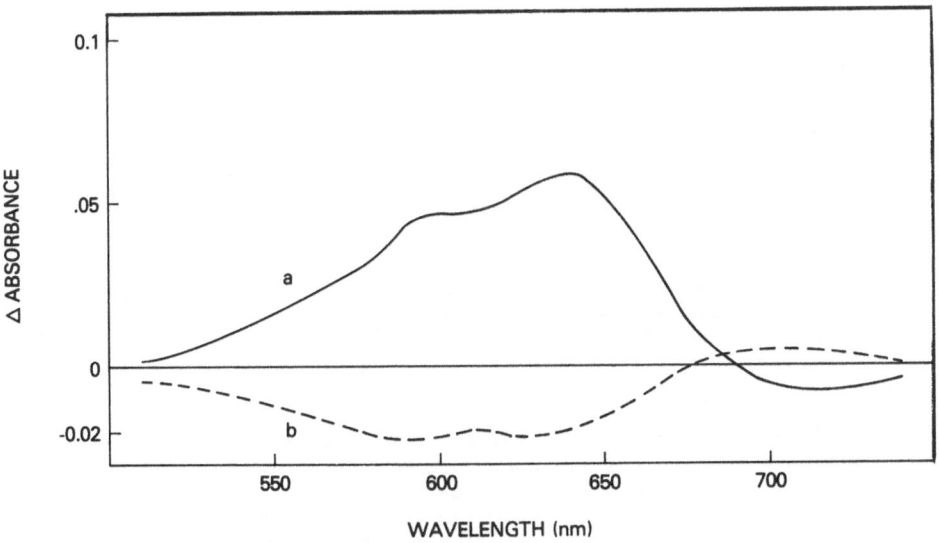

Figure 2. Difference spectra of Cibacron blue in binary solvent and salt solutions. Curve a, difference spectrum of 15uM dye in 50% dioxan-50% water (V/V) versus in 0.1M phosphate buffer; curve b, difference spectrum of 15uM dye in 1.5M NaCl versus in 0.1 M buffer.

The difference spectrum obtained upon mixing Cibacron Blue with spermine in water at a pH of 6.5 in the sample cell versus the dye only at the same concentration and same pH in the reference cell has a spectrum similar to that obtained by the interaction of NaCl with the dye. As shown in Figure 3, spermine-dye interaction produces a positive peak at 685 nm, a negative shoulder at 630 nm, and a negative trough at 585 nm. Such a spectrum is also obtained for the interaction of the dye with oligolysines, polylysine, polyarginine, and protamine.

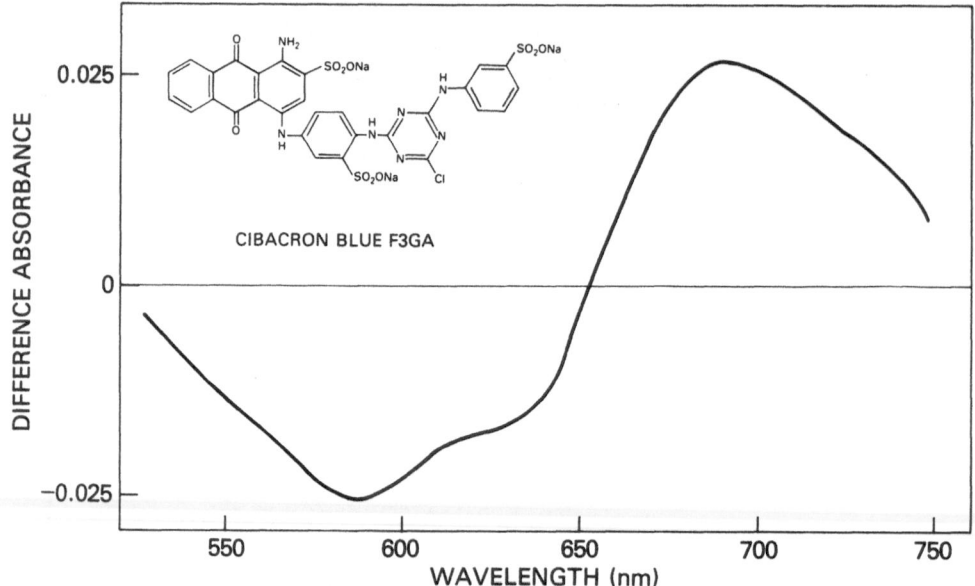

Figure 3.   Difference spectrum obtained for complex formation between Cibacron Blue F3GA and spermine. The final concentration of the dye was 8.14uM and that of spermine was 6.5 uM. The pH of both sample and reference cuvette solutions was kept at 6.5. Temperature, 25 C. Inset structure of the dye.

Spermidine, putrescine, and several other diamines produce similar spectra although the spectral intensity is much less at comparable concentrations. Figures 4-6 show molecular models of the dye, spermine, and the dye-spermine complex. In the dye-spermine complex, with appropriate positioning, the positively charged amino (or imino) groups of spermine can neutralize the negative charges on the sulfonate groups of the dye molecule. The difference spectrum obtained by mixing 1 ml of 30uM of the dye in buffer with 1 ml of 17.7uM of chicken liver dihydrofolate reductase in the same buffer in the partitioned sample cuvette versus mixing 1 ml of 30uM dye in buffer with 1 ml of buffer in the reference cuvette is shown in Figure 7A, curve 1. It is largely characterized by a major positive peak at 700 nm, a negative trough centered at 640 nm, and a positive peak at 610 nm. When NADPH Is added to a final concentration of 50 uM to both cells (about 6-fold ex'ess with respect to the enzyme) there are only minor perturbations of the spectrum (curve 2) with the general features of curve 1 maintained intact. Further additions of NADPH up to 175 and 250 uM changed neither the intensity nor the shape of curve 2. When methotrexate is now added (40uM) to both cells the spectral features are totally eliminated (curve 3) indicating that methotrexate but not NADPH is capable of displacing the dye f from its complex with the enzyme. Folate and dihydrofolate also can displace the dye from the enzyme almost as efficiently as methotrexate.

Figure 4.   Space filling molecular model of Cibacron Blue F3GA

The addition of Cibacron Blue (15uM) to Lactobacillus casei enzyme (9.35uM) is marked by a difference spectrum characterized by a negative trough at 690 nm with positive peaks at 645 nm, 595 nm, and 550 nm (Figure 7B, curve 1). This is almost a mirror image of the spectrum obtained with the chicken liver enzyme. Addition of 25 uM NADPH produces curve 2. Further additions of NADPH (up to 250 uM) do not cause any further changes in curve 2. Addition of even stoichiometric concentration of methotrexate eliminates the vestigial features of curve 2 producing a flat curve 3. This indicates that even in the presence of excess NADPH blue dye can still be bound to the L. casei enzyme (although weakly) and can only be eliminated by methotrexate or folate dihydrofolate. When only methotrexate is added to the blue dye-enzyme complex, the dye is completely displaced from the chicken liver enzyme but there is evidence of trace binding of the dye to the L. casei enzyme (figures not shown). Addition of NADPH to the solutions after methotrexate has been added does not cause any further spectral change with either enzyme.

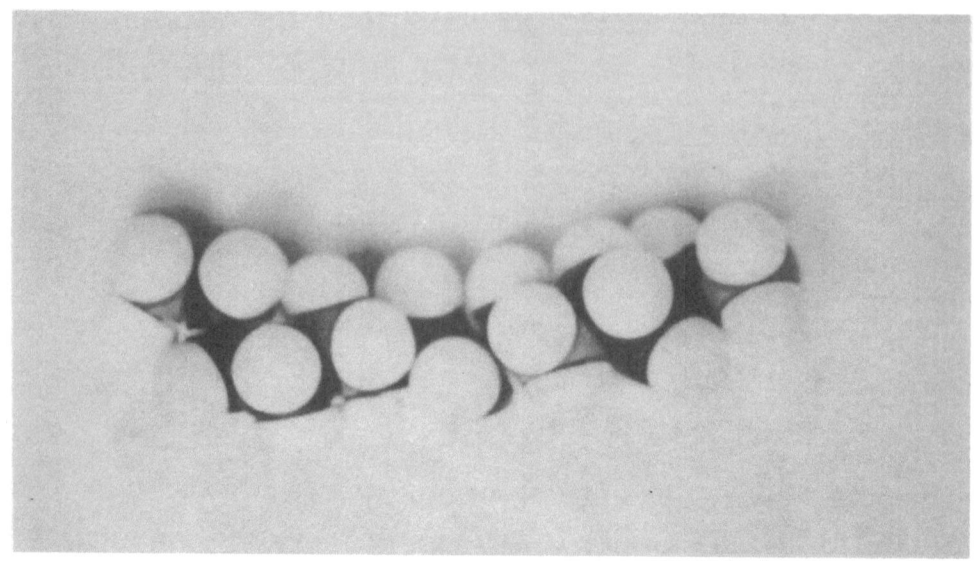

Figure 5.   Space filling molecular model of spermine

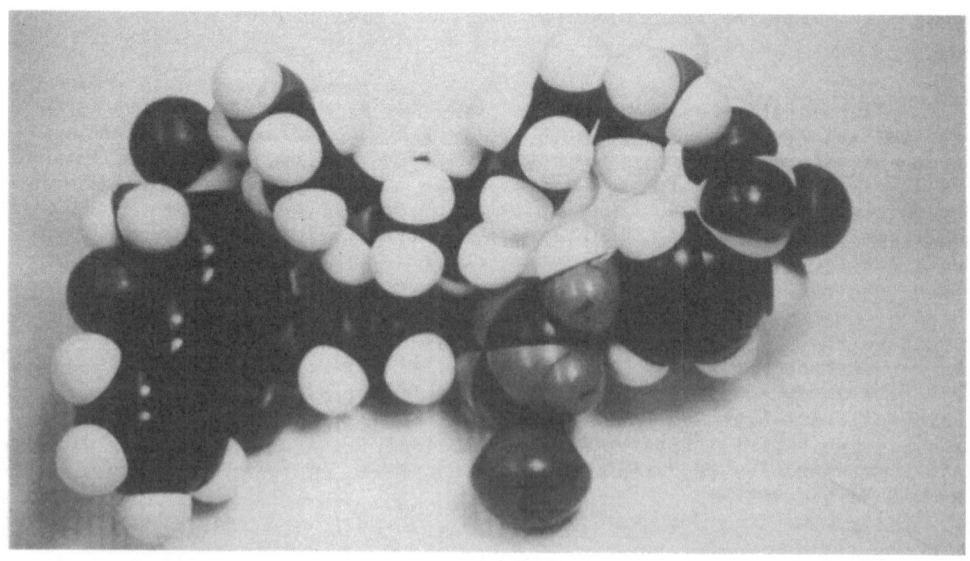

Figure 6.   Model of spermine-dye complex

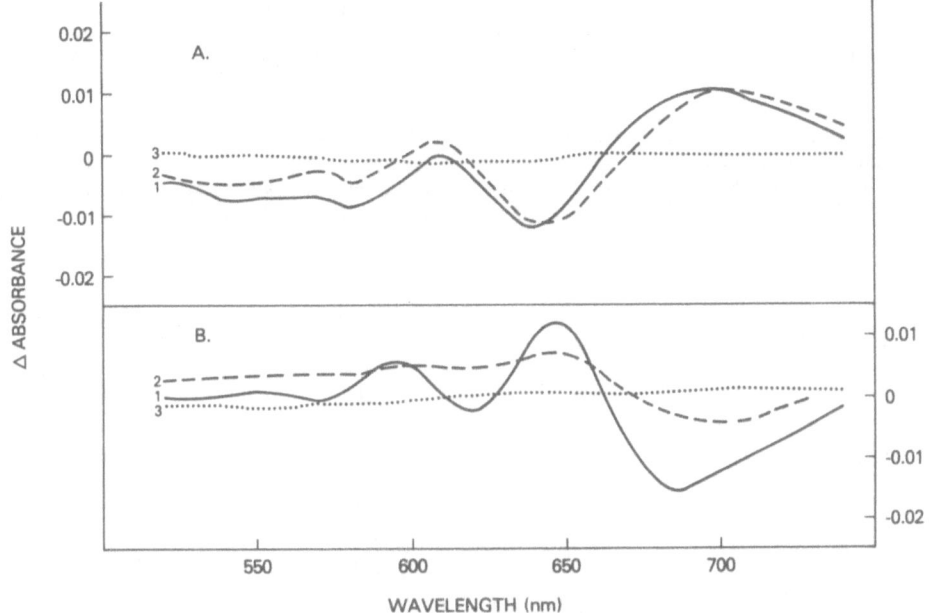

Figure 7. Difference spectra of Cibacron blue bound to chicken liver (A)
and L. casei (B) dihydrofolate reductase. A. Curve 1 sample cell,
dye plus chicken liver enzyme. reference cell, dye plus buffer.
Final dye concentration 15uM; final enzyme concentration 8.85 uM.
Curve 2, spectrum after addition of 50uM NADPH (final) to both
cells. Curve 3, 40uM (final) methotrexate added to both cells.
B. Curve 1, difference spectrum obtained upon mixing dye and
L. casei enzyme. Final dye concentration 15 uM; final enzyme concentration 9.35uM. Curve 3, after addition of methotrexate 22uM
final concentration.

The resemblance of curve a in Figure 2 to curve 1 of Figure 7B is
striking; i.e., the difference spectrum of the dye bound to the L. casei
enzyme is similar in shape to the difference spectrum of the dye in dioxan-
water mixture. In a parallel fashion the spectrum of the dye bound to the
chicken liver enzyme is very similar in shape to that in 1.5M NaCl. In other
words, the difference spectra of the dye bound to the two enzyme species are
mirror images of each other as are the spectra of the dye in dioxan-water
and salt solutions.

When calf rennet was dialyzed in 0.025M citrate at pH 5.5 and applied
to the blue agarose column equilibrated with the same buffer, the passthrough fraction (with absorbance at 280 nm) had some milk-clotting activity
(peak A) as shown in Figure 8. After washing with buffer to basal level of
280 nm absorbance and milk clotting activity, application of 10% (1.7M)
NaCl in the same buffer resulted in elution of milk clotting activity (peak
B). The fractions under peaks A and B were pooled separately. In terms of
clotting activity peak A accounted for 23% and peak B 67% of the total activity applied (accounting for 90% recovery).

It was also found by analysis on a DEAE-cellulose column that peak A consis
ted of 96% pepsin and peak B was 96.3% chymosin.This indicates that pepsin
passes through unadsorbed while chymosin is bound on the blue gel column
and eluted by NaCl. The clotting to proteolytic ratios (C/P) of the rennet,
peak A and peak B were found to be 5.9, 1.5, and 10.8 respectively, thereby
adding further proof for the separation of chymosin and pepsin. The C P ratios for pure pepsin, and chymosin were found to be 1.12 and 9.78 respectively. In a variation of the elution pattern, 50% (V/V) ethylene glycol in
0.025M citrate buffer at pH 6.2 also elutes the bound chymosin.

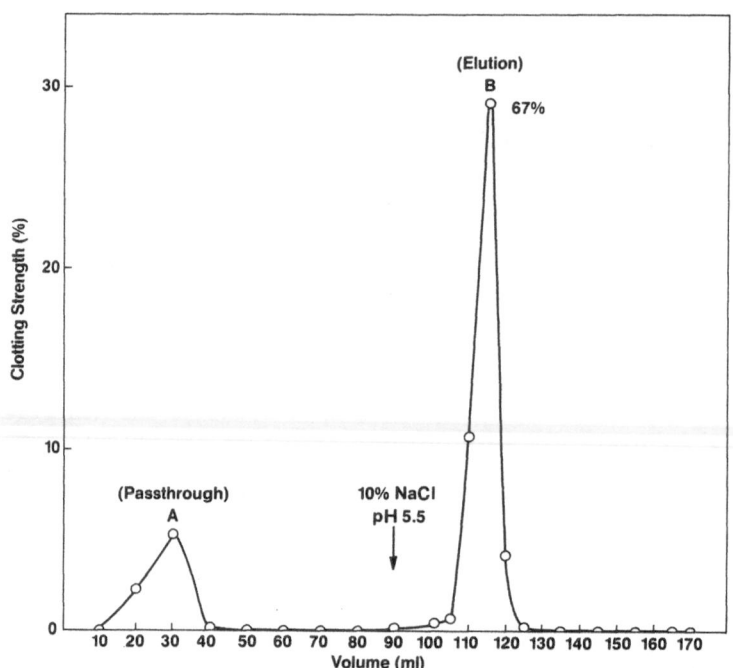

Figure 8.   Calf rennet dialysed in 0.025 M citrate buffer at pH 5.5 was
            applied to Cibacron Blue agarose column (1.5 cm ✕10 cm). Elution
            was effected by 0.025 M citrate buffer pH 5.5 containing 1.7 M
            NaCl. Peak A had 23% and peak B 67% of the clotting activity ap-
            plied.

A salt gradient was applied to elute the bound chymosin as shown in
Figure 9 to determine the strength of the interaction of chymosin with the
dye ligand. The concentration of NaCl was 3.17% (0.54M) at the peak of elution and elution was complete at 6.8% §1.16M). As before the chymosin peak
had a C P ratio of 11.2 while the applied rennet had a ratio of 3. Recovery
of activity was 92%.

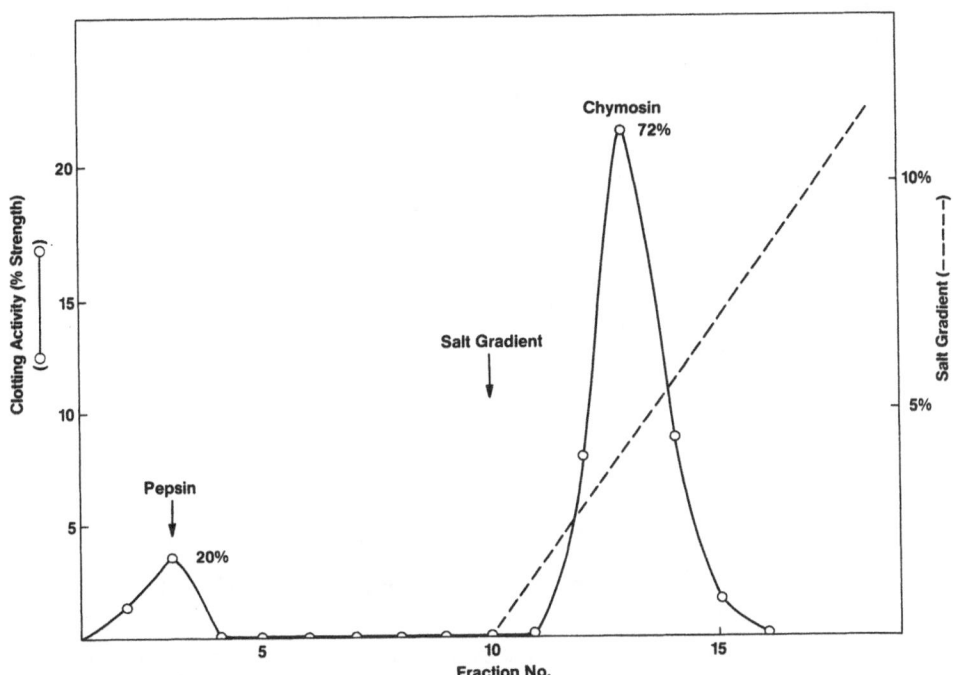

Figure 9.   Same conditions of application as in Figure 8. Elution was effected by 0-18% NaCl gradient. Pepsin (20% of total clotting
activity) passed through unadsorbed. Peak chymosin activity was
eluted at 3.17% (0.54 M) sodium chloride. Clotting activity from
chymosin peak was 72% of the total applied.

The chromatographic behavior of rennet mentioned above is essentially
unchanged if the incubating buffer and the rennet contained 0.75M sodium
sulfate. The feasibility of a large scale separation of rennet was tested
noting that sodium sulfate helped sharpen the separation of chymosin. Calf
stomachs were extracted with 11% (0.77M) sodium sulfate, filtered to remove
debris, and 250 ml of the rennet at 19.4% activity were loaded onto a large
blue gel column equilibrated with 0.025M citrate buffer at pH 5.5 containing 0.77M sodium sulfate. After washing, the bound chymosin was eluted

with 0.85M NaCl. The results are shown in Figure 10. The passthrough fraction (peak A in Figure 10) and the elution fraction (peak B) gave C/P values of 0.99 and 16.8 respectively, indicating the homogeneity of the separated proteases.

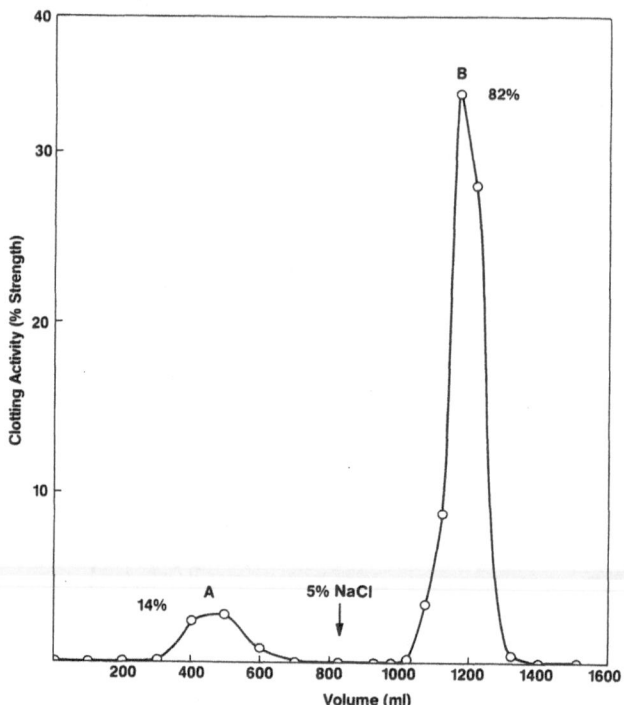

Figure 10. A large scale isolation of chymosin and pepsin. Calf rennet was extracted from calf stomachs using 11% sodium sulfate and applied to the Blue agarose column after clarification. Column was equilibrated with the same medium. After washing, bound chymosin was eluted with 0.025 M citrate containing 5% NaCl at pH 5.5. Recovery of activity was 93%.

DISCUSSION

The interaction of several dyes with proteins have been studied extensively in the past but no singular dye has attracted so much attention as Cibacron Blue, in terms of the applications it has in separating and purifying various classes of enzymes and other proteins. Among the proteins and enzymes purified using Blue agarose are several dehydrogenases and kinases, serum proteins, muscle proteins, endonucleases, interferons, beta lactamases, proteases and a whole class of totally unrelated (in terms of structure or function) proteins.

The construction of a space-filling model of the dye indicates that the dye is a fairly rigid molecule with very little conformational freedom and that the rings are not coplanar (see Figure 4). The visible spectral transitions of the dye are localized on the anthraquinone ring and to a smaller extent the adjacent phenyl ring (8). Since there is very little delocalization of electrons between the rings, only those interactions with the anthraquinone ring, the contiguous phenyl ring and the two attached sulfonate groups would be expected to show changes in absorption spectra. Accordingly cations will interact with the sulfonate groups inducing absorbance changes in the chromophore and nonpolar moieties will do so by interacting with the aromatic rings.

Electrostatic Interactions

As represented in Figure 1 and Figure 2B the difference spectrum in 2.8M NaCl with a positive peak in the 690 nm region and negative double minima in the region 630-585 nm appears to be characteristic of the sodium ion-sulfonate interaction. This "ionic" spectrum of the dye in electrolyte solutions probably arises from a specific interaction of the cations of the salts with the anionic sulfonate groups of the dye. In addition, the intensity of the "ionic" spectrum increases in the order Li<Na<ammonium<K=Rb=Cs for the alkali cation series. This order may represent the relative proclivities of these cations to induce aggregation of the dye and consequent charge screening of the negative charges in the dye molecule.

The effect of the anions needs an explanation in this context. While the positive maximum at 690 nm does not change, the depth of the negative minimum at 585 nm diminishes in the order chloride>bromide>iodide>perchlorate for the sodium salts. Anions are known to disrupt the water structure, in a reverse order of the series shown above. The electrostatic interaction between the cations and the sulfonate groups of the dye is expected to cause a stacking of the dye molecules by virtue of the shielding of the charges and consequent desolvation of the ionic groups. Such a stacking causes a hypochromic effect, partially contributing to the minimum in the difference spectrum. If the anions of the added electrolytes disrupt the water structure, then the tendency to stack will be minimized and correspondingly the hypochromic effect will be quenched. This is the effect seen with the anions, with the perchlorate ion being the most chaotropic anion, causing the least hypochromic effect.

It is interesting to note that ionic interactions predominate in forming the spermine-dye complex. The magnitude of the spectral signal produced by interaction of 6.5uM spermine with 8.2uM dye is comparable to the effect produced by 3M NaCl on 8.2uM dye (9). Spermine is a symetrical molecule with the structure,

$$H_3N^+-(CH_2)_3-N^+H_2-(CH_2)_4-N^+H_2-(CH_2)_3-N^+H_3$$

(see Figure 5) which can stereoselectively interact with the sulfonate groups of the dye through the positively charged nitrogens. A space-filling model suggests (see Figure 6) that the two aminium groups in the middle and a terminal ammonium group in spermine could effectively interact with the three sulfonate groups of the dye. This would produce effective charge shielding and could cause the complex to precipitate, as observed. The titration of the spectral signal for the spermine-dye complex with a pK_a = 8.2 (9) indicates that either an aminium or ammonium group interacts with the sulfonate group on the anthraquinone ring so as to produce the spectral signal.

Spermidine lacks the terminal trimethylene ammonium moiety of spermine and consequently binds rather weakly (9) to the dye presumably due to the loss of one ionic interaction. Putrescine (1,4-diaminobutane), the central segment of spermine binds more weakly than spermidine; nevertheless it produces a spectral signal several fold larger than a comparable NaCl concentration by virtue of bidentate interactions of the two positively charged nitrogens with the sulfonate groups of the dye. It is clear that the interaction is ionic in all these cases, since neutralization of the positive charges on the nitrogens eliminates the spectral signal completely. Addition of salt (at high concentrations) dissociates the spermine-dye complex confirming the electrostatic nature of the complex.

The interaction of arginine, lysine, oligolysines, polylysine, polyarginine, and protamine (which has an arginine content of 66%) with the blue dye produces a characteristic ionic spectrum (similar to that in sodium chloride or spermine) with a maximal spectral signal at very low concentrations of the polymer corresponding to stoichiometric ratios of the positively charged side chains to the sulfonate groups. The amplitude of the signal A(685-585 nm) either increases with increasing chain length of the oligolysines or the same effect is produced at a lower concentration of the higher oligomer (10). The magnitude of the spectral signal is indicative of cooperative effects and the binding strengths of the ionic interactions. It appears fromctra that these polycations interact exclusively with the sulfonate groups of the dye and not much else. It is thus seen that cations or cationic groups can polarize the sulfonate groups of the dye to cause a red shift in the absorption of the blue dye chromophore.

Nonpolar (hydrophobic) Interactions

The difference spectra of the dye in 91% t-butyl alcohol-9% water (V/V) and 50% (V/V) aqueous dioxan (Figure 1) with a positive peak at 655 nm, a shoulder at 610 nm, and a small negative contribution below 550 nm are very similar. Similar spectra are also obtained at several other dioxan-water and t-butyl alcohol-water compositions as well as in neat tetramethyl urea, a partially nonpolar solvent. This is suggestive of a nonspecific hydrophobic interaction of the organic solvent component with the aromatic moieties of the dye chromophore, inducing a red shift in the absorption spectrum which manifests as a positive peak at 655 nm in the difference spectrum. Since nonpolar solvents can also disrupt aggregation of the dye molecules, a hyperchromic effect will result from this and this is seen as a shoulder at 610 nm in the difference spectra of dioxan-water, and t-butyl alcohol-water mixtures.

Thus, the difference spectrum of the dye in binary aqueous solvents characterized by a positive peak at 655 nm and a shoulder at 610 nm is assigned a tentative nomenclature of "nonpolar" or "apolar" spectrum . Such "nonpolar" spectra have also been oberved in ethylene glycol-water system (4, 11).

Dye-Protein Interactions

It may be useful, at this point, to examine the binding of the dye to many proteins studied by several workers (4, 12-17) in the light of the spectral characteristics described here for electrostatic and nonpolar (hydrophobic) interactions of the dye. An X-ray study of the binding of Cibacron Blue to horse liver alcohol dehydrse liver alcohol dehydrogenase (12) showed that the anthraquinone ring and the attached phenyl ring bound to the enzyme at the ADP locus of the coenzyme binding domain but the remaining segments of the dye were bound in a mode unlike that of the nicotinamide-ribose segment of NAD. However, irrespective of whether the proteins to which the dye binds contain a nucleotide binding domain (3) or not and also whether the dye binds at or away from the cucleotide binding domain in those proteins which contain the NAD binding domain, the spectra obtained upon blue dye binding to the proteins can be analysed in terms of electrostatic and nonpolar interactions of the dye with its surroundings. The spectra of the bound dye in several cases resemble the difference spectrum of the dye either in high salt or nonpolar-aqueous solvent mixtures to permit the classification of the interaction of the dye with the protein as either electrostatic or hydrophobic. For example, in the case of chicken liver dihydrofolate reductase, the dye finds a highly aqueous and "ionic" binding environment by encountering a cluster of positively charged amino acid side chains while the interaction of the dye with the Lactobacillus casei dihydrofolate reductase produces an "apolar" difference spectrum indicative of an encounter with nonpolar side chains (Figure 7 and reference 14).

However, caution must be exercised in interpreting the data. In some cases, the dye can interact with the protein both electrostatically and hydrophobically. Correspondingly the spectrum will be a composite of both types. It is obvious when the interaction is classified as electrostatic, the protein has arginine or lysine side chains at the dye binding site interacting with the sulfonate groups of the dye. Likewise when the interaction is classified as hydrophobic, the dye binding site must be replete with the side chains of hydrophobic amino acids. A further aid in identifying the interactions would be the medium used to elute the bound proteins from Blue agarose affinity columns. If salt can elute a protein but not ethylene glycol, it is indicative of an electrostatic interaction and a hydrophobic interaction can be inferred if ethylene glycol or glycerol elutes the protein when salt cannot (18).

Protein Purification

A complete separation of chymosin and pepsin in crude calf rennet was effected by Blue agarose chromatography. Both proteases are highly negatively charged at pH 5.5 but yet chymosin is bound by the gel while pepsin is not. The bound chymosin is eluted either by salt or aqueous ethylene glycol (7). It appears, then, that both electrostatic and hydrophobic interactions occur in the adsorption of chymosin by the blue gel. The two kinds of interactions may be cooperative, i.e., they reinforce each other. If so, the disruption of one kind should lead to the weakening of the other too. Sodium chloride disrupts electrostatic interactions between the dye chromophore and chymosin. This may lead to a weakening of the hydrophobic interactions thereby causing the dissociation of chymosin from the immobilized ligand. Ethylene glycol, on the other hand, disrupts hydrophobic interactions between the enzyme and the dye ligand with concomitant weakening of the electrostatic interactions and elution of the enzyme. Obviously such interactions are not possible between the blue chromophore and pepsin.

It was also noted (7) that in the presence of 0.7M sodium sulfate the binding of chymosin was enhanced. This behavior was attributed (7) to the "kosmotropic" (order-enhancing) nature of sodium sulfate. Chymosin binds to the blue gel at low ionic strength and also in the presence of 0.7M sodium sulfate but does not bind in the presence of even as low as 0.1M sodium chloride. Taken together with the facts that: a) pepsin is not retained by the blue gel, b) the bound chymosin is eluted either by sodium chloride or aqueous ethylene glycol, c) the blue chromophoric ligand can function only as a cation exchanger, and d) both chymosin and pepsin are highly negatively charged at the operating pH of 5.5 we can conclude that the preferential binding of chymosin by the blue gel should be due to a combination of electrostatic and hydrophobic interactions. The primary forces appear to be electrostatic interactions between the positively charged groups in chymosin and the sulfonate groups in the dye. In addition, hydrophobic interactions contribute to a further enhancement of binding.

It is likely that the surface hydrophobicity of chymosin is greater than that of pepsin to cause this differential effect. Chymosin with a higher isoelectric point (pI=4.6) than pepsin (pI=1.2) naturally has a greater number of positively charged side chains (chymosin 17; pepsin 4) at pH 5.5 (19). Proper geometric placement of these positively charged side chains in chymosin could enable chymosin to engage in electrostatic interactions with the sulfonate groups of the dye ligand augmented by favorable hydrophobic interactions. The chymosin separated from calf rennet by Blue gel chromatography displayed an intense difference spectrum in the 550-700 nm region upon mixing with the free dye Cibacron Blue while the separated pepsin or porcine pepsin produced only a feeble effect (spectra not shown). This provides further proof of the differential affinities of chymosin and pepsin to the dye ligand.

CONCLUSION

A knowledge of the basic interactions between Cibacron Blue and other molecules, ionic and nonpolar, has been shown to be helpful in characterizing the interactions between the dye and the proteins. However, caution must prevail in any serious interpretation. In some cases, the dye can interact with the protein in a composite mode. Additionally, the medium used to release the bound protein from Blue agarose, such as salt, change of pH, chaotropic salts, or ethylene glycol should be of help in determining the type of interactions. By an appropriate choice of the loading and eluting conditions, the protein of particular interest could be purified using Blue agarose, in many instances in one step. The dye draws its diversity and versatility by a judicious mix of hydrophobic and ionic groups within the structure of the molecule.

REFERENCES

1. Ryan, L.D., and Vestling, C.S., Rapid purification of lactate dehydrogenase from rat liver and hepatoma a new approach. Arch. Biochem. Biophys. 1974, 160, 279-284

2. Bohme, H.J., Kopperschlager, G., Schulz., and Hoffman E., Affinity chromatography of phosphofructokinase using Cibacron Blue F3GA. J. Chromatogr., 1972, 69, 209-214

3. Stellwagen E., Use of Blue Dextran as a probe for the nicotinamide adenine dinucleotide domain in proteins. Acc. Chem. Res., 1977, 10, 92-98

4. Thompson S.T., and Stellwagen E., Binding of Cibacron Blue F3GA to proteins containing the dinucleotide fold. Proc. Natl. Acad. Sci. U.S.A., 1976, 73, 361-365

5. Kaufman B.T., and Kemerer V.F., Characterization of chicken liver dihydrofolate reductase after purification by affinity chromatography and isoelectric focusing. Arch. Biochem. Biophys., 1977, 179, 420-431

6. International Dairy Federation Standard, 110 1982

7. Subramanian S., Separation of chymosin and pepsin in calf rennet by dye-ligand affinity chromatography. Prep. Biochem. 1987, 17, 297-312

8. Edwards R.A. and Woody R.W., Induced circular dichroism as a probe of Cibacron Blue and Congo Red bound to dehydrogenases. Biochem. Biophys. Res. Commun., 1977, 79, 470-476

9. Subramanian S., Specific interaction of spermine with Cibacron Blue F3GA. Arch. Biochem. Biophys., 1982, 217, 388-391

10. Subramanian S., Spectral changes induced in Cibacron Blue F3GA by salts, organic solvents, and polypeptides: Implications for blue dye interaction with proteins. Arch. Biochem. Biophys., 1982, 216, 116-125

11. Pompon D., and Lederer F., Binding of Cibacron Blue F3GA to Flavocytochrome b2 from Baker's yeast. Eur. J. Biochem. 1978, 90, 563-569

12. Biellmann J-F., Samama J-P., Branden C-I., and Eklund H., X-ray studies of the binding of Cibacron Blue F3GA to liver alcohol dehydrogenase. Eur. J. Biochem., 1979, 102, 107-110

13. Barden R.E., Darke P.L., Deems R.A., and Dennis E.A., Interaction of phospholipase A2 from cobra venom with Cibacron Blue F3GA. Biochemistry, 1980, 19, 1621-1625

14. Subramanian S., and Kaufman B.T., Dihydrofolate reductase from chicken liver and Lactobacillus casei bind Cibacron Blue F3GA in different modes and at different sites. J. Biol. Chem. 1980, 225, 10587-10590

15. Appukuttan P.S., and Bachhawat B.K., Separation of polypeptide chains of ricin and the interaction of the A chain with Cibacron Blue F3GA. Biochem Biophys. Acta, 1979, 580, 10-14

16. Bull P., MacDonald H., and Valenzuela P., The interaction of yeast RNA polymerase I and Cibacron Blue F3GA. Biochim. Biophys. Acta, 1981, 653, 368-377

17. Pompon D., Guiard B. and Lederer F., Binding of Cibacron Blue F3GA to the flavin and NADH sites in cytochrome b5 reductase. Eur. J. Biochem. 1980, 110, 565-570

18. Jankowski W.J., von Muenchhausen W., Sulkowski E. and Carter W.A., Binding of human interferons to immobilized Cibacron Blue F3GA the nature of molecular interaction. Biochemistry, 1976, 15, 5182-5187

19. Foltmann B., Prochymosin and chymosin (Prorennin and Rennin) in Methods in Enzymology, Vol. XIX, Ed. G.E. Perlmann and L. Lorand, 1970, pp 421-436

A QUANTITATIVE STUDY OF THE INTERACTION OF LACTATE DEHYDROGENASE AND IMMOBILISED CIBACRON BLUE F3GA IN THE ABSENCE OF SOLUBLE LIGANDS

ROBERT J. YON and MARK J. EASTON

School of Biological Sciences and Environmental Health, Thames Polytechnic, Wellington Street, London SE18 6PF, U.K.

ABSTRACT

Experimental data for the binding of rabbit muscle lactate dehydrogenase to immobilised Cibacron Blue F3GA (30-300 µM in total dye) in the absence of soluble ligands have been analysed in terms of three theoretical models, here termed the monovalent, equivalent-sites and concerted-cluster models. By the least-squares criterion, equally good fits of the data to each model were obtained, and no evidence of co-operative clusters of matrix-ligands was found. Fifteen estimates of the parameter K_M (intrinsic constant for binding of matrix-ligand) were all in the range $0.6-2.7 \times 10^6 M^{-1}$; 10 were clustered in the range $1.0-1.9 \times 10^6 M^{-1}$. The values of [M] (concentration of accessible matrix-ligand) predicted by the equivalent-sites theory were consistently lower than predicted by the other two theories, and did not show a clear dependence on total dye concentration. The values of [M] predicted by the other two theories were an approximately constant fraction (0.45%) of the total dye concentration. It is concluded that, for the present data, analysis in terms of a monovalent theory is entirely adequate.

INTRODUCTION

Theoretical treatments of affinity chromatography or partitioning under equilibrium conditions lead to equations in which elution volume or partitioning coefficient is expressed in terms of concentrations of soluble and matrix-bound ligand [1-6]. Such equations may or may not include the protein concentration. In quantitative experiments based on such equations, an important consideration is the selection of the appropriate species concentration (i.e. of soluble or matrix-bound ligand, or of protein) to be varied. It can be argued that, if the aim is understanding of the interaction of protein and matrix-ligand, then either or both of these species should be varied, rather than the soluble ligand as used in most published experiments. As pointed out by Hogg & Winzor [7], use of the soluble ligand may give a reliable estimate of the binding constant between protein and soluble ligand, but not necessarily reliable estimates of parameters affecting the protein-matrix interaction.

This communication describes experiments that re-assess the interaction of lactate dehydrogenase and immobilised Cibacron Blue by quantitative studies in which the protein and matrix-ligand concentrations are varied in the absence of soluble ligands. Three theoretical models are compared for their ability to fit the experimental data generated in equilibrium-binding experiments.

THEORY

For present purposes, only equations that explicitly include the protein concentration will be discussed. In the absence of soluble ligand, the relevant theories may all be reduced to a form that expresses the experimental variable R ([total protein]/[free protein]) (dependent variable) as a function (explicit or implicit) of the total concentration of protein $[P_t]$ (independent variable) and of two parameters: the accessible concentration of matrix-ligand [M] and the intrinsic binding constant for the immobilised ligand, K_M. A summary of the relevant models and equations (adapted for the absence of soluble ligands, and for a 4-site protein molecule) follows.

(1) Monovalent theory

$$R = 1 + \frac{4K_M[M]}{1 + 4K_M[P_t]}$$

This equation expresses the condition in which only one of the 4 enzyme sites is involved in binding to matrix-ligand. For multi-site proteins, this implies a relatively low concentration of accessible ligand. The equation above is derived from equations 9 and 17 of Nichol et al. [2], by use of the substitution $R = 1+(V_A-V_A^*)/V_s$.

(3) Equivalent-sites theory

$$K_M = \frac{1 - (1/R)^{1/4}}{(1/R)^{1/4} \{ [M] - 4[P_t](1 - (1/R)^{1/4}) \}}$$

This theory allows multiple (multivalent) binding of dye molecules to protein molecules. Subject only to statistical factors [8,9] it allows unrestricted combination of dye and enzyme molecules and postulates no co-operativity, i.e. equal access by equivalent protein sites to all ligand groups. The model is described by Nichol et al. [4], and has seen several applications, including an analysis of the interaction of lactate dehydrogenase and Cibacron Blue Sepharose in the presence of NADH [10].

(2) Concerted-cluster theory

$$ R = 1 + \frac{4K_M[M]}{1 + 4K_M[P_t]} + \sum_{i=2}^{4} \frac{K_i[X_i]}{1 + K_i[P_t]} $$

This equation also allows multiple (multivalent) binding of dye molecules to protein molecules. It assumes that, in addition to isolated single-dye encounters, protein molecules also encounter the dye as discrete, immovable clusters (i dye-molecules per cluster), to each of which binding is concerted, i.e. highly co-operative. [X_i] are the cluster concentrations and K_i the concerted-binding constants. Both can be expressed in terms of K_M and [M]. The model is presented in full [11] and in summary [12] by the present author. As formulated above, the monovalent equation is seen to describe the limiting behaviour at low dye concentrations.

EXPERIMENTAL

Blue Sepharose with various concentrations of immobilised dye were prepared as described by Heyns & DeMoor [13]. Partitioning experiments were performed at room temperature in 0.067M sodium phosphate buffer, pH7.2. Aliquots of a vigorously-stirred slurry of blue Sepharose were added to aliquots of a stock solution of rabbit muscle lactate dehydrogenase, to give the required packed-gel volume and total enzyme concentration. After 3 minutes of gentle shaking, the gel was separated by centrifugation and the supernatant assayed for enzyme activity. The enzyme concentration in the stock solution was found spectroscopically. The concentration of free enzyme was found by comparison of the activities of the supernatant and stock solutions.

In comparing the fit of experimental data to the three models described above, the best-fit values of the parameters K_M and [M] were found by an empirical non-linear least-squares search procedure. Parameter values were initially sampled over a wide range to avoid entrapment in local minima. The search was then refined in the region of the parameter-values giving the smallest sum-of-squares of residuals. The search was terminated when a change of 0.01% in either K_M or [M] decreased the sum-of-squares by less than 0.001%.

RESULTS

Fig. 1 shows R-vs.[P_t] plots for lactate dehydrogenase (10 concentrations in the range 10-100 nM) binding to blue Sepharose (5 total-dye concentrations in the range 30-300 μM). The lines are theoretical best-fit curves for the various models, obtained by optimising the values of K_M and [M]. The fit to each line (i.e. each dye-concentration) was conducted independently. Except for slight differences at the two highest dye concentrations, the lines for the three models are indistinguishable on the scale of Fig.1. Significance tests, or inspection of residual distributions, did not permit choice of the best model.

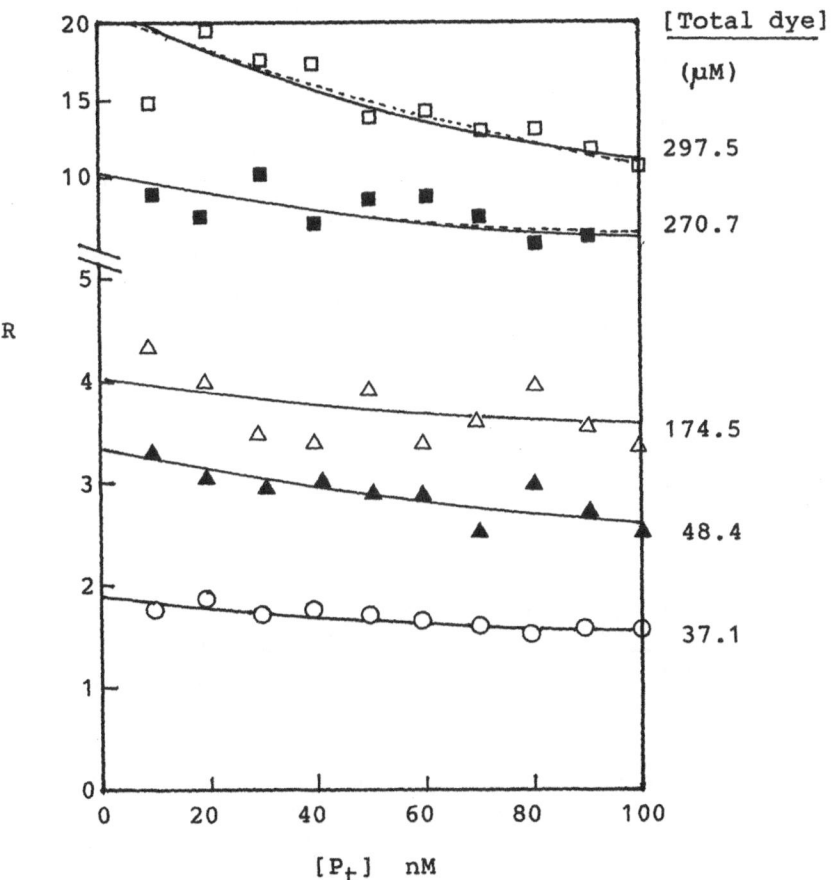

Fig. 1. Dependence of R on the total enzyme concentration. A single line indicates identical fits for the three models, except where a broken line, representing the equivalent-sites model, is shown.

The fitted parameter-values, however, were not the same for all models. Fig. 2 shows these values plotted against total-dye concentration determined by published procedures [14].The values of K_M generated by the three models at each of the five total-dye concentrations (15 values in all) were all in the range $0.6-2.7 \times 10^6$ M^{-1}, with 10 values clustered in the range $1.0-1.9 \times 10^6$ M^{-1}. Since a single intrinsic value is to be expected, the models are reasonably consistent in this respect.

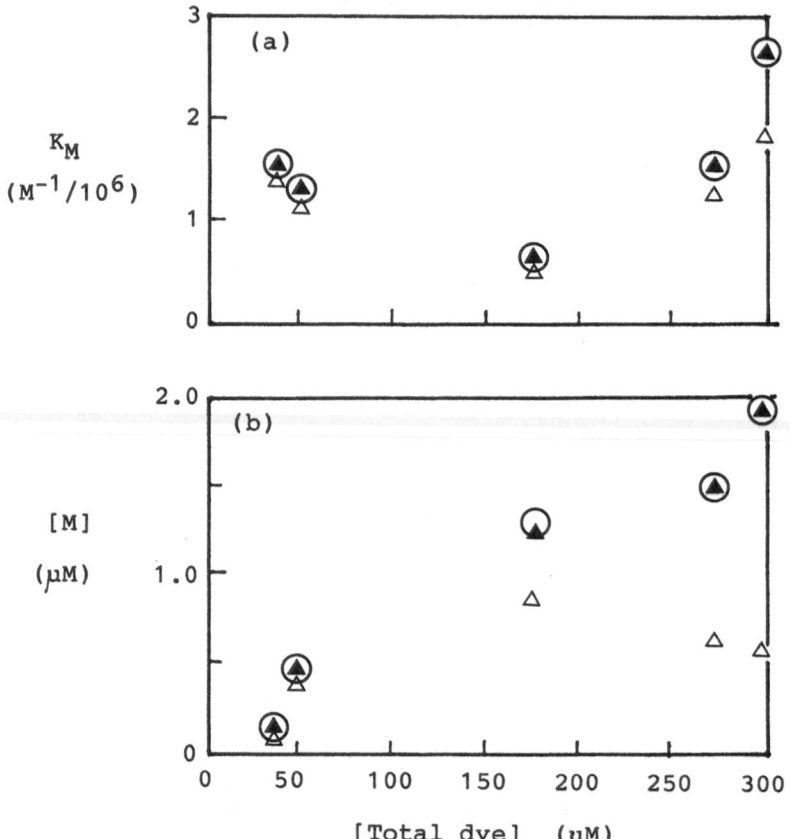

Fig.2. Dependence of (a) K_M and (b) [M] on the total dye concentration.
◯ monovalent theory
▲ concerted cluster theory
△ equivalent-sites theory

This was not the case for the other parameter, [M], however. The monovalent and cluster models gave indistinguishable values for [M] while the equivalent-sites model gave a value that was consistently lower. Moreover the

[M]-values for the former two models clearly increased, in an approximately linear manner, with increase in total-dye concentration. In contrast, the equivalent-sites model generated [M]-values that appeared to reach a plateau, and even decreased slightly, at higher total-dye concentrations.

The linear behaviour of the monovalent and cluster models suggest that a constant fraction of the total dye is accessible to the enzyme. This permits a simultaneous fit of all 50 data-pairs on the assumption of a single value of K_M and a constant fraction of each of the known total-dye concentrations. This fit generated values of 0.45% for the accessible fraction and 1.50×10^6 M^{-1} for K_M.

DISCUSSION

Since the theoretical lines fitted to the data in Fig. 1 were essentially indistinguishable, all models appeared to fit the data equally well, albeit with different parameter-values in some cases. Application of statistical significance tests, therefore, would be unable to select a 'best' model for these data. The presence of any low-concentration, high-affinity adsorption sites, such as are predicted by the cluster theory, would generate an upturn in the R-vs-$[P_t]$ plots if the values of $[P_t]$ were of the same order as the concentration of such sites [11,12]. In terms of the cluster model, therefore, the range of protein concentrations in these experiments was too high to detect such effects. Consequently, as predicted by the theory, the behaviour is indistinguishable in all respects from that predicted by the monovalent model. The monovalent model is preferred to the equivalent-sites model for another reason. It seems reasonable to expect the concentration of accessible dye groups to increase uniformly as the total immobilised dye is increased. This is approximately true for the monovalent model but not for the equivalent-sites model (Fig. 2(b)). We conclude that the monovalent model is adequate for the interpretation of these data.

The values of K_M and the small fraction of accessible dye-groups are compatible with results from Stellwagen's group [15] using various methodologies on a matrix with total dye concentration of 120 uM, and relying on use of soluble ligand (NADH) as the main independent variable and analysis in terms of monovalent theory. They obtained a dissociation constant of 0.23 uM, equivalent to $K_M = 4.3 \times 10^6$ M^{-1}, and an accessible fraction of 1.7% of the total dye, both higher, but of the same order, as the present results for the monovalent model. A later study by the same group [16], using a much higher total dye concentration (11.6 mM) found evidence of multivalency in the interaction of lactate dehydrogenase; analysis in terms of a single intrinsic association constant gave a K_M value of 2.2×10^6 M^{-1}, close to that obtained in the present study. It is interesting to note that Stellwagen's group and ours have worked with similar phosphate buffers near pH7. In contrast, a study by Hogg & Winzor [10] used an imidazole buffer containing a high concentration of NaCl(0.39M), and commercial blue Sepharose. They estimated K_M at the much lower value of

2.3×10^4 M^{-1}. This apparent discrepancy is probably explained
by the salt effect; it has been shown [16] that the
enzyme-ligand affinity is strongly affected by the nature of
the anions present, chloride weakening the interaction relative
to phosphate.

It will be of considerable interest to repeat the present
analysis on matrices with a much higher level of
dye-substitution. In particular we are interested in whether
evidence of clustering may then be found.

REFERENCES

1. Dunn, B.M. and Chaiken, I.M., Quantitative affinity
    chromatography. Determination of binding constants by
    elution with competitive inhibitors. Proc. Natl. Acad. Sci.
    USA, 1974, 71, 2382-2385

2. Nichol, L.W., Ogston, A.G., Winzor, D.J. and Sawyer, W.H.,
    Evaluation of equilibrium constants by affinity
    chromatography. Biochem. J., 1974, 143, 435-443

3. Chaiken, I.M., Quantitative uses of affinity
    chromatography. Anal. Biochem., 1979, 97, 1-10

4. Nichol, L.W., Ward, L.D. and Winzor, D.J., Multivalency of
    the partitioning species in quantitative affinity
    chromatography. Evaluation of the site-binding constant for
    the aldolase-phosphate interaction from studies with
    cellulose phosphate as the affinity matrix. Biochemistry,
    1981, 20, 4856-4860

5. Kyprianou, P. and Yon, R.J., A quantitative study of the
    biospecific desorption of rat liver (M4) lactate
    dehydrogenase from 10-carboxydecylamino- Sepharose.
    Biochem.J., 1982, 207, 549-556

6. Winzor, D.J., Quantitative characterisation of interactions
    by affinity chromatography. In Affinity Chromatography: a
    Practical Approach, eds. P.D.G. Dean, W.S. Johnson and F.A.
    Middle, IRL Press, 1985, pp.149-168

7. Hogg, P.J. and Winzor, D.J., Effects of solute multivalency
    in quantitative affinity chromatography: evidence for
    cooperative binding of horse liver alcohol dehydrogenase to
    blue Sepharose. Arch. Biochem. Biophys., 1985, 240, 70-76

8. Klotz, I.M., Protein interactions with small molecules.
    Accts. Chem. Res. 1974, 7, 162-168

9. Flory, P.J., Molecular size distribution in three
    dimensional polymers. I. Gelation. J. Am. Chem. Soc., 1941,
    63, 3083-3090

10. Hogg, P.J. and Winzor, D.J., Quantitative affinity chromatography: Further developments in the analysis of experimental results from column chromatography and partition equilibrium studies. Arch. Biochem. Biophys., 1984, 234, 55-60

11. Yon, R.J., A cooperative cluster model for multivalent affinity interactions involving rigid matrices. J. Chromatog. IN PRESS

12. Yon, R.J., Computer modelling of multivalent affinity interactions. Biochem. Soc. Trans. IN PRESS

13. Heyns, W. and DeMoor, P., A 3(17)-β-hydroxysteroid in rat erythrocytes. Conversion of 5κ-androstane-3β,1β-diol and purificatioon of the enzyme by affinity chromatography. Biochem. Biophys. Acta, 1974, 358, 1-17

14. Lowe, C.R. and Pearson, J.C., Affinity chromatography on immobilised dyes. Methods Enzymol., 1984, 104, 97-113

15. Liu, Y.C., Ledger, R. and Stellwagen, E., Quantitative analysis of protein:immobilized dye interaction. J. Biol. Chem., 1984, 259, 3796-3799

16. Liu, Y.C. and Stellwagen, E., Accessibility and multivalency of immobilised Cibacron Blue F3GA. J. Biol. Chem., 1987, 262, 583-588

INTERACTION OF AZO DYES WITH CATIONIC POLYPEPTIDES : POLY–L–ARGININE – 4 –DIMETHYLAMINO AZOBENZENE-4' – CARBOXYLIC ACID SYSTEM

ATREYI, M., RAO. M.V.R. AND SCARIA, P.V.
DEPARTMENT OF CHEMISTRY,
UNIVERSITY OF DELHI
DELHI– 110 007, INDIA.

ABSTRACT

The interaction of 4 – dimethylamino azobenzene – 4' – carboxylic acid (DAAC) with poly–L– arginine (PLA) was followed at pH 9.0 by spectrophotometric and circular dichroic measurements. The dye induces a conformational transition of the polypeptide from a random coil state to beta form through an intermediate α – helical state, as the peptide to dye ratio was varied from 6 to 1. In this dye-polypeptide system, besides coulombic interaction between the cationic side chains and the ligand, H–bonding between carboxylate ion of DAAC and guanidinium group of the side chains of PLA appears to play an important role. The differences in the nature of interaction of DAAC, which has a COOH group instead of the sulphonic acid group of methyl orange (MO), and MO with cationic polypeptides of arginine, lysine and ornithine are discussed.

INTRODUCTION

Synthetic polypeptide-dye systems are useful for understanding the interaction of proteins with a variety of ligands. Historically, the first study of dye-polypeptide interaction was that between the cationic dye acridine orange (AO) and the anionic polyglutamic acid (PGA) by Styrer and Blout [1,2]. Extensive studies of this system led to the conclusion that the binding of AO to PGA results in a blue shift of the visible absorption maximum of AO, and that it is the helical conformation of the polypeptide which is the favoured conformational state for dye binding. Further, an induced circular dichroism (ICD) was observed in the visible absoprtion region of the dye. The ICD was attributed to the stacking of dimers of AO in a superhelical array about the α –helical polypeptide, and the concept of the helical conformation as a prerequisite for observing ICD started gaining ground [3,4]. However, later studies on the interaction of the cationic polypeptide, poly-L-lysine (PLL), with the anionic dye methyl orange (MO), showed that PLL– MO complex exhibits ICD, even though the basic polypeptide has a random coil conformation [5,6]. Detailed experimental work done in our laboratories on the interaction of the anionic azo dye, 4 – dimethylamino azobenzene-4' -carboxylic acid (DAAC) [7,8], and also MO, with the three cationic polypeptides poly-L-lysine (PLL), poly-L-ornithine (PLO) [9] and poly-L-arginine (PLA) revealed that factors such as basicity of the cationic group of the polypeptide, the nature of dye anion, pH, ionic strength and finally the length of the side chain have all a role to play in the dye-polypeptide complexation.

4-(Dimethylamino)azobenzene-4-carboxylic acid

(DAAC)

4-(Dimethylamino)azobenzene-4-sulfonic acid

(Methyl orange)

In what follows, results obtained from the investigations of the interaction of poly-L-arginine with DAAC are presented and compared to those obtained with the cationic polypeptides PLL and PLO.

MATERIALS AND METHODS

Poly-L-arginine was obtained from Sigma Chemical Co., USA. Details of the synthesis of 4 – dimethylamino azobenzene– 4' – carboxylic acid (DAAC) were published earlier [10].

Details of procedures used for obtaining spectrophotometric and circular dichroism data were the same as in our earlier publications [7,8,9]. The extinction coefficients are expressed in terms of moles of total dye in the system. Similarly, the ellipticity of the induced circular dichroic bands is given in terms of deg. cm.^{2} decimole $^{-1}$ of total dye. DAAC concentration, unless otherwise specified, in all experiments with PLA was 8×10^{-5} M and the solution pH was 9.0.

RESULTS

Absorption studies

The absorption spectrum of the dye DAAC in aqueous solution, above neutral pH in presence of poly-L-arginine, showed a new band at 372 nm and this was accompanied by a decrease in the intensity of the characteristic 464 nm absorption band of the dye, as can be seen from the absorption spectra of the dye and its mixture with PLA, having a polypeptide to dye ratio (P/D) equal to six, shown in Fig. 1. Actually, the intensity of the 372 nm band was found to be a complex function of the polypeptide to dye ratio; it increased upto a P/D of 2 and then decreased until P/D = 3, and then again increased with P/D with the maximal value being attained at P/D ≥ 6 (Inset, Fig. 1).

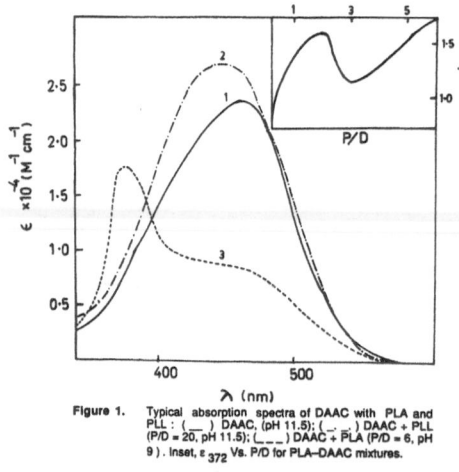

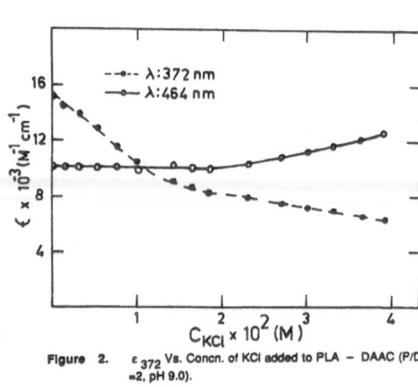

Figure 1. Typical absorption spectra of DAAC with PLA and PLL : (___) DAAC, (pH 11.5); (_._) DAAC + PLL (P/D = 20, pH 11.5); (___) DAAC + PLA (P/D = 6, pH 9). Inset, ε_{372} vs. P/D for PLA–DAAC mixtures.

Figure 2. ε_{372} vs. Concn. of KCl added to PLA – DAAC (P/D =2, pH 9.0).

On addition of KCl to the PLA-DAAC mixture with a P/D of 2 at pH 9.0, the intensity of the 372 nm band was reduced (Fig. 2) and the intensity of 464 nm band was not completely restored even with 0.04 M KCl. On the other hand, on progressive addition of dioxane to PLA-DAAC system (P/D = 2, pH 9.0), the 372 nm band is weakened, and in 40% (V/V) dioxane the band disappeared (Fig. 3); the stength of the 464 nm band was more or less restored to its original intensity. Dioxane itself has marginal effect on the absorption characteristics of the dye.

Circular dichroic studies

PLA in aqueous solutions at pH 9.0 has a random coil conformation, as can be concluded from its circular dichroic (CD) spectrum (Fig. 4), which exhibits a small +ve band around 217 nm and a strong negative band at about 197 nm. On addition of DAAC to PLA, the features of the CD spectrum change (Fig. 4) from that of a random coil to that of a α –helix, characterised by two extrema at 207 and 222 nm (P/D = 2.6). On further increase in DAAC concentration, the α–helical CD pattern remarkably changes over to one with a single minimum at 222 nm, and a +ve maximum at about 200 nm; this CD pattern corresponds to that of a beta conformation.

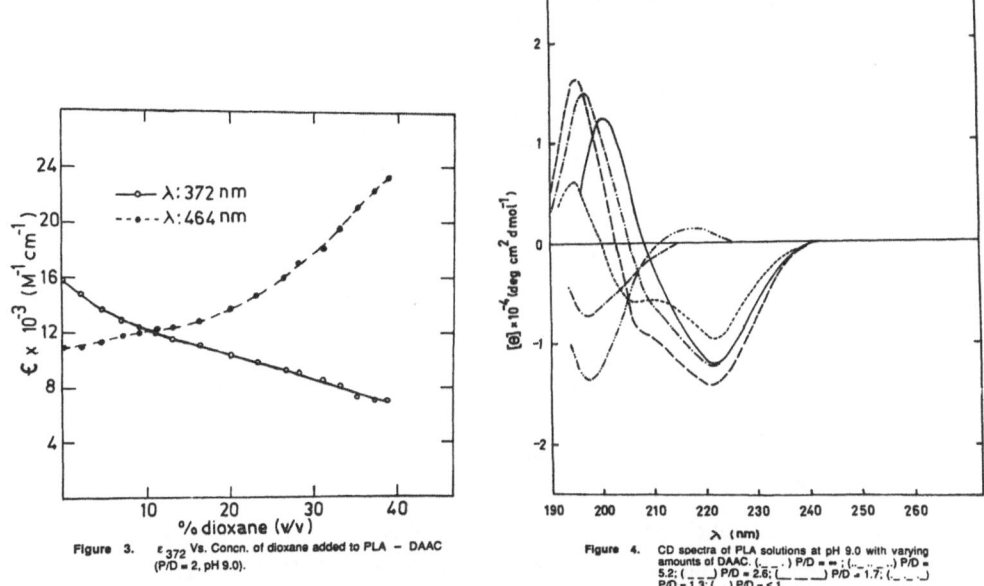

Figure 3. ε_{372} Vs. Concn. of dioxane added to PLA – DAAC (P/D = 2, pH 9.0).

Figure 4.   CD spectra of PLA solutions at pH 9.0 with varying amounts of DAAC. (._ _ _.) P/D = ∞ ; (.._.._..) P/D = 5.2; (_ _ _) P/D = 2.6; (_ _ _ _) P/D = 1.7; (._ _ _.) P/D = 1.3; (___) P/D ≤ 1.

The new absorption band (372 nm) of the dye in presence of PLA was found to be associated with Cotton effects (Fig. 5). The ICD spectrum was conservative with a +ve long wavelength band (385 nm) and a negative band at 360 nm; the two bands were almost of equal intensity. However, the features of the induced circular dichroism in this system were markedly different from those observed with poly-L-lysine-DAAC system (curve 2, Fig. 5) ; not only is the location of the ICD bands considerably red shifted, but the rotational strength of the bands is much higher than that for the arginyl-dye system.

The intensitiy of the induced circular dichroism was also found to be a complex function of polypeptide to dye ratio (Fig. 6), as was also true of the intensity of the absorption band of the peptide bound dye at 372 nm (Fig. 1, inset). An increase in P/D, at constant concentration of DAAC, enhances the intensity of ICD upto a P/D of about 2 and thereafter, there was a slight fall in the rotational

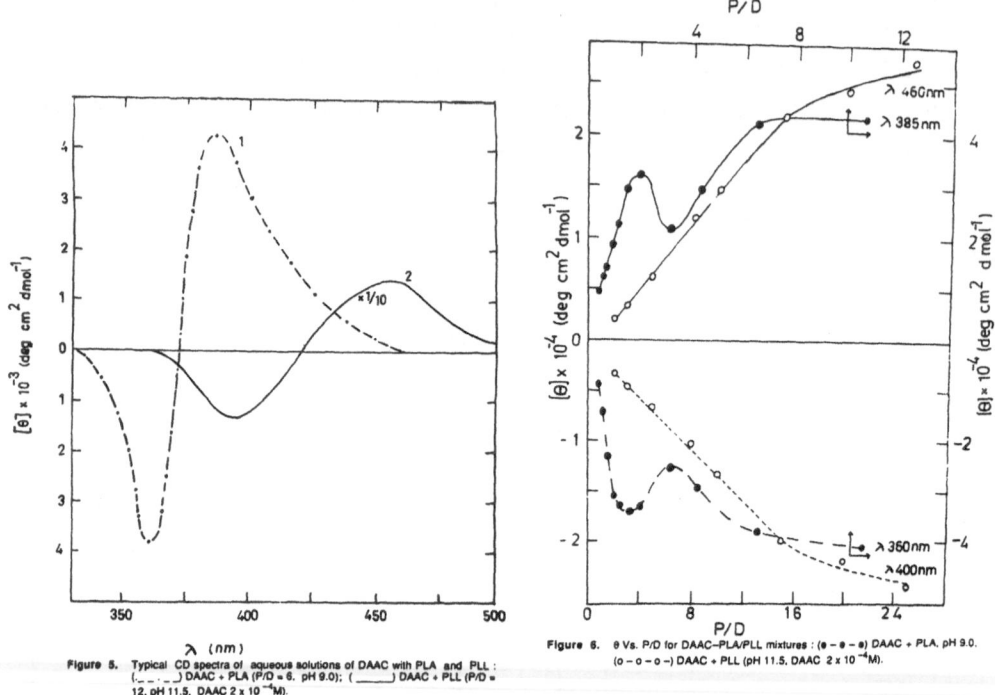

Figure 5. Typical CD spectra of aqueous solutions of DAAC with PLA and PLL : (.._.._) DAAC + PLA (P/D = 6, pH 9.0); (_____) DAAC + PLL (P/D = 12, pH 11.5, DAAC 2 x 10⁻⁴M).

Figure 6. θ Vs. P/D for DAAC–PLA/PLL mixtures : (•–•–•) DAAC + PLA, pH 9.0. (o–o–o–) DAAC + PLL (pH 11.5, DAAC 2 x 10⁻⁴M).

strength of both the bands till a P/D of 3. And above a P/D of 3, the intensity of the bands again increased till a levelling off was seen at a P/D ≥ 6. The changes in the ICD band intensities for the PLL–DAAC system, also depicted in Fig. 6, on the other hand, show a continuous increase in the intensity with P/D.

An elucidation of the role of electrostatic interactions in PLA–DAAC binding was sought from a study of the effect of KCl on the induced CD bands. The plot of θ_{385} and θ_{360} as a function of added KCl (Fig. 7) showed a decrease in the intensities with an increase in the concentration of KCl. The induced CD, however, persisted even when the concentration of KCl was 0.05M in a PLL–DAAC mixture with P/D = 2 and pH = 9.0.

Fig. 8 shows the effect of dioxane on the induced CD of PLA–DAAC, at pH 9.0 and P/D = 2. Addition of dioxane resulted initially in a small decrease in the ellipticity values followed by more or less a constant ellipticity upto a concentration of 20% (V/V) dioxane. Further increase in dioxane concentration caused a decrease in the θ values and the induced CD bands disappeared in 40% (V/V) dioxane. This decrease in ellipticiy for the induced CD bands was accompanied by a decrease of θ_{222} from − 14,500 to −2500, indicating an order–disorder transition of the PLA backbone (spectrum not shown).

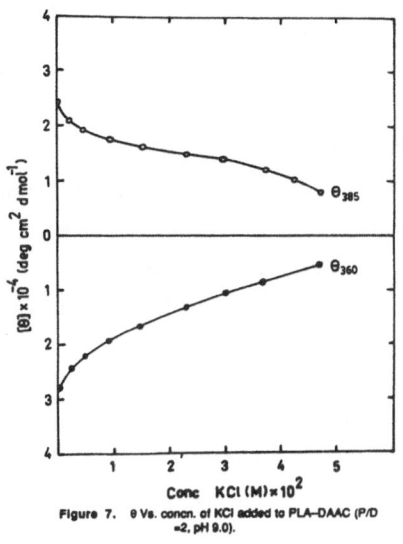

Figure 7. θ Vs. concn. of KCl added to PLA–DAAC (P/D =2, pH 9.0).

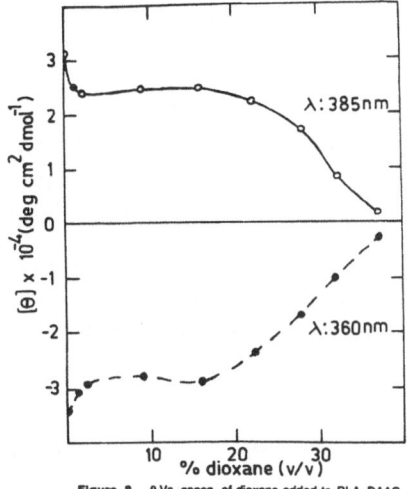

Figure 8. θ Vs. concn. of dioxane added to PLA–DAAC (P/D =2, pH 9.0).

DISCUSSION

The azo dye DAAC has a planar carboxylate group but MO has instead a pyramidal sulphonate group and one would therefore expect differences in the interaction of the two dyes with cationic polypeptides. While MO forms [5,6,11] complexes with all of the positively charged polypeptides (PLL, PLO & PLA) with a large blue shift of its absorption band ($\Delta \lambda = 100$ nm), DAAC does so only with PLA ($\Delta \lambda = - 94$ nm); it does not interact with PLO, but weakly couples to PLL ($\Delta \lambda = -18$nm) [7]; however, when DAAC is covalently linked to side chains of PLL [13] one does notice a larger blue shift ($\Delta \lambda = - 64$ nm).

The azo dye DAAC induces conformational transitions of the PLA backbone, and interestingly, the conformational state stabilized depends on P/D. At high P/D values (> 3, pH = 9.0), when the dye molecules are widely dispersed on the polypeptide backbone, the random coil conformation of PLA is not altered, but on increasing the dye concentration, the polypeptide undergoes a transition first to a helical state (P/D = 3 to 2) and then to a β conformation (P/D = 1). These conformational changes are also reflected in the ellipticity of ICD bands; initially, the ICD (at constant DAAC concentration) increases upto a P/D of 2 and then decreases (P/D = 2 to 3) and then again increases (Fig. 6). The data for PLL–DAAC system given in the same figure shows a near linear variation of θ with P/D; this behaviour is so becuase the transition in this system occurs from α to β state.

It is interesting to note that DAAC transforms the helical conformation of PLL to β sheet form in pH = 11.5 solutions, while MO binding at pH 7.0 does not cause any change. MO binds to cationic poly-L-arginine (pH =4 − 11) and transforms it to helical state when P/D = 1 [11], but DAAC stabilizes β form of PLA at this P/D. The PLO-MO complex is also in β state. The above data clearly indicate differences in the mode of binding of the azo dyes to the cationic peptides.

Salt as well as dioxane affect the PLA–DAAC complex. 0.05 M salt does not disrupt the complex completely as evidenced by the persistence of the 372 nm band as well as ICD (Figs. 2 and 7), indicating the importance of non-electrostatic forces in the interaction. On the other hand, in 40% dioxane medium, the ICD is totally lost, the 372 nm band disappears and the PLA backbone also retains the random coil conformation (Fig. 3 and 8).

Guanidinium group of the side chains of PLA being planar has two H-bond donor sites and can make a pair of H-bonds with a COO⁻ group as was seen in the crystal structure of the dipeptide Arg − Glu [14]; such a binding has also been proposed for the complex of PLA with polyglutamic acid [15]. It is significant that when PLA is unionized, DAAC does not bind to it, indicating the necessity of the positive charge of the side chain for complex formation. It has been shown earlier [7] that PLL–DAAC interaction is predominantly hydrophobic, but in PLA it appears that H − bonding is more crucial for complexation. The fact that DAAC fails to interact with $-(CH_2)_3 NH_3^+$ side chains of PLO shows clearly that electrostatic interactions play only a minor role when the anionic dye is a carboxylate dye; MO with a sulphonic acid group binds to all the three polypeptides more or less in a similar manner through predominantly electrostatic interactions. It is interesting to note that replacing the N − dimethyl amino group by $- NH_2$ or $- NO_2$ or changing the acidic group from the para position to any other position in the azobenzene moiety decreases or completely destroys binding [16]. The above results highlight the complexities involved in the understanding of the factors responsible for specificity in ligand-protein interactions.

REFERENCES

1. Styrer, L. and Blout, E.R., J. Am. Chem. Soc., 1961, **83**, 1411 − 18.

2. Blout, E.R. and Styrer, L., Proc. Natl. Acad. Sci. USA 1959, **45**, 1591–93.

3. Sato, Y. and Hatano M., Bull. Chem. Soc. Japan 1973, **46**, 3339–44 and references therein.

4.   Ikeda, S., Yoshida, T. and Imae., T. Biopolymers 1981, **20,** 2395 and references therin.

5.   Hatano, M., Yoneyama, M., Sato, Y. and Kawamura, Y., Biopolymers. 1973, **12,** 2423–30.

6.   Murakami, K., Sano, T. Kuru, N., Ishii K. and Yasunaga, T. Biopolymers, 1983, **22,** 2035–44.

7.   Scaria, P.V., Atreyi, M. and Rao, M.V.R. Biopolymers 1986, **25,** 2349–58.

8.   Scaria, P.V., Atreyi, M. and Rao, M.V.R., Int. J. Biol. Macromol 1988, **10,** 60–62.

9.   Atreyi, M. Scaria, P.V. and Rao, M.V.R., J. poly. Sci.(C) polym. lett. 1987, **25,** 249–55.

10. Atreyi, M., Rao, M.V.R. and Scaria, P.V., J. Macromol. Sci. Chem. 1984, **A 21,** 15–19.

11. Yamamoto, H. and Nakazawa, A., Biopolymers 1984, **23,** 1367–77.

12. Quadrifoglio, F., and Crescenzi, V., J. Coll. and Int. Sci. 1971, **35,** 447–459.

13. Authors' unpublished results.

14. Lancelot , G., Mayer, R. and Helene, C., J. Am. Chem. Soc.. 1979, **101,** 1569–76.

15. Mita, K., Ichimura, S. and Zama, M., Biopolymers, 1978, **17,** 2783–2798.

16. Hiroyuki, H., Nakazawa, A and Hayakawa, T., J. Poly. Sci. (C) poly Lett. , 1983, **21,** 131– 8.

KINETIC AND THERMODYNAMIC OF THE INTERACTION
OF DYES WITH BIOLOGICAL MACROMOLECULES

J. AUBARD, M.A. SCHWALLER, A. ADENIER and G. DODIN
Groupe de Dynamique des Interactions Macromoléculaires
Institut de Topologie et de Dynamique des Systèmes
de l'Université Paris 7, associé au C.N.R.S. (UA 34)
1, rue Guy de la Brosse, 75005 Paris, France.

The investigations of the interaction of small organic molecules (effector with biological macromolecules (proteins and nucleic acids) or organized structures like phospholipid membranes constitute one of today's most active fields in biochemistry. The goals of this research effort are manifold and range from understanding the basic mechanisms and forces involved in the interaction, to the development of compounds endowed with particular properties (pharmaceuticals with antimicrobial, antiviral and antitumoral activities, dyes for medicinal diagnosis, dyes for foodstuffs, etc...).

Though it does not permit to readily infer what will be the fate of an effector in a living system, the "in vitro" approach is a necessary step to the knowledge of the interaction modes with the various components of the biological system. These "in vitro" studies are usually oriented to the determination of the properties of the effector - macromolecule complexes in terms of their, structures, thermodynamics of formation, biochemical properties such as their inhibitory effects on metabolic chain etc...

More seldom, however, the dynamics of the formation of the complexes themselves is investigated though it may lead to valuable information. This situation arises from the fact that short-lived steps are involved in the formation of the final stable adduct which can be observed only with particular time-resolving techniques.

We have developed fast kinetic experimental set-up based on a rapid perturbation of the equilibria. We shall present Temperature-Jump (T-Jump spectroscopy and show how it has been used to investigate the binding of a membrane dye, MC540, and of DNA-acting drug from the ellipticine series (NMHE). Prior to the time-resolved studies, the thermodynamics of these systems have been investigated.

THERMODYNAMIC ANALYSIS

The changes of light absorption and fluorescence of the effectors are

commonly used to monitor the binding process. Fig. 1 shows an example for
MC540 (a cyanine dye) bound to liposomes, here used as a membrane model.
When increasing amounts of liposomes are added to an aqueous solution of
MC540, the absorption spectrum of the free dye is progressively red shifted,
leading to the formation of a new band which can be assigned to the bound
form of the dye (1).

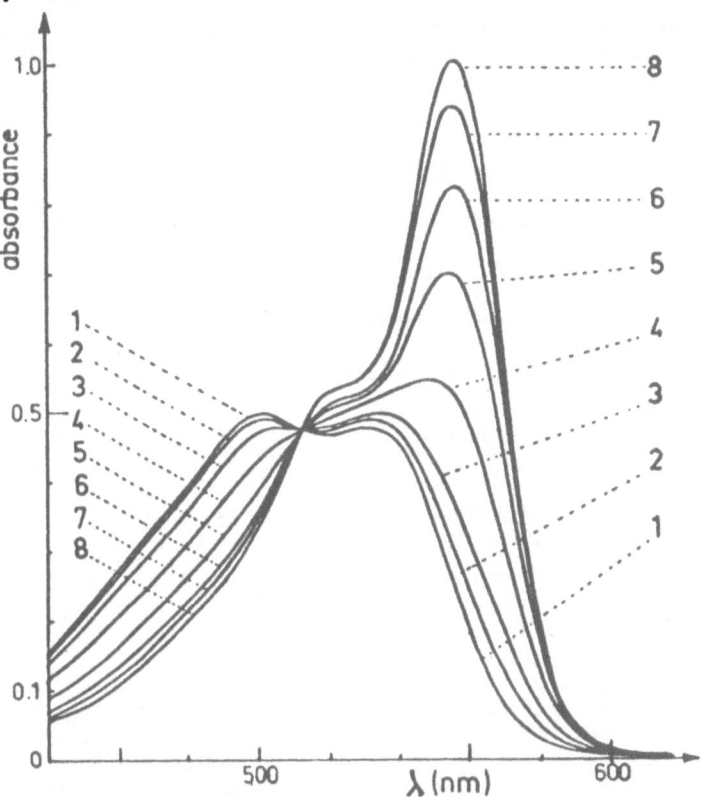

Figure 1. Titration of MC540 by liposomes. Spectrum 1 is the absorbance
    of the free aqueous dye; spectrum 8 of the fully complexed dye.

Analog spectral behaviours are also observed for drugs bound to DNAs and
these optical variations (absorption or fluorescence) may be explained by
considering that essentially strong hydrophobic interactions occuring
between the aromatic effectors and macromolecules lead to the formation of
tightly bound complexes (1,2).

    The interaction between effectors and biological macromolecules can
thus be viewed as arising from the binding of the drug, D, to a set of n
identical and independent binding sites, S, leading to the formation of an
hydrophobic complex, C

$$S + D \underset{}{\overset{K_i}{\rightleftharpoons}} C \qquad\qquad (Scheme\ 1)$$

with an intrinsic association constant K_i

$$K_i = \frac{[\overline{C}]}{[\overline{S}][\overline{D}]} \qquad (1)$$

where $[\overline{C}]$, $[\overline{S}]$ and $[\overline{D}]$ are equilibrium concentrations; generally macromolecular concentrations being expressed with respect to a basic entity, B (base pair in DNA, phospholipid in vesicle, protein in membrane..), the preceding equation rewrittes

$$K_i = \frac{[\overline{C}]}{n[\overline{B}][\overline{D}]} \qquad (2)$$

where $n[\overline{B}]$ now represents the equilibrium concentration of binding sites.

Binding parameters, K_i and n, can be determined with the aid of the Scatchard binding isotherm. Substituting in equation (2) $r = [\overline{C}] / [B]_o$, the site occupation parameter, yields after simple rearrangement to the Scatchard equation

$$r/[\overline{D}] = K_i (n - r) \qquad (3)$$

where $[\overline{D}]$, the concentration of the free drug and r can be computed from the spectral data (2). A typical Scatchard plot, $r/[\overline{D}]$ versus r, for the binding of an ellipticine derivatives (NMHE) to DNA is given in Fig.2. The slope of the line provides K_i whereas the intercept in the r axis allows to determine n.

The major source of errors in these experiments comes from the tendancy of the aromatic organic molecules to aggregate by stacking in aqueous solutions. In some cases, with dyes that aggregate severily, it is no longer possible to assume in the Scatchard equation (3), that $[\overline{D}]$ is now equal to $[\overline{D_1}]$ the concentration in monomer which is the genuine binding species (2,3). Then prior any binding experiments it is important to determine on which structure the dye exists in solution. Generally at the concentrations used for the titration experiments, solutions of aqueous dyes are mixtures of monomeric, D_1, and dimeric D_2, species

$$D_1 + D_1 \xrightleftharpoons{\ K_D\ } D_2 \qquad \text{(Scheme 2)}$$

and one has to determine precisely the dimerization constant, K_D, in order to estimate the proportion of the free monomeric species in solution. The kinetic approach to this problem, which gives reliable measurements of K_D will be presented further in this paper.

The thermodynamic aspect, just developed, only gives a coarse description of the binding process since it neither provides information about the mechanisms nor as concerns the number and life-times of transient species involved in the reaction. Therefore the present paper is now devoted to the kinetics of the binding by fast T-Jump relaxation methods.

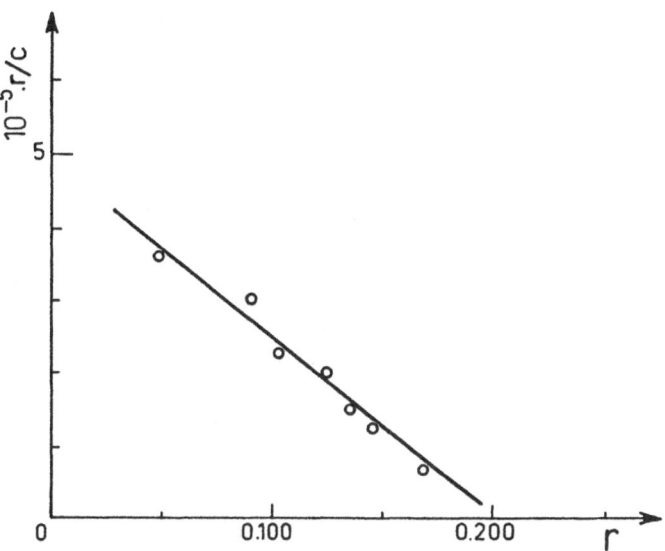

Figure 2. Fluorescence Scatchard plot for NMHE with DNA. The full line is
the best fit of the experimental data according with equation (3)

KINETIC STUDIES BY T-JUMP METHODS

Chemical relaxation methods are the best tool avalaible today to study
very fast chemical or biological equilibria in solution. The basic principle of these methods (4) is to perturb a chemical system at equilibrium by
a rapid change (transient or periodic) of an external parameter such as
temperature, pressure, electric field... .   Following the rapid variation
of the reactants' concentrations to their new equilibrium value allows the
determination of the reaction rates. In transient chemical relaxation
techniques, such as temperature-jump (T-Jump), the concentration change
following the perturbation (i.e. the relaxation process) is usually
monitored by means of absorption, emission or diffusion of the light.
However today time-resolved optical absorption spectroscopy is the most
popular and reliable detection method used for monitoring the relaxation.
Relaxation measurements from nanosecond to second are currently obtained
and allow to investigate a large number of chemical events occuring in the
course of a reaction. Thus as concerns the dimerization process evoked
above, this reaction is known to be always rapid and its study sometimes
requires using the *Laser T-Jump technique* which is today the fastet
transient relaxation technique (5). This is the case for MC540 (Fig. 3)
for which relaxation times fall within the microsecond range (6).

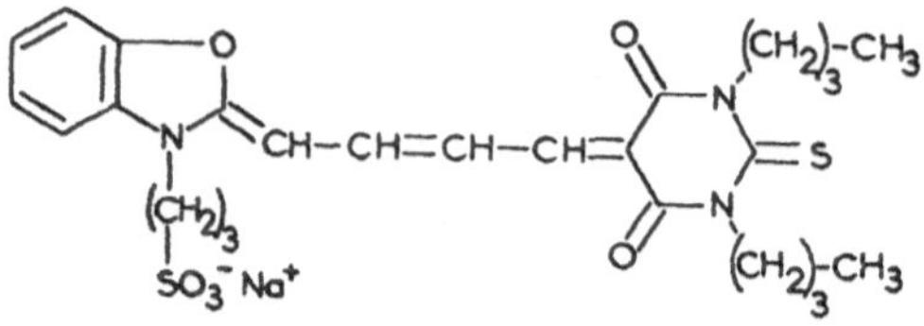

Figure 3. Merocyanine 540, Dimerization process in neutral aqueous solution

For the dimerization process, plotting the relaxation times as a function of the overall dye concentration, D_o

$$\tau^{-2} = 8 \, k_1 k_{-1} D_o + (k_{-1})^2 \qquad (4)$$

allows the determination of the rate constants k_1 and k_{-1}. The ratio k_1/k_{-1} gives the dimerization constant K_D thus allowing a reliable approach to this problem, whereas the spectroscopic equilibrium studies often failed in this analysis (6).

The dye-binding equilibria are not so fast, relaxation times generally range from 10^{-3}s to 10^{-1}s and the *Joule T-Jump* instrument is well suited to study this process. With this system a 5°C temperature perturbation is obtained within a few microseconds by discharching a large capacitor (0.05μF), charged up to 20 kV, across a specially design optical cell (7). Relaxation signals detected by an extremely sensitive absorbance spectrometer (ca. 5.10^{-4} O.D unit sensitivity) are digitized and processed by an on-line minicomputer. A typical relaxation curve, for the binding of NMHE to DNA, is shown in figure 4a. Two successive slow phases ($\tau_1 = 4$ ms and $\tau_2 = 30$ ms) are observed in the relaxation signal (a third one, falling in the μs time range is not apparent in the picture) and can be assigned to two distinct drug-DNA complexes, probably arising from two distinct DNA sites (8).

When the same T-Jump experiments are undertaken on the binding of dyes (MC540, ellipticine - derivatives) to artificial membranes, only one slow relaxation time is observed in the signals (Fig.4b). From these two similar binding experiments to different biological macromolecules, the

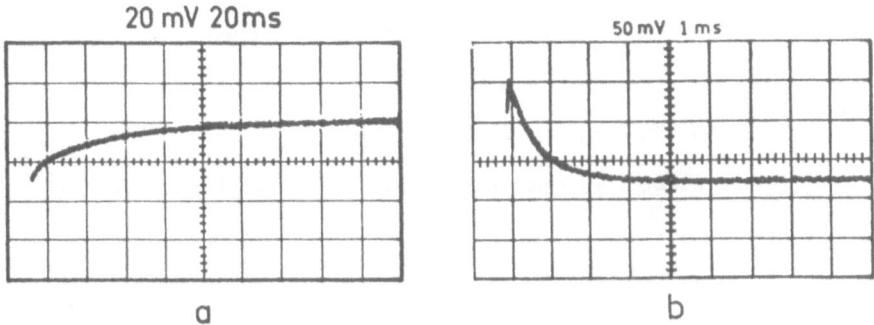

Figure 4. T-Jump relaxation signals; (a) NMHE-DNA solution, recorded at
λ = 310nm; (b) MC540-Liposome solution, recorded at λ = 560nm.

kinetic approach allows now to precise the mechanism involved in the
binding process, whereas the thermodynamic approach gave the same picture.

Thus for dye-membrane binding, a good fit of our kinetic data (1)
were obtained according to scheme   3

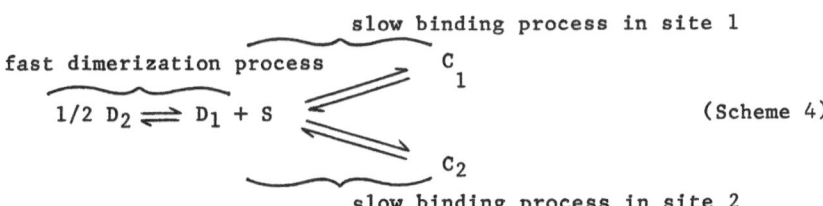

$$\text{fast dimerization process}$$
$$1/2 \; D_2 \rightleftharpoons D_1 + S \rightleftharpoons C \qquad \text{(Scheme 3)}$$
$$\text{slow binding process}$$

where C the fully complexed dye has a life-time of $\underline{ca}$ 1 ms.

In the case of NMHE-DNA binding (and for other numerous drugs) our
kinetic data were satisfactorily analysed according to scheme   4

$$\text{slow binding process in site 1}$$
$$\text{fast dimerization process} \qquad C_1$$
$$1/2 \; D_2 \rightleftharpoons D_1 + S \qquad\qquad \text{(Scheme 4)}$$
$$C_2$$
$$\text{slow binding process in site 2}$$

where C_1 and C_2 are two hydrophobic complexes, probably arising from drug
intercalation between base pairs and drug bound in the DNA grooves (8).
Furthermore from the detailed analysis of the relaxation times, in this
latter case, it has been possible for the first time, to detect some
selectivity of the drug for specific DNA sequences (8).

Finally, it has been state recently that the knowledge of the
dynamics of the drug - macromolecule interaction could have some biomedical
importance if we consider that long-residence times of complexed drugs
(into or onto the macromolecule) are probably associated with enhanced

"in vivo" activities. In that sense, the present results could thus give an usefull contribution to future more rational drug-design.

REFERENCES

1. Dodin, G. and Dupont, J., J. Phys. Chem., 1987, **91**, 6322-6326.

2. Dodin, G., Aubard, J. and Falque, D., J. Phys. Chem., 1987, **91**, 1166-1172.

3. Chaires, J.B., Dattagupta, N. and Crothers, D.M. Biochemistry, 1982, 21, 3927-3932.

4. Eigen, M. and De Maeyer, L., in Techniques of Organic Chemistry, Wiley-Interscience, 1963, Part.2, pp. 895-1054.

5. Aubard, J., Meyer, J.J. and Dubois, J.E., Chem. Instrum., 1977, **8**, 1-16.

6. Adenier, A. and Aubard, J., J. Chim. Phys., 1987, **84**, 921-927.

7. Dreyfus, M., Dodin, G., Bensaude, O. and Dubois, J.E., J. Amer. Chem. Soc., 1975, **97**, 2369-2376.

8. Schwaller, M.A., René, B., Paoletti, C., Aubard, J. and Dodin, G. VIIIème Forum de Cancérologie, Faculté de Médecine Xavier Bichat, Paris, June 1988.

Chapter 3

Dye–Ligand Adsorbents in Chromatographic Systems

THE USE OF BIOSPECIFIC ELUTION IN PURIFYING PROTEINS FROM DYE-LIGAND ADSORBENTS

ROBERT K. SCOPES
Centre for Protein and Enzyme Technology,
La Trobe University, Bundoora, Vic., 3083, Australia

ABSTRACT

Elution of protein from dye-ligand adsorbents can be carried out by non-specific means (salt or pH changes), in which case generally poor resolution is achieved. Biospecific elution using natural ligands to the protein can, in the appropriate conditions, achieve a much higher degree of purification. For biospecific (affinity) elution to succeed, the dye must bind at the natural ligand's site; screening of a range of dyes to find such an interaction may be necessary to achieve optimum results from affinity elution. The use of differential column adsorption with affinity elution can result in a one-step purification of 20 to 100-fold at the first step in many instances.

INTRODUCTION

The introduction of dye-ligand adsorbents for protein isolation nearly twenty years ago added an additional mode of adsorption to the others available at the time, i.e. ion-exchange, hydrophobic, hydroxyapatite, and biospecific affinity. Whereas the initial application of Cibacron Blue adsorbents was for isolating dehydrogenases and kinases, later findings indicated that many other types of protein would bind [1,2,3]. Indeed, we now think of dye adsorbents as being selective rather than specific, since out of a mixture of many proteins they may select the one we want, but also bind many others. Because of the fact that many proteins bind, together with the observation that gradient elution (salt, pH) does not achieve high resolution, the degree of purification obtainable using a dye adsorbent may not be great.

Affinity chromatography, in which a known ligand of the protein is attached by chemical means to the inert matrix, has since its introduction used biospecific behaviour at two stages - during adsorption, and at the elution step. The latter, which has variously been termed affinity elution, biospecific elution or substrate elution [4], involves the incorporation of the natural affinity ligand into the elution buffer, thereby competing with the immobilized ligand for the specific proteins. This process depends on the partitioning of the protein between adsorbed and non-adsorbed state, and operates ideally when the partitioning ratio between these is around 10. However, it does not depend on the nature of the adsorption; all that is required is that the binding of free ligand to the protein forms a complex which itself has little or no affinity for the adsorbent. Consequently, affinity elution procedures have been used successfully from non-biospecific adsorbents, such as ion-exchangers [4-6] and dyes (which are not "biospecific" in the strictest sense). Using affinity elution, especially when coupled with differential adsorption columns [7], the potentials of high resolution using dye-ligand adsorbents can be realized.

ELUTION OF PROTEINS FROM DYE-LIGAND COLUMNS

1.   Ionic Strength. Dyes can be regarded as multi-functional in that they can interact with proteins electrostatically (most dyes used being highly negatively-charged), hydrophobically through the chromatophoric rings and linkage rings in their structure, through hydrogen bonding, and in particular by van der Waals forces between π-electron orbitals and dipolar stuctures such as α-helices in the proteins. Nevertheless, the predominant interactions seem to be those that are weakened by increasing salt concentrations; most proteins can be eluted from dyes by 1-1.5 M NaCl, and better by chaotropic salts such as NaCNS. On the other hand, stucture-forming salts such as $(NH_4)_2SO_4$ may elute fewer proteins, as the hydrophobic

interactions are increased more effectively by sulphate anions. Indeed, the more hydrophobic proteins may require conditions of elution such as very low ionic strength (even water), slightly alkaline buffers, or the presence of ethylene glycol, as would be used from a true "hydrophobic" adsorbent.

Partly because of the conflicting effects of the salts on the balance between polar and non-polar interactions, gradient elution with salts rarely achieves the sort of resolution expected from an ion-exchanger. Gradient elution is often less satisfactory than a stepwise programme of salt concentration increases, which will elute the desired proteins in a smaller volumes, through perhaps slightly more contaminated with unwanted material.

2. Effect of pH. The importance of pH in both adsorption and elution on dye-ligand columns is not as great as with ion-exchangers, but nevertheless for many proteins it is essential to use the correct pH for adsorption. The negative charges on dyes suggest that ion-exchange behaviour should mean more protein-binding at lower pH, and this is found to be true [8,9]. It is also an observation that non-electrostatic interactions are strengthened by lower pH, due possibly to the more positively-charged proteins interacting more strongly with electron dense surfaces on the adsorbents. Nevertheless, many negatively-charged proteins will bind to the negatively-charged dyes at all pH's between 5 and 10, but can be eluted by salt. Other proteins may be eluted by quite small changes (increases) in pH, exhibiting the expected cation-exchange behaviour. Thus, there is a range of behaviour which is not initially predictable. Elution by an increasing pH gradient can be more successful than using a salt gradient, and more successful (chromatofocussing excepted) than using pH gradients with ion-exchangers. Because the adsorption behaviour on dyes changes little in the ionic strength range 0.02 to 0.2, a strong buffer solution can be used to generate the pH gradient at, say $I = 0.15$. In this way buffering effects of bound proteins, and

any titratable groups on the dyes can be overcome and a smooth pH elution gradient obtained. However, pH gradient elution from dye adsorbents has not been widely used to date.

3.   Affinity Elution. As indicated above, elution of proteins from dye adsorbents by inclusion of a biospecific ligand in the buffer has frequently been successful. The implication of a successful affinity elution process is that the interaction between protein and dye is at the ligand's binding site, and this interaction is thus prevented by free natural ligand binding to the protein [10]. However, there are other possible explanations, including conformational changes of the protein when forming the protein-ligand complex, or an increase in overall negative charge by binding a negatively-charged ligand which might lessen electrostatic interactions with the dye. Most evidence points to the first possibility, biospecific interactions. In many cases it has been shown that the dye in question can act as a competitive inhibitor to the ligand in free solution, confirming the concept that the dye binds at the biospecific ligand site [11, 12]. Direct demonstrations of dyes binding to substrate sites in dehydrogenases have been made using X-ray crystallography [13]. However we must not assume that dye-protein interactions are always of this type. Nor is the inability of natural ligand to elute a protein from a dye adsorbent necessarily a proof that the interactions are not through the natural ligand binding site. Conditions must be found, for successful affinity elution, in which the protein is already partly desorbed (see Figure 1), and the added ligand's concentration should be at least 10 X K_d (K_d as measured in those particular buffer conditions) [4] Only if these conditions are met, and the protein is still not eluted, can we assume that the dye-protein interaction is through parts of the protein surface other than the ligand binding site.

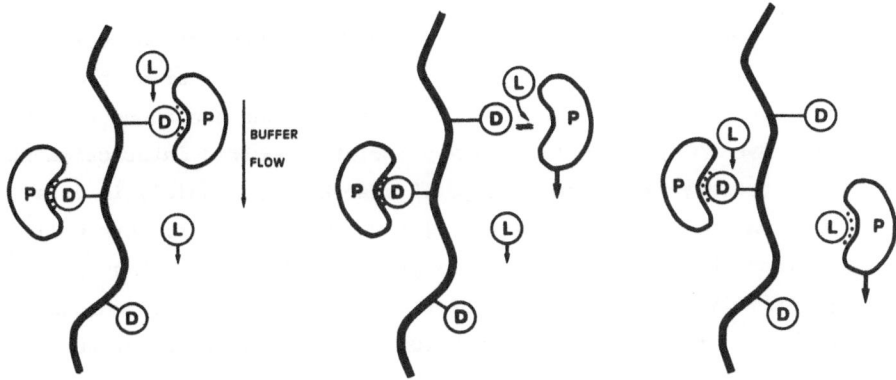

<u>Figure 1.</u>

(a) Protein   (b) Protein   (c) Natural
     bound   dissociates   ligand
     to dye.   from dye.   prevents
                                                 recom-
                                                 bination.

    The procedure for successful biospecific elution of a
protein from a dye-ligand adsorbent column involves, after
adsorption, washing with a buffer which results in a slight
weakening of the interaction. This causes the protein to
commence moving slowly down the column, though still spending
most of the time immobilized in the adsorbed state. Specific
ligand is than added to the buffer at 10-20 x K_d concentration,
whereupon the protein binds to the ligand and is eluted.
Decreasing the flow-rate at this stage can improve the sharpness
of elution. The process is illustrated in Figure 1. Only
proteins binding to that particular ligand (which were
appropriately weakly interacting with the column) should appear
in the elution peak.

SELECTION OF APPROPRIATE DYE ADSORBENT

1.   Screening of Columns. A biospecific interaction, between
a natural ligand's binding site and a dye, may occur with only a
few dyes, and the strength of the interactions are bound to vary
from dye to dye. To find an appropriate specific interaction it
is necessary to screen a range of adsorbents, possibly in more
than one buffer condition [7,14]. The most selective dye will
be that which binds the desired protein well, but with a
relatively small proportion of the other proteins applied also
binding. For a full investigation, a quantitative assay for the
desired protein should be available,.and the fraction passing
through the columns tested. This fraction should also be tested
for total protein content. It is then possible that results
such as illustrated in Table 1 might be found. Potentially
useful dyes are I, G and D. The latter, although letting
through some of the desired protein, binds a relatively small
proportion of the total protein. Also of potential use is dye
E, which allows all of the desired protein through (a "negative"
selectivity") but still binds the other proteins in comparable
amounts. It is most unlikely then that dye E would have any
significant biospecific interaction with our desired protein,
but D, G and possibly I are likely to. This screening is
specifically selecting for interactions that are, for a
particular dye, stronger with the desired protein than with the
other proteins present.

TABLE 1
Possible Results of a Screening of Dye-ligand Columns

Dye (in Order of Protein-Binding	% of Total Protein in Non-absorbed Fraction	% of desired Protein in Non-absorbed Fraction
A	72	100
B	68	100
C	67	90
D	62	30
E	59	100
F	59	60
G	54	10
H	48	50
I	44	0
J	43	20
K	40	10
L	38	0
M	38	0

2.   Differential Adsorption Techniques. In the example given in
Table 1, if dye G were chosen, nearly all the desired protein
would bind, but also 46% of the applied protein in general.
This would imply a 2-fold purification at the adsorption stage.
If on the other hand two columns, E and G, were coupled together
in series, much of that 46% would in fact be removed on E,
leaving as little as 5-10% binding to G, perhaps a 20-fold
purification at the adsorption stage. Because of the
differential selectivity of the dyes, not all of the 41%
expected to bind to E would have bound to G, but in general the
nature of the proteins binding is very similar. Thus for
maximizing the degree of purification, a two-column procedure is
generally the best; biospecific elution from the second
"positive" absorbent then maximizes the purification [4,7]. In
many cases we have achieved 50-fold plus purification from a
crude extract using differential column adsorption followed by
affinity elution; recovery is usually at least 80% (Table 2).

TABLE 2

Glycolytic Enzymes of *Zymononas mobilis*: Elution from Dye Adsorbents

Enzyme	Adsorbent	pH	Elution
Glucokinase	Scarlet MX-G	6	increase pH **
Fructokinase	Red H-E3B	6	MgATP
Glucose 6-P dehydrogenase	Scarlet MX-G	6	NADP$^+$
6-P gluconolactonase	Green H-E4BD	6	glucose 6-P
Gluconate kinase	Brown MX-5BR	6	gluconate + ATP
6-P gluconate dehydratase	Blue H-EGN	6.5	glycerol 1-P
KDPG aldolase	Yellow MX-GR	6	pyruvate
Glyceraldehyde-P dehydrogenase	Violet H-3R	6.5	NADH
Phosphoglycerate kinase	Red 3-BA	6	1,3-P$_2$-glycerate
Phosphoglycerate mutase	Red 3-BA	6	3-P-glycerate
Enolase	Not adsorbed on any dye at pH 6		
Pyruvate kinase	Yellow MX-4R	6	P-enolpyruvate
Pyruvate decarboxylase	Yellow H-E4R	5.5	increase salt **
Alcohol dehydrogenase-1	Scarlet MX-G	6	NAD$^+$
Alcohol dehydrogenase-2	Blue H-4R	6.5	NAD$^+$
Gluconolactonase	Brill. Blue R	6	increase pH **
Glucose: fructose oxidoreductase	Brill. Blue R	6	increase salt **

** not affinity eluted

CONCLUSIONS

Not all dye-protein interactions are "biospecific", and when
not, the technique of affinity elution is unlikely to be successful.
But with many enzymes, especially those with negatively-charged
substrates, biospecific elution has been successful [10]. It is
probable that dye interactions involving positively-charged groups in
the enzyme's binding site (which normally interact with the true
substrate) are important. In particular, sulphonate groups are likely
to mimic phosphate groups on, for example, ATP, phosphorylated sugars,
etc. In our experience, uncharged ligands have not successfully
eluted enzymes, whereas negatively-charged ligands have almost always
been successful [10]; but there are exceptions to this rule both

ways. It is probable that a different range of absorbents, not necessarily related to dyes, may be more appropriate for proteins with uncharged ligands.

REFERENCES

1. Kopperschläger, G., Freyer, R., Diezel, W. and Hofmann, E., Some kinetic and molecular properties of yeast phosphofructokinase. FEBS Letters, 1968, 1, 137-141.

2. Swart, A. and Hemker, H., Separation of blood coagulation factors II, VII, IX, and X by gel filtration in the presence of dextran blue. Biochimica et Biophysoca Acta, 1970, 222, 692-695.

3. Lowe, C.R., An Introduction to Affinity Chromatography, Series, Laboratory Techniques in Biochemistry and Molecular Biology, ed., T.S. Work and E. Work, Amsterdam, North-Holland, 1979, Vol. 7, pt. 2, pp. 269-522.

4. Scopes, R.K., Protein Purification: Principles and Practice, 2nd. ed., Springer-Verlag New York Inc., 1987, 1-329.

5. Pogell, B.N., Enzyme purification by selective elution with substrate from substituted cellulose columns. Biochem. Biophys. Res. Commun., 1962, 7, 225-230.

6. Scopes, R.K., Purification of glycolytic enzymes by using affinity elution chromatography. Biochem. J., 1977, 161, 253-263.

7. Scopes, R.K., Multiple enzyme purifications from muscle extracts by using affinity elution chromatographic procedures. Biochem. J., 1977, 161, 265-277.

8. Scopes, R.K., Use of differential dye-ligand chromatography with affinity elution for enzyme purification: 2-keto 3-deoxy 6-phosphogluconate aldolase from *Zymomonas mobilis*. Anal. Biochem., 1984, 136, 525-529.

9. Baird, J.K., Sherwood, R.F., Carr, R.J.G. and Atkinson, A., Enzyme purification by substrate elution chromatography from Procion dye-polysaccharide matrices. FEBS Letters, 1976, 70, 61-66.

10. Angal, S. and Dean, P.D.G., The use of immobilized Cibacron Blue in plasma fractionation. FEBS Letters, 1978, 96, 346-348.

11. Scopes, R.K., Dye-ligands and multifunctional Adsorbents: an empirical approach to affinity chromatography. Analytical Biochemistry, 1987, 165, 235-246.

12. Ashton, A. and Polya, G., The specific Interaction of Cibacron and related dyes with cyclic nucleotide phosphodiesterase and lactate dehydrogenase. Biochem., J., 1978, 175, 501-506.

13. Subramanian, S. and Kaufman, B.T., Dihydrofolate reductase from chicken liver and Lactobacillus casei bind Cibacron Blue F3G-A in different modes and at different sites. J. Biol. Chem., 225, 10587.

14. Biellmann, J.-F., Samama, J.-P., Bränden, C.-I and Eklund, H., X-ray studies of the binding of Cibacron Blue F3GA to liver alcohol dehydrogenase. Eur. J. Biochem., 1979, 102, 107-110.

15. Qadri, F., and Dean, P.D.G., The use of various immobilized triazine affinity dyes for the purification of 6-phosphogluconate dehydrogenase from Bacillus stearothermophilus. Biochem., J., 1980, 191, 53-62.

THE INFLUENCE OF MATRICES ON DYE-PROTEIN INTERACTION

A.A. GLEMZA, B.B. BAŠKEVIČIUTE, V.A. KADUŠEVICIUS,
J.H.J. PESLIAKAS, O.F. SUDZIUVIENE
ESP "Fermentas", Vilnius, USSR

Adsorbents with immobilized dyes are employed for purification of enzymes and other proteins in affinity chromatography. Adsorbents are useful, because the ligands are chemically and biologically stable. Its immobilization on polymer is simple and the interaction with certain enzymes i.e. proteins is sufficiently specific. That allows us to look at them as affinity (or pseudoaffinity) adsorbents.

In All-Union Research Institute of Applied Enzymology "Fermentas", in the course of 10 years, investigations to create large scale processes obtaining physico-chemically and funcionally pure enzymes have been performed. They are applicable in gentle modification of structure of nucleid-acids in immuno-enzymatic and photometric analysis. There are some technological processes in which dye adsorbents are used.

Dye-adsorbent consists of 2 components: matrix-polymer and a dye, which is a specific ligand of some enzyme. They are bound covalently. Interaction of numerous commercial dyes with various enzymes and blood proteins has been investigated by us.

Dye-protein interaction has been analysed in solutions applying pure dyes by means of differential spectroscopy and circular dichroism and inhibitory analysis. Let us make some suppositions of possibility to use selected dye as a ligand of adsorbent. Finally the application of dye-adsorbents have been tested by chromatography. At this stage the yield of main activity and the yield of contaminant activity have been studied.

Presently, our experimental data is not sufficient to make any conclusive remarks of dye-enzyme interaction mechanisms, though some of the conclusions being quite appropriate.

TABLE 1
Adsorption-desorption of human serum albumin on sepharose-dyes

Dye	Adsorption capacity, mg ml	Percentage of protein eluted by agent, %	Total desorption yield, %
Orange 5K	12	1,4 M NaCl - 76 1 M KCNS - 0 8 M urea - 7	83
Orange 4K	16	1,4 M NaCl - 60 1 M KCNS - 6 8 M urea - 1	67
Scarlet 2Z	13	1,4 M NaCl - 59 1 M KCNS - 8 8 M urea - 2	69
Red 6S	7	1,4 M NaCl - 55 1 M KCNS - 4 8 M urea - 3	62
Cibacron Blue F3GA	19	1,4 M NaCl - 51 1 M KCNS - 23	74
Red-violet 2KT	19	1,4 M NaCl - 48 1 M KCNS - 37 8 M urea - 4	89
Red-Brown 2K	9	1,4 M NaCl - 34 1 M KCNS - 11 8 M urea - 16	61
Bordo ST	7	1,4 M NaCl - 33 1 M KCNS - 67	100
Blue KX	6,4	1,4 M NaCl - 30 1 M KCNS - 13 8 M urea - 16	59
Yellow 5Z	3,8	1,4 M NaCl - 24 1 M KCNS - 7 8 M urea - 26	57
Violet 5K	12	1,4 M NaCl - 16 1 M KCNS - 50 8 M urea - 1	67
Red-Brown 2KT	20	1,4 M NaCl - 13 1 M KCNS - 55	68
Yellow light resistant 2KT	10	1,4 M NaCl - 10 1 M KCNS - 25 8 M urea - 15	50

A group of dyes has been discovered to interact preferently only with dehy-drogenases. Another group has been observed to possess lower specificity, forming complexes with dehydrogenases, kinases, non-specific nucleases and hormonal contaminants of human albumin. Finally, it was established that dye scarlet 4 ZT did not form stable complexes with any of the enzyme under investigation.

Data given in Table 1 enable us to evaluate the contribution of va-rious interactions in the formation of albumin-dye complexes./1, 2/. Investigation results showed that to a certain degree hydrophobic and elec-trostatic forces are involved in the binding of dyes to proteins. The in-fluence of these forces in respect to individual dyes differ within a wide range. This might explain the discrepancy in data, occuring between the dis-sociation constants of protein-dye interaction and the chromatographic be-haviour of the some proteins on dye adsorbents. In the future elaboration of concrete enzyme interaction in solution after the combination with chro-matography is required. This is the way of most effective ligand selection. It was spoken of a certain matrix, which was the basis of any adsorbent and had an influence upon ligand (dye)-protein interaction. The choice of matrix is determined by many factors – its physico-chemical, mechanical, hydrody-namical properties and stability in chromatography. Presently, the majority of dyes have been immobilized on agarose, in many cases being unsurpassed when creating biospecific adsorbents. Today, however, I shall present some data, concerning the use of inorganic matrixes for cretion of dyes contai-ning adsorbents. We used aminosilochrome S-80 as well as macroporous silica gels MSA-1500 and MSA-2500 containing high specific surface and large pores. However, their application in chromatographic fractionation of proteins is impossible because of hydrolytic instability and non-specific adsorption of neutral and basic proteins. We needed to elaborate efficient modification methods, which would decrease non-specific adsorbent-protein interaction and increase hydrolytic stability. For these experiments inorganic matrixes were coated by soluble dextran and glutaric aldehyde /3/. As can be seen from our experiments on hydrolytic stability, the modification of aminosilo-chrome S-80 considerably decreased the rate of bond dye hydrolysis. The re-sults of hydrolytic stability of modified Aminosilochrome S-80 you can find in Table 2.

Modification of inorganic matrixes allowed us to decrease non-specific adsorption to a considerable extent. As can be seen from Table 3, irreversi-ble, non-specific adsorption has been entirely excluded after amino silica

TABLE 2

The dependence of stability of affinity adsorbent
on matrix modification

Matrix	Red-Brown 2K conc. M ml	Hydroliza tion time	Released dye quantity, M ml	Dye releasing rate $\ln c$	Dye releasing ratio matri modified matrix
Aminosilochrome S-80	13,7	17	1,62	0,1260	–
Aminosilochrome S-80, modified by glutaric aldehyde	6,8	17	0,10	0,0154	8,2
Aminosilochrome S-80, modified by dextran	6,9	17	0,17	0,0262	4,8

TABLE 3

Adsorption-desorption of proteins on aminosilochrome
S-80, silica gel MSA-2500 and their derivatives

Matrix	Dextran quantity, mg g	Protein	Protein adsorption, mg g	quantity desorption, mg g	desorption %
Aminosilochrome S-80	–	Albumin Hoemoglobin Cytochrome c	37,7 7,0	23,6 2,6	62,6 37,1
Aminosilochrome S-80 dextran 40 000	63,9	Albumin Hoemoglobin Cytochrome c	14,4 2,8 3,0	14,4 1,3 2,8	100 46,5 93,3
Aminosilochrome S-80 dextran 40 000 acetylated	63,9	Albumin Hoemoglobin Cytochrome c	14,2 1,7 0	14,2 1,3 –	100 70,5 –
Aminosilica gel MSA-2500	–	Albumin Hoemoglobin Cytochrome c	15,4 15,6 3,2	14,7 7,0 1,0	95,6 44,8 31,3
Aminosilica gel MSA-2500 dextran 60000	13,1	Albumin Hoemoglobin Cytochrome c	11,9 2,7 0	11,9 2,7 –	100 100 –

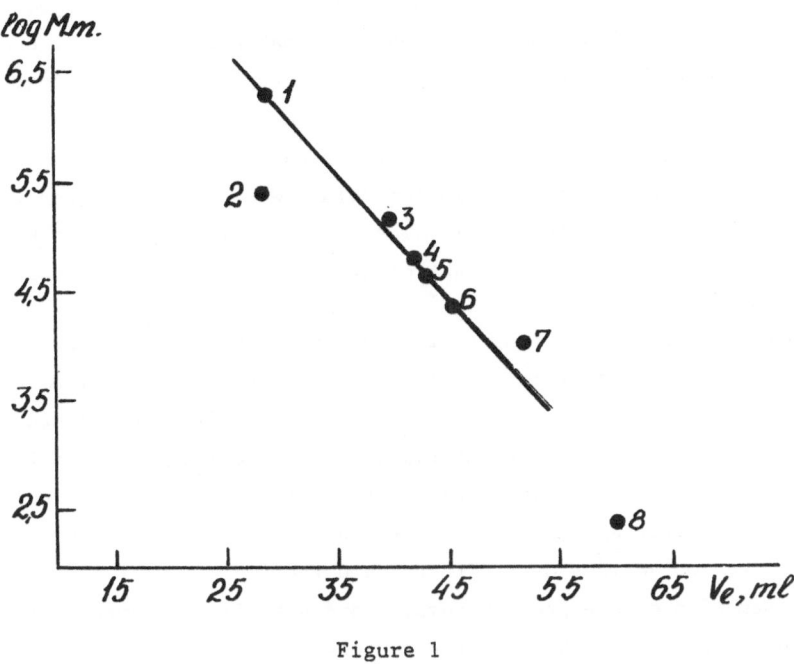

Figure 1

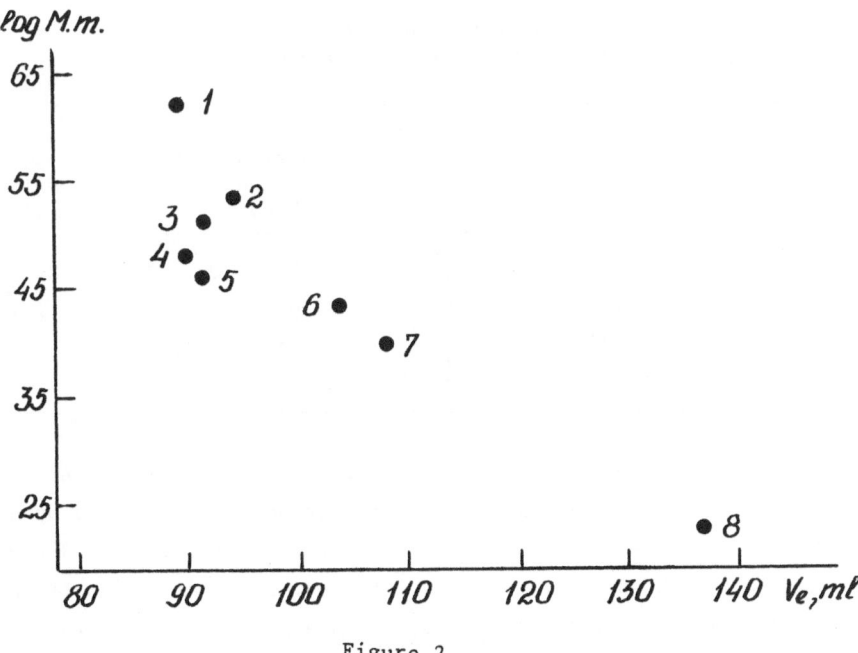

Figure 2

gel MSA-2500 modification by dextran 60 000. These results have not been obtained with aminosilochrome S-80. Data, summarized in Table 3, confirmed investigated behaviour of modified inorganic matrixes at chromatographic conditions. Dextran derivative of amino silica gel MSA-2500 proved to be a good gel-filtrating material on which practically no effect of matrix-protein interaction was observed. Elution volume was linearly proportional to their molecular weight. (Figure 1). There was no linear relation between elution volume and molecular weight of proteins on modified aminosilochrome S-80 (Figure 2).

When analyzing the results of batch adsorption of albumin, we came to a conclusion, that dye-protein interaction has been very complicated. In the formation of dye-protein complexes, forces of different nature have been involved. This reflects in the possibility of albumin to elute different reagents. As can be seen from Table 1 at desorption of albumin from different dye-sepharoses NaCl, KSCN and urea have been used. Different results were obtained on various dyes. It approved the influence of different forces on formation of dye-protein complex and significant contribution towards concrete dye-protein (albumin) interaction.

TABLE 4
Comparison of adsorption-desorption of human serum
albumin on different adsorbents

| Dye | Eluted agent | Percentage of eluted protein | | | |
		Sepharose	Modified silochrome S-80	FSL-gel	Spherone 100000 LC
Orange 5K	NaCl	76	67	20	30
	KCNS	0	33	66,5	66,8
	urea	7	0	0	0
Yellow 5Z	NaCl	24	58	7,5	
	KCNS	7	30	77,5	n.d..
	urea	26	7	–	

Table 4 represents data of interaction of albumin with dye, immobilized on different matrixes sepharose, silochrome S-80, modified by dextran, gel-FSL and spherone. As it is seen from data, there are great differences in the same dyes bond to the various matrixes. It shows a significant contribution of matrix towards the nature of dye-protein complex formation.

Now we have a possibility to compare the results, obtained by chromatographic fractionation of yeast glucose-6-phosphate-dehydrogenase on ad-

TABLE 5

Chromatographic purification of yeast G6PDH

Dye	Sepharose CL-6B			Modified aminosilochrome S-80			Dextran silica gel MSA-2500		
	Capacity, U/M	Activity yield %		Capacity, U/M	Activity yield %		Capacity, U/M	Activity yield %	
		NaCl	NADP		NaCl	NADP (03 MNaCl)		NaCl	NADP
Orange 5K	3,0	32	35	3,8	76	28	9,0	51	62
Red 6S	4,7	70	17	1,7	57	79	4,0	80	55
Red-Brown 2K	2,8	74	68	2,0	66	26	2,0	92	66

sorbents, created by immobilization of 3 dyes on sepharose CL-6B, modified aminosilochrome S-80 and amino silica gel MSA-2500 modified by dextran. (Table 5).

Obtained results are the basis to conclude that matrix can change the nature of dye-protein interaction. Aminosilochrome S-80 practically in all cases interferes with enzyme desorbed non-specifically (NaCl) as well as specifically (NADP), in comparison either with sepharose or modified amino silica gel MSA-2500.

In all probability, it was effect of "non-specific adsorption", which was not completely excluded by means of aminosilochrome S-80 modification. Thus, a better correlation can be observed between sepharose CL-6B and dextran-silica gel MSA-2500, in which non-specific adsorption practically is not observed.

REFERENCES

1. Baškevičiūte B.B., Sūdžiuviene O.F., Pesliakas J.H.J., Glemza A.A., Mygunov V.N., Liachova T.D., Pozina I.M. Study of human albumin interaction with adsorbents, containing immobilized dyes. Prikl. Biochim. Mikrobiol., 1987, 22, 584-590.

2. Mygunov V.N., Pesliakas J.H.J., Liachova T.D., Sudžiuviene O.F., Pozina I.M., Glemža A.A. Comparative analysis of protein composition of raw materials of human albumin, obtained by the aid of silochrome adsorbent. Gematologiya Transfuziologiya, 1985, 30, 58-62.

3. Kaduševičius V.A., Sūdžiuviene O.F., Pesliakas J.H.J. Modification of inorganic matrixes for synthesis of biospecific adsorbents. Bioorg. Khim., 1983, 9, 1128-1135.

4. Kaduševičius V.A., Sudžiuviene O.F., Pesliakas J.H.J. Chromatography of yeast glucose-6-phosphatedehydrogenase on adsorbents with immobilized dyes. Prikl. Biokhim. Mikrobiol., 1986, 22, 291-296.

EVALUATION OF CHROMATOGRAPHY SUPPORTS PREPARED BY GRAFTING REACTIVE DYES ONTO DEXTRAN COATED SILICA BEADS.

YOLANDE KROVIARSKI, XAVIER SANTARELLI°, SYLVIE COCHET, DANIEL MULLER°,
THIERRY ARNAUD*, PIERRE BOIVIN and OLIVIER BERTRAND.
INSERM U160, Hôpital Beaujon, 92118 Clichy Cedex
LRM (°), CSP Ave J.B. Clément 93430 Villetaneuse
IBF Biotechnics (*) 92390 Villeneuve la Garenne FRANCE

ABSTRACT

Procion Blue HE-GN was grafted onto silica beads coated with DEAE Dextran, and evaluation of the supports was performed. It was shown that the same chemistry used for grafting dyes onto agarose (involving treatment in hot alkaline conditions) could be used. The prepared supports could be used for purification of 6-phosphogluconate dehydrogenase using the affinity of the enzyme for the dye. Capacities of dextran coated silica based supports were found much lower than those of agarose based supports although very high levels of dye substitution could be obtained. Comparison of supports prepared with dextrans differing by DEAE content showed that at least in the model system used (6PGD and Procion Blue HE-GN), satisfactory interaction between enzyme and immobilized dye could only be obtained with a low DEAE groups content. Interaction between enzyme and immobilized dye seemed to be in some way kinetically limited : capacities of the dextran coated silica based dye columns decreased significantly with increasing mobile phase velocity.

INTRODUCTION

Dyes immobilized on agarose have known wide acceptance as chromatographic supports because among other virtues immobilization chemistries are very simple. On the other hand agarose is not a perfect support because it cannot withstand high pressures and flow rates.

Immobilisation of dyes onto silica through silane chemistry is not so simple, moreover this approach does not always fully protect against unwanted interactions between proteins and silica backbone.

Coating of silica beads with DEAE dextran has been proposed previously (1-2) as a way to circumvent the problem of unwanted interactions between proteins and silica, as well as a way to increase chemical resistance of silica to alkaline conditions.

Some trials have been made with dextran coated silicas grafted with Procion Blue HE—GN a dye which was previously used for purification of 6 Phosphogluconate dehydrogenase (6PGD) from human hemolysate (3).

MATERIALS AND METHODS

Procion Blue HE—GN was a much appreciated gift of I.C.I. France (Clamart). Spherodex X—015 silica beads (pore diameter, 1250 angstroms, particle size, 40 to 100 micrometers) were obtained from IBF Biotechnics (Villeneuve la Garenne France). Unsubstituted Dextran (T 500) and one sample of DEAE Dextran were purchased from Pharmacia (Uppsala Sweden).

Preparation of DEAE Dextran and coating of silica beads:
Preparation of Dextran substituted with a defined and low level of DEAE groups was performed as described in reference 2. Technique for coating of Silica beads with DEAE Dextran was taken from reference 2.

Grafting of dye onto Dextran coated silica beads:
Basically the same technique which was used for attachment of dyes onto agarose (3) has been used with dextran coated Silica : Dry dextran coated Silica was suspended in 2% NaCl (w:v) and put under vacuum. Support was then rinsed on a fritted disc with 0.2 M NaOH containing 2% NaCl and transferred in a stoppered vessel containing dye dissolved in the same solution (ratio of silica volume to total volume was 1:2). Vessel was then tumbled at 60°C for one hour. Then silica was rinsed by large volumes of 0.2 M NaOH 6M Urea and water on a Buchner funnel. The procedure was repeated twice or thrice in order to raise level of dye substitution, (number of treatments is indicated later in the text by the postcript I, II or III).

Analytical methods applied to the characterization of the diverse preparations of modified silica beads.
Dye incorporation in silica beads was evaluated by elemental analysis for Sulfur. Pore size diameter distribution was analysed by mercury porosimetry in a Carlo Erba apparatus.

Chromatography experiments performed with the various modified silica beads.

   Frontal chromatography: Frontal chromatography experiments have been performed with 1.6 ml volume columns (0.8 cm internal diameter). Before hemolysate loading, columns were equilibrated in buffer A : 10 mM KOH adjusted to pH 6.5 with solid MES containing 20 mM NaCl and 2mM MgCl$_2$ (4). Flow rate was 3 ml/h, enzymatic activity was assayed in 0.5 ml fractions, collected from the beginning of hemolysate loading. Hemolysate was prepared with same buffer used for equilibration of columns as detailed in reference 3.

   Zonal chromatography: Modified silica beads were slurry packed in 0.46 cm internal diameter and 10 cm high stainless steel columns and equilibrated in buffer A. After hemolysate loading, columns were rinsed till return of the recorder trace to baseline with buffer A followed by buffer B (of same composition as buffer A but without MgCl$_2$ and adjusted to pH 7.5). Buffer B did not elute any enzymatic activity but unwanted proteins. Thereafter column was developed with 6 ml of buffer A then with nucleotides dissolved in 6 ml of the same buffer at the concentrations indicated later in table 2. Column was finally rinsed with buffer A. Yields of eluted enzymatic activity were evaluated by taking into account all enzymatic activity found in the eluate after injection of the pulse of nucleotide containing eluent.

RESULTS AND DISCUSSION

Although the alkaline conditions used for grafting of the dyes were rather harsh no dissolution of the silica beads was observed. Light microscopic examination of the beads did not reveal any alteration of the beads. Dextran coated silica was found in fact so resistant to alkaline hydrolysis that quantitative evaluation of dye fixation on the beads had to rely on elemental analysis, hydrolysis method used with silica beads on which dye had been grafted through silane chemistry (5) was found inapplicable.

Dye incorporation in modified silica beads were 2.9 milligrams of dye per one milliliter of support (Procion Blue HE-GN Dextran coated silica I), 16.5 mg/ml ·(Dye DCS II), 31.2 mg/ml (Dye DCS III). Dye incorporation in Ultrogel A4 (III) and Sepharose CL 4B (III) which have been used for comparison purposes were respectively 6.0 mg/ml and 6.2 mg/ml.

Pore size analysis was conducted on the Procion Blue HE-GN Dextran coated silica (II) and gave results similar to those given by the manufacturer for the bare silica suggesting that neither dextran coating nor dye grafting could induce a change in the pore sizes distribution.

Frontal chromatography of hemolysate on the various supports.

Results of frontal analysis of 6PGD retention from hemolysate with three silica based supports and with agarose based supports are shown in figure 1.

Examination of the breakthrough curves obtained with the various supports reveals that our immobilized dyes on dextran coated silicas have much lower capacity than agarose based supports with the same grafted dye.

Much higher levels of substitution were obtained on Dextran coated silica than on agarose supports. However frontal chromatography did show that capacity for the enzyme of the modified silica supports was much lower than with agarose based supports (the differences in the capacities of immobilized dyes on different agarose based matrices, at a similar level of substitution, has been already noted -3-, also by others -6-). Of the three levels of substitution obtained on modified silica supports, it is the intermediate value (Dye DCS II) which gave the better result.

It has to be stressed also that immobilized dye support prepared with silica beads coated with commercial DEAE Dextran did not retain any 6PGD from hemolysate. This fact suggests that presence of many positive charges onto the surface of the coated silica has a deleterious effect on interaction between enzyme and support. 43 % of glucose units of the commercial DEAE Dextran were carrying a DEAE substituent and only 7.5 % in the lab prepared DEAE Dextran.

118

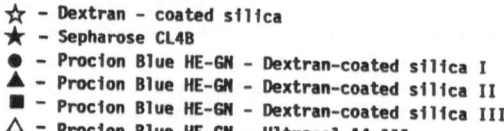

Figure 1. Elution profiles of 6PGD activity obtained with various supports. The value of 100 % on the ordinate axis corresponds to the 6PGD activity value present in hemolysate loaded onto the column.

Zonal chromatography experiments

Results of zonal chromatographies conducted in a somewhat overload mode are reported in table 1.

TABLE 1

Results of zonal experiments performed by loading hemolysate onto immobilized Procion Blue HE-GN Dextran coated Silica II.

Dilution of hemolysate	Flow rate	6PGD activity deposited	6PGD activity retained	
1/36	12 ml/h	0.288 IU	0.153 IU	
				53%
1/36	60 ml/h	0.359 IU	0.077 IU	
				21%
1/6	60 ml/h	0.357 IU	0.193 IU	
				55%

Chromatographies were conducted at two different flow rates. Results of table 1 show unequivocally that fixation of enzyme on the immobilized support is in some way kinetically limited, amount of retained enzyme being twice lower when flow rate is increased by a factor of five. For comparison purposes chromatographies have been conducted with Dye Sepharose CL 4B III sedimented in a column of same volume (but of different geometry because of compressibility of the agarose gel) at the same flow rates. It was observed that capacity was the same at 12 and 60 ml/h.

Increasing 6PGD concentration in loaded hemolysate sample (third line of the table) does allow a greater amount of enzyme to be retained on the column.

Results of table 2 show that affinity elution is efficient on the silica based support as it is on agarose based supports. The observation (made earlier and discussed elsewhere -3-) that NADP hydrolysed by phosphodiesterase is more efficient than unhydrolysed NADP to elute the enzyme holds true for the silica support as it did with agarose based supports.

TABLE 2

Affinity elution of 6PGD from Procion Blue HE-GN Dextran coated silica II
(Flow rate was 60 ml/h)

6PGD activity deposited	6PGD activity retained	6PGD activity eluted by NADP
0.357 IU	0.193 IU 55%	10 mM — 0.166 IU — (86% of retained activity)
0.347 IU	0.211 IU 60%	5 mM — 0.166 IU (78%)
0.347 IU	0.197 IU 56%	2.5 mM — 0.096 IU (48%)

CONCLUSIONS

These experiments demonstrated that dextran coated silica beads could be used in place of agarose based supports for preparation of immobilized dye chromatography supports. At least in the model system used, amount of DEAE groups present in the coating polymer seems to be an important point as it is when Dextran coated silica is used for size exclusion chromatography (2). High levels of dye incorporation are easily attained, dextran coated silica beads can support harsh alkaline conditions needed for the coupling of dye. Nevertheless it appears that some problems are not solved : enzyme interaction with dye seems to be kinetically limited, capacity is much lower than obtained on agarose supports. Future work will appreciate the usefulness to graft dyes using a spacer arm and evaluate the interest to use other coating polymers.

ACKNOWLEDGEMENTS

Sincere thanks are expressed by the authors to ICI France for the generous gift of dye used in this study.

REFERENCES

1. Tayot, J.L., Tardy, M. and Gattel, P., Ion exchange and affinity chromatography on Silica derivatives. In Methods of Plasma Proteins Fractionation, ed. J.M. Curling, Academic Press, New York, 1980, pp. 149-60 .

2. Santarelli, X., Muller, D. and Jozefonvicz, J., Dextran coated silica supports for HPSEC of proteins. J. Chromatogr., 1988, 443, 55-62.

3. Kroviarski, Y., Cochet, S., Vadon, C., Truskolaski, A., Boivin, P., Bertrand, O., Simultaneous purification of human 6-phosphogluconate dehydrogenase from human hemolysate with chromatography on an immobilized dye as the essential step and extensive use of automation (simultaneous purifcation of lactate dehydrogenase). J. Chromatogr., 1988 (in press)

4. Scopes, R.K., Strategies for enzyme isolation using dye-ligand and related adsorbents. J. Chromatogr., 1986, 376, 131-40.

5. Lowe, C.R., Glad, M., Larsson, P.O., Ohlson, S., Small, D.A.P., Atkinson, T. and Mosbach, K., High performance liquid affinity chromatography of proteins on cibacron blue F3G-A bonded silica. J. Chromatogr., 1981, 215, 303-16;

6. Dean, P.D.G., Qadri, F., Jessup, W., Bouriotis, V., Angal, S., Potuzak, H., Leatherbarrow, R.J., Miron, T., George, E. and Morgan M.R.A., Design faults in affinity chromatography, In Affinity Chromatography and Molecular Interactions ed. J.M. Egly. Editions INSERM, Paris, 1979, pp.321-44.

FAST ANALYTICAL DYE-LIGAND CHROMATOGRAPHY IN PROCESS DEVELOPMENT

L. JERVIS
Department of Biology, Paisley College,
High Street, Paisley, PA1 2BE,
Scotland, U.K.

ABSTRACT

The use of Macrosorb KAX-Cibacron Blue as a dye-ligand adsorbent for both analytical and preparative protein purification is described. Although designed for process-scale chromatography, the adsorbent shows excellent chromatographic behaviour on an analytical scale. It is an ideal material for use in process development, allowing rapid optimisation of conditions and direct scale-up.

INTRODUCTION

Dye-ligand chromatography has been used widely for protein purification (1,2). The dyes used can be attached easily to a range of matrix materials including those suitable for laboratory-scale, process-scale, or high performance analytical applications(3). The conditions required for the above applications have resulted in the development of a variety of matrix materials. None of these are entirely suitable for the range of conditions encountered from analytical through to process-scale work.

Protein purification is an essential feature of biotechnological industries and process development involves considerable attention to purification scale-up problems. Many of these arise from the need to change adsorbent matrix materials during scale-up. Ideally, the same adsorbent should be used throughout process development so that conditions developed in the laboratory can be transferred with confidence to the production plant.

The new high performance Macrosorb matrix materials are adaptable to either small or large-scale chromatography. Their rigid, inert,

Kieselguhr-agarose composite structure is highly porous, has excellent flow properties, and allows very high flow rates. The Matrix material is easily modified by the covalent attachment of ion-exchange groups, affinity ligands, or dyes such as Cibacron Blue. Macrosorb KAX-Cibacron Blue has been used in the development of a process for the preparative-scale purification of lactate dehydrogenase (LDH).

MATERIALS AND METHODS

Development work was carried out using Macrosorb KAX- Cibacron Blue packed in an analytical column (5mm x 50mm). The column was run at a linear flow rate of $0.5cm.min^{-1}$. Equilibration and washing buffer was K.phosphate, 50mM, pH7.5. Sample was whiting muscle extract prepared in the same buffer. Eluant was 2mM NADH in the same buffer. After conditions had been optimised, a direct scale-up was carried out using a preparative column (50mm x 50mm) packed with the same adsorbent and run at the same flow rate.

RESULTS AND DISCUSSION

The results obtained under both analytical and preparative conditions are summarised in Table 1.

TABLE 1
Performance of Macrosorb under analytical and
preparative conditions

Scale	Column Volume (cm^3)	Volume Flow Rate ($cm^3.min^{-1}$)	Bound Enzyme (μ.Katals)	Recovered Enzyme (μ.Katals)	(%)
Analytical	0.98	0.1	66.7	65.7	98.5
Preparative	98	10	6370	6050	95

All results, including cycle time, were identical.

The maximum performance of the analytical system was tested by carrying out runs under progressively shorter cycle times. Even with a cycle time of 5 minutes, corresponding to a linear flow of $10cm.min^{-1}$, performance was good. Complete removal of LDH from the sample occurred and enzyme recovery by specific elution was greater than 95%.

The recovered enzyme was judged to be greater than 95% pure by SDS-PAGE.

The results obtained show that Macrosorb KAX-Cibacron Blue, although designed for process-scale chromatography, gives excellent performance under analytical conditions. It is, therefore, an ideal dye-ligand adsorbent for use in protein purification process development.

REFERENCES

1. Subramanian, S., Dye-ligand affinity chromatography: the interaction of Cibacron Blue F3GA with proteins and enzymes. CRC Crit. Rev. Biochem., 1984, 16, 169-205.

2. Qadri, F., The reactive triazine dyes : their usefulness and limitations in protein purifications. Trends in Biotechnol., 1985, 3, 7-12.

3. Groman, E.V. and Wilchek, M., Recent developments in affinity chromatography supports. Trends in Biotechnol., 1987, 5, 220-224.

A COMPARISON OF MACROSORB-CIBACRON BLUE UNDER FIXED-BED, BATCH
ADSORPTION AND FLUIDISED-BED CONDITIONS

Ewan R. Robertson
Biology Department, Paisley College of Technology,
High Street, Paisley, Renfrewshire, PA1 2BE, Scotland.

Macrosrob-Cibacron Blue is a chromatographic medium which, in being a
composite of kieselguhr and agarose [1] with the triazine dye Cibacron Blue
F3GA immobilised to the agarose component, combines a very rigid structure
with a wide specificity for protein adsorption[2]. The high density
(1.16g/ml) of the material gives fast packing characteristics when used in
fixed-bed columns and also makes it ideal for use in fluidised bed systems
[3]. The adsorbent was assessed under fixed-bed, batch adsorption and
fluidised bed conditions in relation to the partial purification of potato
tuber aldolase.

In all three systems a 50ml volume of adsorbent was used with a
quantity of clarified potato extract containing approximately 150nKatals
of aldolase activity. The amount of aldolase adsorbed, the time taken to
reach adsorbent saturation and the percentage of the bound activity which
could be recovered by elution with fructose-1,6-bisphosphate, was
determined for each of the three systems. Aldolase activity was assayed
by the method of Gracey et al (1969) [4].

Table 1

Linear flow rate	total activity adsorbed	time to adsorbent saturation	elution with fructose-1,6-bisphosphate
Fixed-bed			
260cm/hour	67nKatals	4.2 minutes	91%
Batch adsorption			
not applicable	68nKatals	20.0 minutes	88%
Fluidised-bed			
1200cm/hour	57nKatals	20.0 minutes	80%

The results indicate that, with the high flow rate through the fixed-bed column, adsorbent saturation was achieved in a very much shorter time than with either of the other two systems. This high flow rate was possible thanks to the rigid structure of Macrosorb. Despite using a much higher flow rate in the fluidised-bed system and recirculating the extract continuously through the bed, a much longer time was required to achieve saturation. It has been claimed that fluidised beds can be used with unclarified crude extracts[5] but such material was found to foul the system used here with the immediate loss of fluidation.

REFERENCES

1. Clonis, Y.D., Biotechnology, 1987, 5, 1290-1293
2. Bite, M.G., Eurochem '86-Process engineering today, 1986, 137-142
3. Hammond, P.M., Atkinson, T. and Scawen, M.D., J. Chromatog., 1966, 366, 79-89.
4. Gracey, R.W., Lacko, A.G. and Horecker, B.L., J. Biol. Chem., 1969, 244, 3912-3919
5. Burns, M.A. and Graves, D.J., Biotechnol. Prog., 1985, 1(2), 95-103.

POLYMER DESIGN IN DYE CHROMATOGRAPHY :
From the definition of monomers to the evaluation of
polymeric supports

J. A. MAZZA[a] , P. OUTUMURO[a] , Y. MOROUX[b] & E. BOSCHETTI[b] *

(a)   VILMAX, Santiago del Estero 366 - 1075 Capital
        Federal Buenos Aires, Argentina
(b)   IBF-Biotechnics, 35 av. Jean Jaurès, 92390 Villeneuve
        La Garenne, France

   *   *To whom all correspondence should be addressed*

ABSTRACT

The limitations of immobilized dyes presently available in
large scale chromatography are mainly due to the relatively
high cost of sorbents, which results from expensive and
quite complicated synthesis.
     In this paper, we propose a general approach for the
preparation of a dye-chromatographic support by
copolymerization of special acrylic derivatives of dyes of
interest. The chemical modification of pre-activated dyes
(<u>e.g.</u> Cibacron Blue) with introduction of an acrylic
copolymerizable arm , will be presented in the first part.
Then the preparation of polymers and their characterization
will be discussed.
     These special polymer-dye supports should be of large
interest since their production may be achieved at low cost
and under reproducible conditions. In fact , it is easy to
manage the total composition and properties according to the
nature of the main monomer and the acrylic dye as well as
the concentration of different monomers .

INTRODUCTION

Dye affinity chromatography provides a very powerful
technique for purification of many biologicals of potential
interest in diagnostics and therapeutics. These special
sorbents are obtained by chemical immobilization of triazine
dyes (Cibacron and Procion dyes) (1,2) , pyrimidinyl dyes

and vinyl-sulfone remazol dyes (3,4,5) on a hydroxyl-containing solid matrix. The dye which is activated by the presence of a reactive chlorine group, can be immobilized in strong alkaline medium (6). Alternatively, the reactive dye can be immobilized after introduction of an amino spacer arm on a carboxyl-containing matrix through a condensation agent (7).

However, those classical techniques of immobilization suffer from two main drawbacks :
- no possibility of predicting the dye density on the matrix polymer,
- unspecific adsorption of dye molecule on the support which makes washing very difficult and generates troubles due to dye leakage.

To overcome these difficulties and thanks to our experience in acrylic copolymerization (8, 9, 10, 11), we started a research program on the triazine dyes modification by introduction of a copolymerizable acrylic moiety. The aim of this work was then to prepare in one single step dye-affinity sorbents in granular or beaded form (12).

In this paper, we describe the complete methodology and present preliminary evaluation results on the synthetized supports.

Materials and Methods

Chemicals : Cibacron Blue F3 GA was purchased from FLUKA (Switzerland) and Reactive Red 120 was from SIGMA (USA) ; all other chemicals and standard proteins were bought respectively from ALDRICH (Belgium) and SIGMA (USA).

Preparation of reactive acrylic dyes : The principle of reactive dyes synthesis (Cibacron Blue F3 GA and Reactive Red 120) is illustrated on Fig. 1 and 3. The methodology is extensively described in the patent application (12). Briefly, for the blue dye, the synthesis was performed in two steps :
- introduction of an amino spacer arm,
- introduction of the acrylic moiety.

80 g of Cibacron Blue F3 GA were dissolved in 2 000 ml of DMF at 40°C. We added to this solution 53 g of hexamethylene diamine under stirring and then 8 g of pure pyridin. After stirring overnight, the pH was decreased to 2 by adding 160 ml of 10 M hydrochloric acid and 1,875 g of sodium chloride. 7,500 ml of water were added to precipitate the modified dye and after 30 minutes stirring, the dye was recovered by filtration. The latter product was then washed with 7,500 ml of water at pH 2 to remove the excess reagent (i.e. DMF) and finally dried at 70-80°C (product 3 ; Fig. 1).

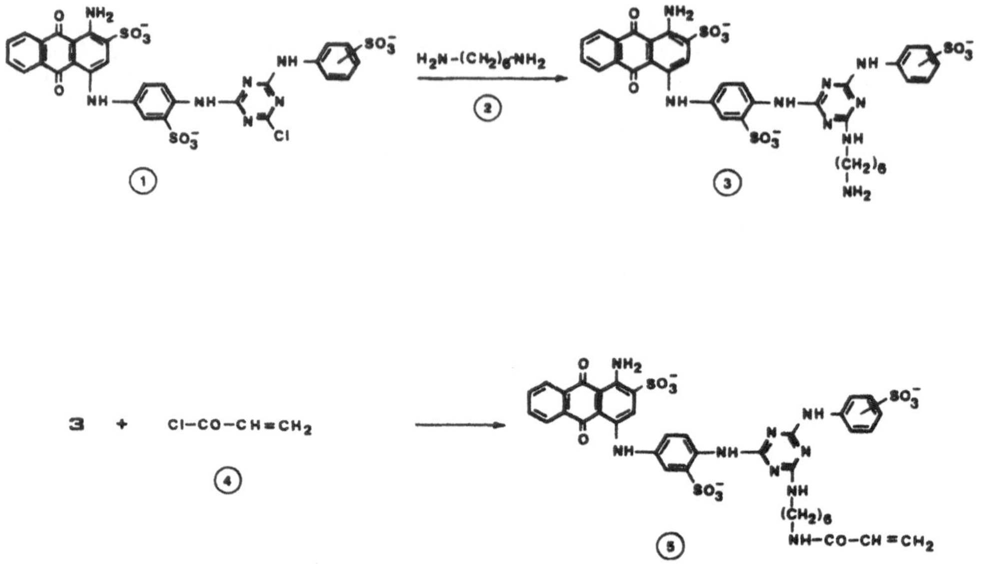

FIGURE 1. Chemical reaction scheme for the synthesis of N-acryloyl-
hexamethylene-N'-cibacron blue (5).
1 : cibacron blue ; 2 : hexamethylene diamine ; 3 : epsilon-
amino-hexamethylene-N'-cibacron blue; 4 : acryloyl chloride
(see the text for details)

The product 3 (Fig. 1) was dissolved in 2,000 ml of
pure dry DMF and then added with 800 ml of triethylamine.
After stirring for complete dissolution of the dye, 17 g of
acryloyl chloride were added and the mixture stirred for two
hours. Once the reaction was achieved, the solution was
added with 7,500 ml of 25 % sodium chloride in water and
stirred for 15 minutes ; then 750 ml of 10 M hydrochloric
acid were added. The precipitate was then recovered by
filtration , suspended again in water and finally dried at
50°C.

The final product (product 5 ; Fig. 1) was
characterized by NMR spectra (Fig. 2) and its purity checked
by thin layer chromatography using the following solvent
mixture : n-butanol, n-propanol, ethylacetate, water
(2:4:1:3). The intermediate 3 (Fig. 1) was nearly absent.

The acrylic derivative of Reactive Red 120 was prepared
in the similar way as per the Blue Cibacron F3 GA (see
Fig. 3). 0.2 moles of Reactive Red 120 were dissolved in
2,500 ml of water at 50°C which contained 20 % of sodium
carbonate. 1 mole of hexamethylenediamne and 0.2 mole of
pyridin were added under stirring. After 1 hour stirring at
80°C, the reaction was achieved (intermediate 3 ; Fig. 3) .
The control was effected by thin layer chromatography as
indicated for the Blue derivative. The temperature was then
decreased to 5-10°C . Hydrochloric acid was added to pH 7
and the precipitate obtained was collected and dried at
75°C. 0.2 mole of intermediate 3 . Fig. 3 was dissolved in
2,500 ml of water and triethylamine added up to pH 9.5 at
0-5°C. The mixture of tosyl-acrylic anhydride and DMF (1:1)

added dropwise. The reaction was controlled by thin layer chromatography. The addition of the mixture was stopped when the reaction was complete . Under stirring, acetic acid was poured into the solution until obtaining pH 5 ; potassium chloride was added up to 5 % concentration and then the precipitate obtained was collected by centrifugation and washed in the presence of acetone. It was finally dried at 60-70°C under vacuum.

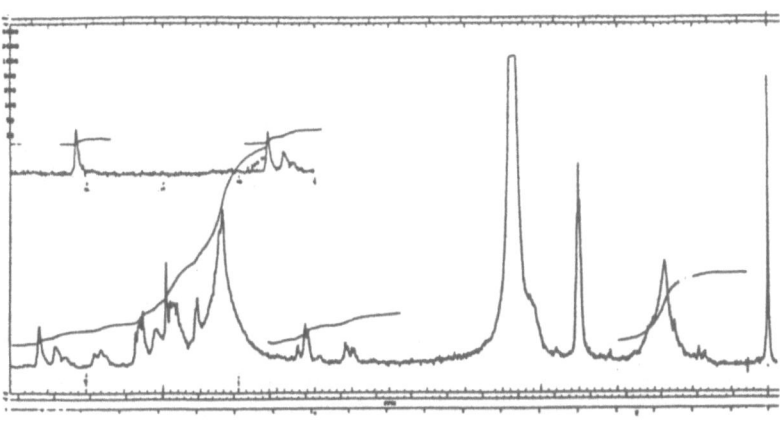

FIGURE 2.   *NMR spectra for the acrylic cibacron blue obtained as indicated in Figure 1*

Block copolymerization studies : 50 ml of demineralized degased water were added with 19 g of N-acryloyl-amino-2-hydroxymethyl-2-propanediol-1,3 containing different quantities of acrylic dye (red or blue) and then with 2 g of N,N'-methylene-bis-acrylamide . After solubilization, the solution was placed in a water bath at 50°C and 125 mg of ammonium persulfate and 100 ulitres of N,N,N',N',-tetramethylethylenediamine were added. The solution was rapidly stirred . A few minutes later, the polymerization occured and the temperature increased of about 10-20 degrees Celsius. The block polymer obtained was passed through a stainless steel net in order to get small gel particles. These particles were extensively washed to remove the dye excess successively with water, 0.15 M sodium chloride, 0.15 M sodium chloride / 25 % ethanol mixture, ethylene-glycol/water mixture and finally with 0.1 M phosphate buffer. The resulting gel was stored at room temperature in a 1 M sodium chloride solution containing 0.02 % sodium azide as bacteriostatic agent.

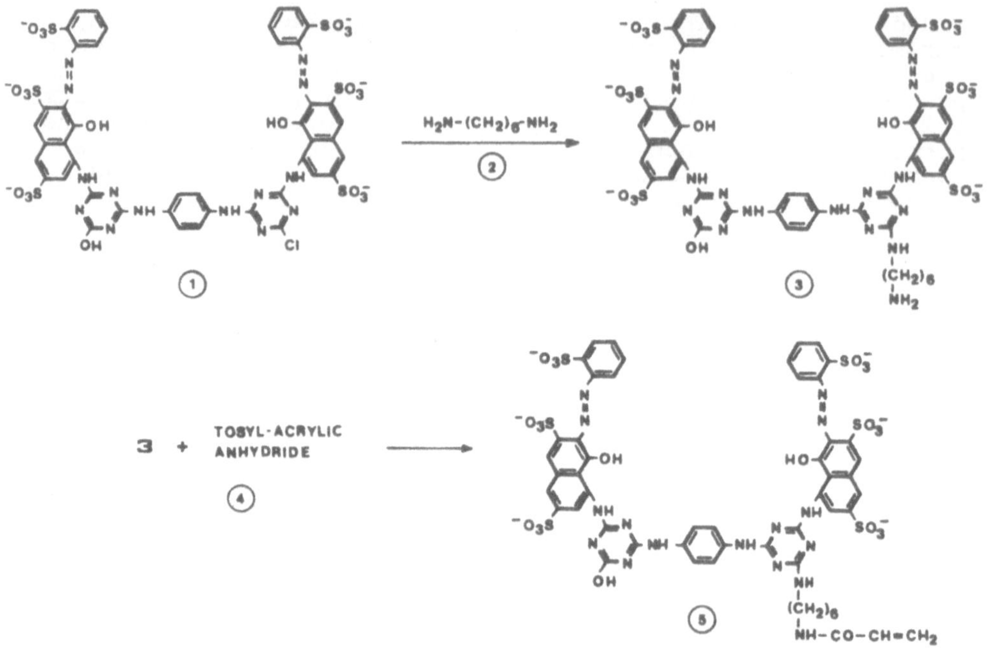

FIGURE 3.   Chemical reaction scheme for the synthesis of N-acryloyl-
hexamethylene-N'-Reactive Red 120 (5).
1 : Reactive Red 120 ; 2 : hexamethylene diamine; 3 : epsilon-
amino-hexamethylene-diamine-N'-Reactive Red 120 ;   (see
the text for details)

Polymerization in bead form : Beads of gel were
obtained by injecting the same aqueous solution of monomers
and the catalysts described above into a stirred paraffin
oil bath before polymerization. The dye microdroplets
obtained were transformed into gel beads during the
copolymerization as described elsewhere (13, 14, 15).

Determination of dye incorporation into the polymers :
The dye content of the different polymers obtained was
determined by measuring the sulfonate group content per ml
of gel assuming that the Cibacron Blue derivative contained
three groups per molecule and that each molecule of Reactive
Red 120 had six sulfonate groups . This determination was
effected by frontal analysis : titration in column of an
acid-regenerated gel was performed with a 50 mM sodium
hydroxide solution.

Evaluation of the polymeric dye sorbent : The
characterization of dye sorbents and evaluation of their
performances were assessed by classical methods such as :
- determination of the sorption capacity by frontal
  analysis,
- preliminary separation trials using protein mixtures.

RESULTS AND DISCUSSION

The results on the characterization of Blue and/or Red polymers are summarized on Tables 1 and 2 .

TABLE 1
Summary of main results obtained with
acrylic cibacron blue copolymers

ACRYLIC BLUE CONCENTRATION (%) BEFORE POLYMERIZATION	UMOLES DYE PER ML SOLUTION	% ACRYLIC DYE ON TOTAL MONOM.*	UMOLE DYE PER ML OF GEL	ACRYLIC DYE INCORPORATION (%)	SORPTION CAPACITY FOR ALBUMIN MG / ML OF GEL **	SORPTION EFFICIEN. MG ALBUMIN PER UMOLE OF DYE
0.10	1.026	0.25	N.D.	N.D.	0.49	N.D.
0.20	2.05	0.50	1.12	54.6	1.48	1.32
0.25	2.56	0.62	N.D.	N.D.	1.74	N.D.
0.30	3.08	0.75	1.39	45.1	1.90	1.37
0.40	4.10	1.00	1.50	36.6	1.78	1.19
0.50	5.13	1.25	N.D.	N.D.	3.18	N.D.
0.60	6.16	1.50	1.63	26.5	2.78	1.70
0.80	8.21	2.00	1.70	20.7	3.50	2.05
1.00	10.26	2.50	2.48	24.2	4.67	1.88
1.25	12.83	3.12	3.80	29.6	5.28	1.39
1.50	15.40	3.75	2.95	19.2	4.73	1.60
1.75	17.96	4.37	5.14	28.6	4.18	0.82
2.00	20.50	5.00	3.40	16.6	6.37	1.87
2.50	25.66	6.24	6.25	24.3	7.05	1.13
4.50	46.20	11.25	10.50	22.7	6.74	0.64

*   TOTAL AMOUNT OF MONOMERS WAS 40 % (W/V)

* *   AVERAGE FROM TWO DETERMINATIONS AT PH 7.4 IN 0.085 M PHOSPHATE BUFFER CONTAINING 0.15 M NACL

The quantity of incorporated blue acrylic dye was proportional to its initial concentration in the aqueous solution. The conversion monomer-polymer was not quantitative (between 20 % and 50 %) ; it decreased regularly with the increase of the concentration of blue acrylic dye . Although this situation is classical when compared to copolymers in which the main monomer is structurally different, it was not similar for both the blue and red acrylic dyes. The incorporation of the latter was in fact practically constant (about 18 %) whatever the concentration of the dye monomer was. This difference could be assigned to the autraquinone nature of Cibacron Blue F3 GA which has a little inhibition activity on the acrylic polymerization process. This inhibition is proportional to the relative concentration of the acrylic reactive antraquinone dye. It is to be noted here that the yield of the chemical immobilization of activated triazine dye on a classical sorbent using one of the methodologies described in the past is also low (e.g. 10 - 20 %) (16).

TABLE 2
Summary of main results obtained with
acrylic reactive red 120 copolymers

ACRYLIC BLUE CONCENTRATION (%) BEFORE POLYMERIZATION	UMOLES DYE PER ML SOLUTION	% ACRYLIC DYE ON TOTAL MONOM.*	UMOLE DYE PER ML OF GEL	ACRYLIC DYE INCORPORATION (%)	SORPTION CAPACITY FOR OVALBUMIN MG / ML OF GEL **	SORPTION EFFICIEN. MG OVALBUMIN PER UMOLE OF DYE
0.1	0.595	0.25	0.11	18.5	0.40	3.64
0.2	1.19	0.50	0.18	14.3	1.22	6.77
0.4	2.38	1.00	0.45	19.3	1.18	2.62
0.8	4.76	2.00	0.88	18.4	4.32	4.91

* TOTAL AMOUNT OF MONOMERS WAS 40 % (W/V)

** SORPTION CAPACITY WAS DETERMINED AT PH 4 IN 0.01 M ACETATE BUFFER

The yield of the polymerization reaction (incorporation of functionalized dye) was not high (between 20 % and 50 %). However, it was clear that the sorption capacity of the supports obtained was proportional to its dye content (see Fig. 4).

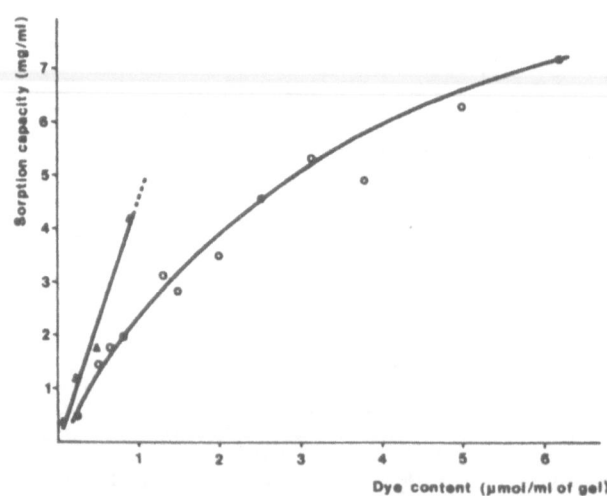

FIGURE 4. *Protein sorption capacity of red and blue matrices as a function of the chromophore content in the polymer. o---o capacity of blue polymer for bovine serum albumin at pH 7.4 (0.025 M phosphate buffer . --- capacity of red polymer for egg ovalbumin at pH 4.0 (0.01 M acetate buffer).*

Since the sorption capacity increased, it would be possible to obtain highly substituted gels with improved characteristics. Unfortunately, we were limited by the aqueous solubility of the functionalized dyes. Nevertheless, the studied supports demonstrated a sorption capacity of the same magnitude as the commercially available immobilized dyes. The sorption efficiency (_i.e._ the quantity of protein adsorbed per umole of immobilized dye) was remarkably constant for the two series of gels. The sorption efficiency of blue acrylic polymer was 1.41 (on average) whereas the average value of same the characteristic was 4.48 for the red acrylic polymer . The higher efficiency of the red acrylic polymer is probably due to the presence of six sulfonate groups per molecule instead of three as in the case of blue polymer (see Fig. 3).

The chromatographic behaviour of these sorbents was undoubtedly very interesting. It demonstrated the possibility of adsorbing and desorbing typical proteins under buffered conditions, and using a salt gradient as an eluent.

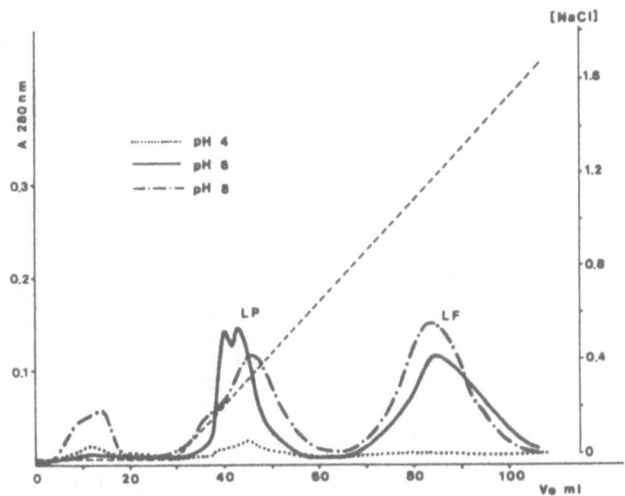

FIGURE 5.   Chromatographic separation of a mixture of crude lactoperoxidase (LP) and lactoferrin (LF) from milk on the polymeric blue sorbent. Column : 1 x 8 cm ; buffers : 0.05 M Tris-HCl , pH 8 or 0.05 M acetate pH 6 and 4 ; elution gradient : sodium chloride from zero to 2 M ; flow rate : 45 cm/hour.

Fig. 5 shows a chromatographic separation of a crude mixture of lactoferrin and lactoperoxidase from milk at different pH on a blue derivative. Impurities were found in the flowthrough whereas lactoperoxidase and lactoferrin were perfectly separated under a salt elution gradient at the same pH. It is interesting to note that the pH plays an important role in the separation efficiency of these two proteins. At pH 4, most proteins were strongly adsorbed. On the red derivative, the same mixture behaved differently : at pH 4 and 6, lactoperoxidase was found in the flowthrough and lactoferrin separated under the salt gradient (Fig. 6).

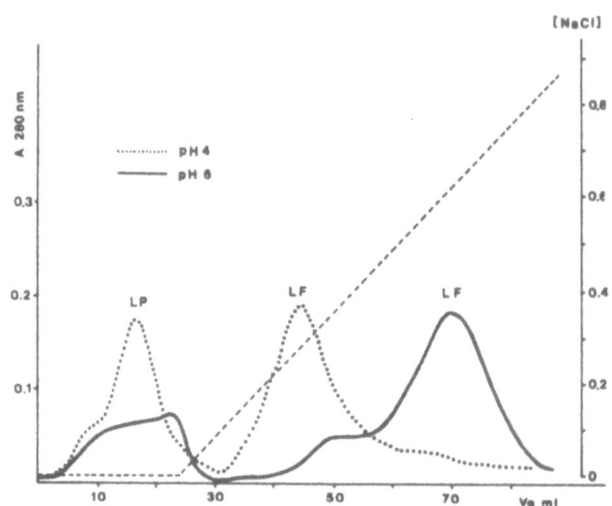

FIGURE 6.   Chromatographic separation of a mixture of crude lactoperoxidase (LP) and lactoferrin (LF) from milk on the polymeric red sorbent. Column : 1 x 6 cm ; buffer : 0.05 M acetate pH 6 and 4 ; elution gradient : sodium chloride from zero to 2 M ; flow rate : 36 cm/hour.

All these preliminary results are encouraging as it will be possible to apply this methodology to the chromatographic synthesis of immobilized dye and perhaps, to extend it to other dyes of interest. This approach associates a number of benefits. The first important advantage over the classical dye immobilization on a neutral support is the possibility of designing the most appropriate polymer with a predictable and controllable substitution degree. As the dye is a real part of the polymer, it is very solidly anchored and does not leach during the chromatography or regeneration steps.

On the chemical point of view, it is also important to point out that the support synthesis is realized in one single step and hence reduces the length of the synthesis process, and saves time and labour.

The principle we describe in this paper (preparation of dye polymer and then copolymerization) is also appropriate for the preparation of electrophoresis plates in order to explore the separation possibilities in analytical affinoelectrophoresis as described in the past , without application of a non-reproducible step of dye reaction on already prepared plates.

Another extension of this copolymerization technique is the preparation of linear soluble chains, permitting to carry out two-phase separations and related techniques.

Finally, it is possible to prepare polymers (in linear or

cross-linked forms) using mixture of different acrylic dyes with associated properties for the concomitant adsorption of some proteins and desorbing them selectively afterwards.

CONCLUSION

One of the methods described for the Chromatographic Support Design (CSD) (11) was applied here in the synthesis of copolymers containing special acrylic derivative of dyes of interest. This approach demonstrated the great potential of creating sorbents in accordance with the suited characteristics and capable of separating proteins mixtures by liquid chromatography.

The polymeric material was stable, easy to obtain and did not give leachables : all advantages particularly appreciated in comparison with the classical immobilized dyes.

The possibility to obtain these polymers in bead form opens another interesting way of support synthesis.

Trials on the extension to other dyes, to application of copolymers at complex protein mixtures and to large scale preparation of sorbents are being in progress in our laboratories.

REFERENCES

1. Lowe, C.R., Small, D.A.D., Atkinson, A., _Int. J. Biochem._, 1981, 13, 33.

2. Turner, A.J., Hryszko, J., _Biochim. Biophys. Acta_, 1980, 613, 256.

3. Birkenmeyer, G., Tschechonien, B., Kopperschlager, G., _FEBS Lett._, 1984, 174, 162 - 66

4. Mislovcova, D., Gemeiner, P., Kuniak, L., Zemak., J., _J. Chromatogr._, 1980, 194, 95 - 99.

5. Byefield., P.G.H., Copping, S., Bartilett, W.A., _Biochem. Soc. Trans._, 1982, 10, 104

6. Atkinson, T., Hammond, P.M., _Biochem. Soc. Trans._, 1981, 9, 290 - 293.

7. Girot, P., Boschetti, E., unpublished results

8. Brown, E., Racois, A., Boschetti, E., Corgier, M., _J. Chromatogr._, 1978, 150, 101.

9. Brown, E, Couturier, M., Touet, J., Boschetti, E., _Bull. Soc. Chim._, 1986, 4, 669

10. Touet, J., Pierre, C., Brown, E., Boschetti, E.,

<u>Makromol. Chem.,</u> 1987, 8, 377.

11. Boschetti, E., Proc. Symp. "Protein Purification", Toulouse, 1988, p. 23.

12. Patent pending (1988)

13. Boschetti, E., Tixier, R., Uriel, J., <u>Biochimie,</u> 1972, 54, 439

14. Brown, E., Joyeau, R., Racois, A., Boschetti, E., <u>Proc. Aff. Chrom. Sym.,</u> Strasbourg, 1979, p. 37.

15. Girot, P., Boschetti, E., <u>J. Chromatogr.,</u> 1981, 213, 389.

16. Girot, P., Boschetti, E., unpublished results.

THE USE OF DYES IN AFFINITY LIGAND
AND PROTEIN IMMOBILIZATION

Russell Rines, A.H.L. Mulder, and William H. Scouten
Chemistry Department
Baylor University, Waco, Texas

Reinhart Loy
Schleicher & Schuell, Inc.
10 Optical Dr., Keene, New Hampshire

ABSTRACT

We have been using heterocyclic dyes to activate hydroxylic matrices, such as paper or Sepharose (agarose), for subsequent covalent attachment of ligands or proteins through a secondary amine (lysine) or thioether (cysteine) linkage. The activation provides a colored matrix whose coupling to ligands can be followed by the loss of color. The dye can also serve as an affinity ligand for the affinity-directed immobilization of a protein. In the latter case, the protein is attached to the matrix in a specific way, due to its affinity for the dye. The affinity-directed immobilization of IgG through its Fab segment is being used as a model system.

As an example of the chromophoric aspect, the activation of paper with FMNI (3-fluoro-2-methyl-5-[2-naphthol-1-azo]-indazolium tosylate) and the coupling of several proteins is discussed.

As an example of affinity-directed immobilization, the coupling of anti-DNP antibodies to DNP-CMP (5-amino-2-chloro-N-[2,4-dinitrophenyl]-1-methylpyridinium tosylate) activated Sepharose is discussed.

INTRODUCTION

The compound FMP (2-fluoro-1-methylpyridinium tosylate) has been used to activate hydroxylic matrices. The covalent attachment of an amine or thiol to

such activated matrices leads to secondary amine or thioether linkages, respectively [1,2]. We have synthesized several 2-fluoro- and 2-chloro-1-methyl-
pyridinium salts and used these compounds to activate paper or agarose
(Figure 1). Each of these derivatives have had a chromophoric and/or affinity
ligand attached to the pyridinium ring. This chemistry is not limited to

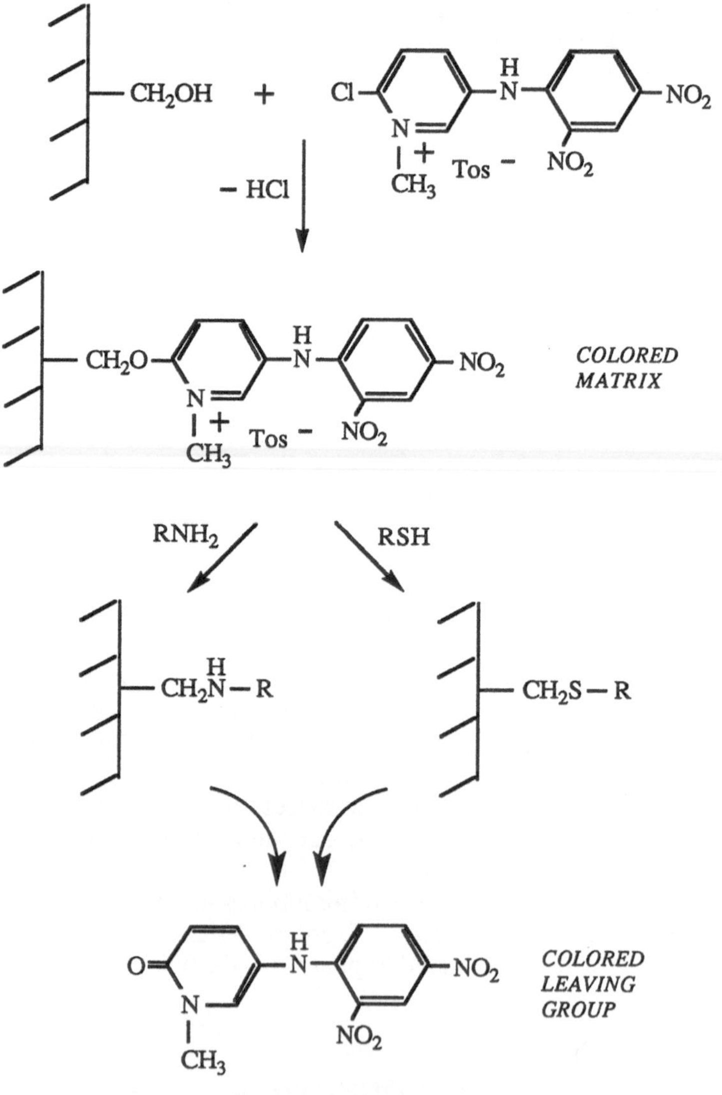

Figure 1. Activation of hydroxylic matrix with dye
and ligand coupling.

pyridinium systems, and we have used other heterocyclic ring systems as well, e.g., FMNI (Figure 2). The purpose of the chromophoric group is to provide a colored matrix whose coupling to ligands can be followed by loss of color. In

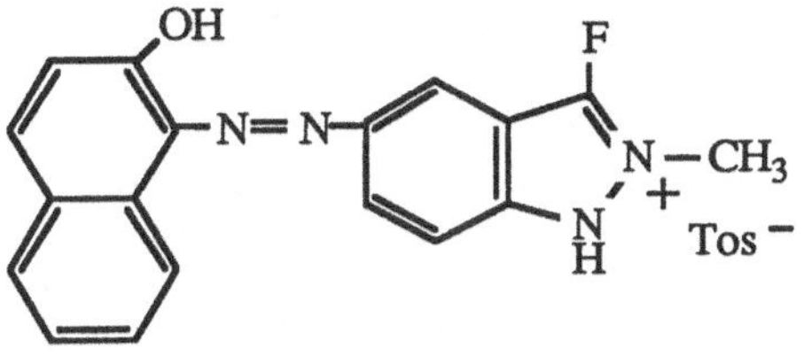

Figure 2. FMNI (3-fluoro-2-methyl-5-[2-naphthol-1-azo]-indazolium tosylate).

the case of the affinity ligand-containing agent, the enzyme (or other protein) to be attached is first attracted to the matrix in a rapid affinity binding mode. A nucleophilic residue near the enzyme active site then displaces the activating agent in a slower step, and the enzyme becomes covalently attached to the matrix (Figure 3). This should lead to a more homogeneous immobilization than current methods allow, as well as to a higher activity of the immobilized material. Affinity-directed immobilization might also eliminate the need to purify the enzyme before immobilization, since in the first step the gel functions as an affinity column.

All of the work to date on affinity-directed immobilization has been with DNP-CMP activated Sepharose. We have attached affinity purified anti-DNP antibodies to this matrix and are currently trying to prove that the immobilization was indeed affinity-directed. To accomplish this, the immobilized antibody is cleaved with pepsin or papain and the eluted fragments obtained compared to those obtained from cleavage of the same antibody in free solution. Since the antibody (IgG) is expected to be immobilized via its Fab fragment, the ratio of Fab to Fc fragments obtained from papain cleavage should be 1:1. If the antibody is cleaved in free solution, this ratio should be 2:1 (Figure 4). Similarly, cleavage of the immobilized antibody with pepsin should give only Fc fragment, and a 1:1 ratio of F(ab)'$_2$ to Fc in free solution (Figure 5).

MATERIALS AND METHODS

5-Amino-2-chloropyridine, 4-aminothiophenol, and 3-chloro-5-nitro-indazole were purchased from Aldrich Chemical Co. The anti-DNP antibody (developed in goat) was acquired from ICN Biochemicals. Centricon 30 microconcentrators were obtained from Amicon Corp. The paper used for immobilizations was 589-WH, a product of Schleicher and Schuell, Inc.

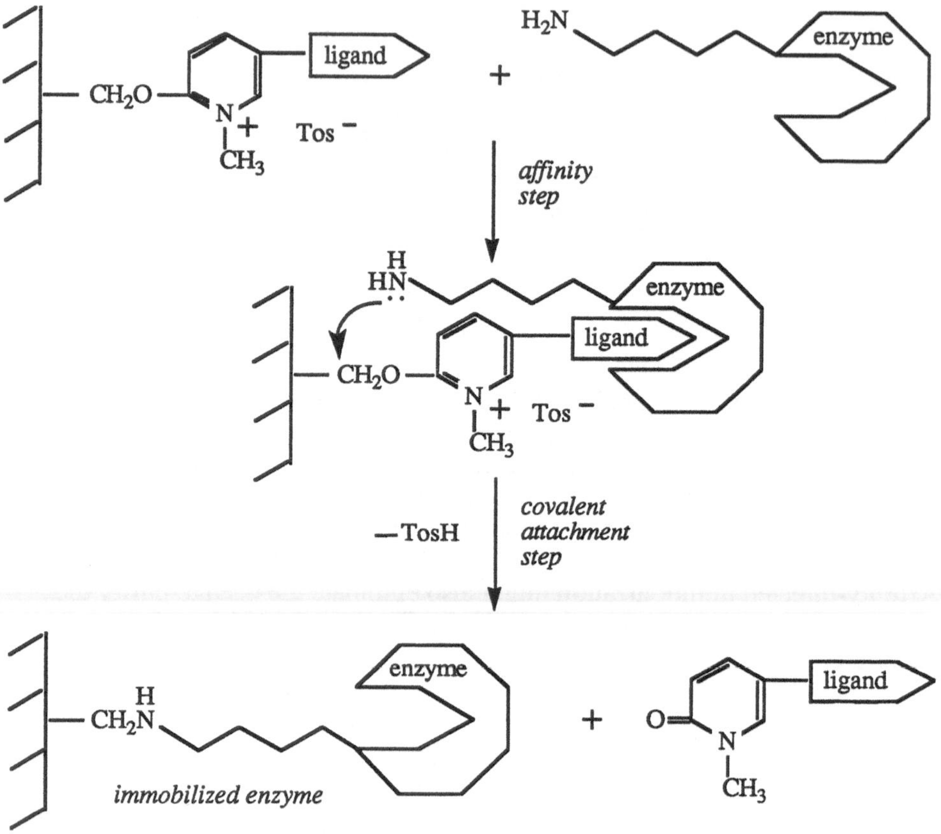

Figure 3. Affinity-directed immobilization scheme.

Synthesis of FMNI (3-fluoro-2-methyl-5-[2-naphthol-1-azo]-indazolium tosylate

3-Chloro-5-nitroindazole (3.9 g) was refluxed in 250 ml o-xylene with 5.0 g AgF for 2 days in the dark with a CaCl$_2$ drying tube to exclude moisture. Silver salts were filtered off and the xylene removed by flash evaporation. The crude 3-fluoro-5-nitroindazole was dissolved in 150 ml 95% ethanol and 10.3 g sodium dithionite in 50 ml H$_2$O added dropwise while stirring under reflux. After refluxing for 2 hours, the solution was filtered after cooling to room temperature, the filter cake washed with 30 ml boiling ethanol, then with 30 ml boiling H$_2$O. The solvents were removed from the filtrate by flash evaporation, then the remaining residue refluxed for 1.5 hours with 1.64 ml concentrated HCl

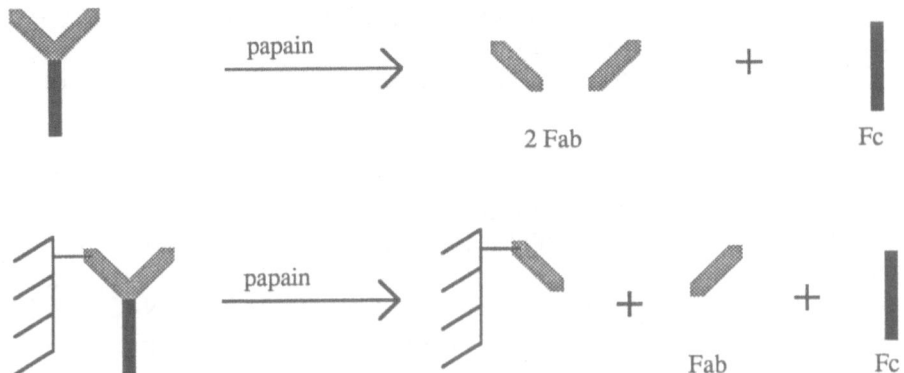

Figure 4. Cleavage of immobilized and free anti-DNP with papain.

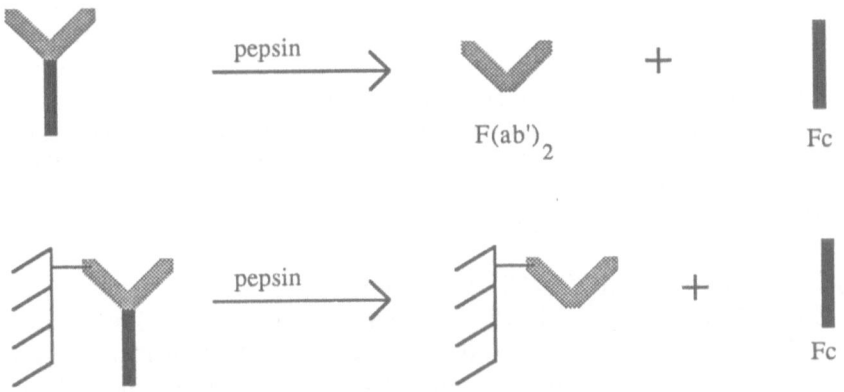

Figure 5. Cleavage of immobilized and free anti-DNP with pepsin.

in 30 ml H_2O. The aqueous HCl was removed by flash evaporation leaving the 5-amino-3-fluoroindazole HCl salt. The amine salt was dissolved in 50 ml H_2O and diazotized at 5° C by dropwise addition into 12 ml H_2O containing 6 ml concentrated HCl and 2.2 g sodium nitrite. After stirring for 20 minutes, the solution was poured into 30 ml 10% NaOH containing 4.3 g β-naphthol and stirred for 1 hour at room temperature. The resulting (3-fluoro-5-[2-naphthol-1-azo]-indazole was extracted into toluene with a continuous liquid/liquid extractor (overnight) then the toluene removed by flash evaporation and the residue dried over P_2O_5. The dye was then methylated by refluxing for 3 hours in 30 ml dry toluene with an equal molar amount of methyl tosylate. The toluene was removed by flash evaporation and 100 ml dry ether added and left at 5° C overnight to crystallize. The FMNI (3.5 g) tosylate was filtered off and stored over P_2O_5.

Synthesis of DNP-CMP (5-amino-2-chloro-N-[2,4-dinitrophenyl]-1-methylpyridinium tosylate)

Five grams of 5-amino-2-chloropyridine was refluxed with 5.0 ml of 2,4-dinitrofluorobenzene and 5.5 ml of dry triethylamine in 75 ml of dry acetonitrile for 2 hours. The acetonitrile was removed by flash evaporation and the remaining orange solid washed with 95% ethanol, then dried in vacuo over $CaCl_2$ (95% yield, m.p. 164-165° C). Ten grams of the 5-amino-2-chloro-N-(2,4-dinitrophenyl)-pyridine was refluxed with 50 grams of methyl tosylate in 55 ml of dry toluene for 30 minutes. The yellow solid was filtered, washed with dry ether, then dried in vacuo over $CaCl_2$. The resulting 5-amino-2-chloro-N-(2,4-dinitrophenyl)-1-methylpyridinium tosylate was used without further purification (50% yield, m.p. 208-212° C, decomposition).

Activation of paper with FMNI

Six 20 x 20 cm 589-WH sheets were washed 3 times with acetone, air dried, then dried over P_2O_5. The sheets were then placed between polyethylene screens and immersed in 200 ml dry acetonitrile containing 15 g FMNI and 4.5 ml triethylamine in a glass dish sealed with tape. The paper was shaken for 4 hours at room temperature on a mechanical shaker, turning once or twice to ensure uniform solution contact. The sheets were then washed successfully with acetone/2 mM HCl solutions having volume/volume ratios of 75/25, 50/50, 25/75, then finally with 2 mM HCl. The pink/orange sheets were then cut into 0.25 inch diameter disks.

Coupling of protein to FMNI activated paper

The FMNI activated disks (5 g) were placed in a glass dish with sufficient 0.2 M $NaHCO_3$ protein solution (100 mg protein per 30 ml buffer) to cover them. The dish was rocked gently for 14 hours at room temperature. The disks were washed with 0.2 M $NaHCO_3$, then immersed in 0.1 M Tris-HCl, pH 8.0 for 2 hours. The disks were then washed with 1 M NaCl, and finally with PBS, pH 7.5. The disks were dried and stored desiccated.

Quantitation of protein immobilized on FMNI activated disks

Each disk was dissolved in 0.3 ml 65% H_2SO_4 for 2 hours. A 10% solution of Na_2CO_3 (5 ml) was added and when foaming stopped, 1.3 ml of Biorad protein reagent dye concentrate (Coomassie Blue from Biorad Corp.) added. The A_{595} was taken and protein concentration determined from protein standards treated exactly as the disks. The average of at least 5 disks were used for the quantitations.

Activation of Sepharose with DNP-CMP

Sepharose was washed with 20 column volumes of water, then with 20 column volumes each of the following water:acetone mixtures (% v:v); 75:25,

50:50, 25:75, 0:100. The gel was then washed with 20 gel volumes of dry DMF (N,N-dimethylformamide), then resuspended in 4 gel volumes of 20 mM DNP-CMP, 144 mM triethylamine in dry DMF. The gel was shaken in a water bath for 12 hours at 55° C. The activated gel was washed with DMF until the filtrate was clear, then once with 20 column volumes of acetone, followed by 20 column volumes each of the following water:acetone mixtures (% v:v); 25:75, 50:50, 75:25, 100:0. Finally, the gel was washed with 20 column volumes of 10 mM HCl and stored in 10 mM HCl.

Preparation of DNP column for purification of anti-DNP

Sepharose was activated with FMP as described by Ngo [1]. Amino-thiophenol was coupled to the activated gel by incubating 50 grams of gel (wet weight) with 50 ml 0.5 M p-aminothiophenol in ethanol:water:acetone (5:4:1, v:v:v%), 0.1 M NaCl and 0.04 M NaHCO$_3$, pH 9.0 for 1 hour at 25° C. After coupling, the gel was washed with 25 column volumes ethanol:water:acetone (5:4:1, v:v:v%); ethanol; ethanol:acetone (1:1, v:v%); acetone; and water.The gel was then resuspended in 4 column volumes of 62 mM 2,4-dinitrofluorobenzene, 144 mM triethylamine in dry acetonitrile. The gel was shaken for 16 hours at 25° C. The DNP-gel was washed with acetonitrile until the filtrate was colorless, then with 20 column volumes of 10 mM HCl and stored in 10 mM HCl.

Affinity purification of anti-DNP antibodies

A 2 cm x 1 cm DNP column was equilibrated with 0.2 M PBS, pH 7.5. The column was loaded with 0.25 ml anti-DNP solution (0.6 mg/ml), then washed with PBS until the A$_{280}$ was less then 0.02. The antibody was eluted with 0.1 M glycine-HCl, pH 2.5 and 1.5 ml fractions collected. The protein containing fractions were pooled and concentrated by ultrafiltration in Centricon 30 microconcentrators.

Affinity-directed immobilization of anti-DNP

A 2 cm x 1 cm DNP-CMP column was equilibrated and loaded with affinity purified anti-DNP exactly as described for the affinity purification step in the previous section.

Proteolytic cleavage of anti-DNP antibodies

The DNP-CMP gel with immobilized anti-DNP was incubated with 2 column volumes of 20 mM NaH$_2$PO$_4$, 20 mM cysteine-HCl, 10 mM EDTA, pH 6.0 with a papain/antibody ratio of 1:100 by weight. The gel was incubated for 5 hours at 37° C, then 6 column volumes of 10 mM Tris-HCl, pH 7.5 added to stop the digestion. The pepsin cleavage was exactly the same as the papain except it was done in 20 mM sodium acetate, pH 4.5 with a pepsin/antibody ratio of 1:50 by weight, and the incubation was carried out at 37° C for 2 hours.

The free antibody was cleaved by papain and pepsin as described for the immobilized antibody, except buffer was added instead of the gel.

HPLC analysis of antibody fragments

The fragments from the proteolytic cleavage of the antibody were concentrated by ultrafiltration in Centricon 30 microconcentrators. The concentrated fragments were injected into a Waters ALC 202 HPLC equipped with a Waters 450 variable wavelength detector set at 280 nm. The mobile phase was 0.067 M NaPO$_4$, 0.3 M NaCl, pH 6.8 at 1.0 ml/minute on a TSK 3000SW exclusion column purchased from Biorad Corp.

RESULTS AND DISCUSSION

Coupling of protein to FMNI activated paper

Paper activated with FMNI was a pink/orange color. Color was removed from the paper in the process of coupling protein, however, not all the color was removed. With all the dyes we have synthesized and used to activate hydroxylic matrices, we have not been able to remove all the color from the matrix under conditions normally employed for protein immobilization. Even a small nucleophile such as ethanolamine (1 M, pH 9.0) will not remove all the color from the matrix. We are not sure if this is due to some physical/chemical property of the matrix, the coupling chemistry, or both.

The maximum amount of protein that can be immobilized on a 0.25 inch diameter paper disk with FMP is approximately 5.6 μg. As shown in Table 1, the amount of protein immobilized to disks activated with FMNI is approximately the same as with FMP. This shows that halogenated heterocyclic salts other than pyridiniums can be used to activate hydroxylic matrices.

Table 1.
Immobilization of protein on S + S 589-WH 0.25 inch diameter disks
activated with FMNI.

Protein	μgm/disk
BSA	3.2
chicken OVA	4.3
pepsin	6.9
trypsinogen	3.2

Affinity-directed immobilization of anti-DNP on DNP-CMP activated Sepharose

The activation of Sepharose with DNP-CMP requires harsher conditions than activation with FMP (12 hours at 55° C compared to 30 minutes at room temperature). We believe the reason for this is the lower reactivity of the 2-chloropyridiniums compared to the 2-fluoropyridiniums. In fact, 2-chloro-1-methylpyridinium does not react with alcohols. The reason DNP-CMP activates hydroxylic matrices is probably due to the electron withdrawing nature of the 2,4-dinitrophenylanilino function.

Anti-DNP antibody was affinity purified before application to the DNP-CMP column to remove any antibody that would recognize epitopes other than DNP. This could occur since the immunogen used was DNP-BSA. Table 2 shows the elution times of the antibody fragments on the HPLC exclusion column from the cleavage of the antibody (immobilized and free) with papain and

Table 2.
The elution of antibody fragments from a TSK 3000SW HPLC column.
The area of the peaks decreases from top to bottom.

elution time (minutes)

	papain		pepsin	
affinity	3.7	[Fc]	4.3	
immobilized	4.5		3.0	[F(ab')$_2$]
antibody	5.5		----	
free antibody	5.5		3.0	[F(ab')$_2$]
	3.7	[Fc]	4.5	

pepsin. If the peak eluting at 3.7 minutes is the Fc fragment from the papain cleavage, then the results are as expected. There is a greater proportion of this fragment released from the immobilized antibody, since half of the Fab fragments are covalently bound to the gel (see Figure 4). Similarly, if the peak eluting at 3.0 minutes from the pepsin cleavage is the F(ab')$_2$ fragment, there should be none of this fragment released from the immobilized antibody since it should be covalently attached to the matrix (see Figure 5). However, it is possible that we have a degree of affinity-directed immobilization, and that some of the antibody is attached randomly.

When antibody is cleaved with either papain or pepsin, the fragments produced are often cleaved again into smaller fragments. The reason that the large fragments are produced is because the hinge region is more exposed and therefore more accessible to the enzymes. There were several peaks that appeared at later times (smaller fragments) that are not shown in Table 2, that could be fragments of the original fragments produced. Attempts to characterize the larger fragments have been unsuccessful so far.

One other piece of data that suggests that the immobilization was affinity-directed is that if the antibody is applied to the DNP-CMP column at pH 6.0, it is not possible to elute it from the column with acidic buffers, as is possible with the DNP column. At pH 6.0, small amine nucleophiles (1 M) displace the DNP-CMP groups exceedingly slowly (less than 10% displacement after 4 days, data not shown). The reason that the antibody is immobilized at pH 6.0 could be due to the higher effective concentration of nucleophilic residues on the antibody from the interaction with the DNP group, i.e., there may be a nucleophilic residue held in the proper position by the noncovalently bound antibody.

We are planning to affinity-immobilize monoclonal anti-DNP antibodies to the DNP-CMP matrix. The advantage of monoclonals is that the fragments produced by proteolytic cleavage will be the same (hence easier to characterize) from a homogeneous population of antibodies. However, one problem with using monoclonal antibodies is that a particular clone may not have a nucleophilic amino acid side chain in the proper position for immobilization. With the polyclonal antibodies we have been using, there is a better chance that at least one population of antibodies will have a suitably positioned nucleophilic residue. We also plan to put different length spacer arms between the heterocyclic ring and the DNP hapten to see what effect this has on the immobilization.

REFERENCES

1. Ngo, T.T., Facile activation of Sepharose hydroxyl groups by 2-fluoro-1-methylpyridinium toluene-4-sulfonate: Preparation of affinity and covalent chromatograghic matrices. Biotechnology, 1986, **4**, 134-137.

2. Ngo, T.T., US Patent 4,582,875 issued to Bioprobe International, Tustin, California

Chapter 4

Dye Ligands in Affinity Partition, Affinity Ultrafiltration and Affinity Precipitation

DYE-LIGAND AFFINITY PARTITIONING - A POWERFUL METHOD FOR STUDYING ENZYME-DYE INTERACTION

G. KOPPERSCHLÄGER, J. KIRCHBERGER, T. KRIEGEL and M. NAUMANN
Institute of Biochemistry, Karl-Marx-University Leipzig
Liebigstrasse 16, Leipzig, 7010, German Democratic Republic

ABSTRACT

Affinity partitioning of proteins in aqueous two-phase systems composed of dextran and poly(ethylene glycol), the latter being partially replaced by dye-substituted polymer, was evaluated for the study of enzyme-dye interaction. Due to its high sensitivity the method turned out excellently suited for the recognition of any interaction between a certain enzyme and a dye-ligand as indicated by determining the actual partition coefficient.
A number of enzymes which bind more or less specifically to triazine dyes has been selected for studying the influence of various chemical and medium parameters on the effect of dye-affinity partition. Thus, chemical modification of the dye molecules turned out appropriate to obtain informations on the chemical nature of the enzyme-dye interaction.
Affinity partitioning also proved applicable to the study of the structural dynamics of enzymes. This was exemplified on the specific interaction of yeast phosphofructokinase with selected triazine dyes.
Finally, the recent knowledge of the application of affinity partitioning for separation and purification of enzymes will be discussed.

INTRODUCTION

Affinity partitioning combines the property of biological macromolecules like proteins to partition in aqueous two-phase systems and their ability to recognize natural and artificial affinity ligands. Basically, two-phase systems are generated by, i.e., dissolving two polymeric substances of different hydrophobicity in water. They have been introduced as analytical and preparative tool in biology and biochemistry by Albertsson and his school [1] twenty-five years ago and become more and more attractive in downstream processing of protein purification today.

Several pairs of polymers have been used for affinity partitioning but the most popular system elaborated contains dextran or hydrolyzed starch and poly(ethylene glycol) (PEG) forming the lower and the upper phase, respectively. In order to obtain two-phase systems the concentrations of the two polymers have to exceed minimum values which mainly depend on the molecular weight of the polymers and on the temperature.

The partition of a protein or of any other macromolecule is characterized by the partition coefficient, K, which is defined as the concentration ratio of the respective substance in the upper and the lower phase. The K-value depends on a large number of external factors that include the concentration and the kind of the polymers, the pH of the buffer, the temperature and other factors. Since the partition of proteins also shows great individual response which is related to a number of intrinsic parameters like molecular weight and surface properties separation can be governed by taking advantage of such distinct parameters. Furthermore, the partition of proteins can be steered both efficiently and selectively by covalent binding of biomimetic ligands to one of the phase-forming polymers providing a special branch of liquid/liquid phase partition, called affinity partitioning [2]. Applying this principle any protein of interest can be extracted from the bulk into the ligand containing phase provided that interaction between the protein and the ligand does occur.

A number of natural ligands has been introduced in affinity partitioning technology [3] but practical application in protein purification failed so far probably due to the expensive preparation of those ligand-polymers. With the discovery of the peculiarity of certain reactive dyes to function as "pseudoligands" for many complementary domains of enzymes and other proteins, however, the diversity in the preparation of liganded two-phase systems has extremely increased due to the commercial availability of the dyes and the potency of this procedure to scale up.

Beside practical aspects affinity partitioning seems to be useful as analytical tool for studying protein-dye interaction. Small changes of ligand binding caused e.g. by a partial chemical modification of the ligand, by adding competing effectors, or by changing the buffer compositions are mostly reflected by an alteration of the corresponding partition coefficient of the protein.

In this contribution examples will be presented in which affinity partitioning was applied to study enzyme-dye interaction in more detail. For doing this alkaline phosphatase (AP), phosphofructokinase (PFK), lactate dehydrogenase (LDH), and glucose 6-phosphate dehydrogenase (G6PDH) were selected as model enzymes. The latter three enzymes are potentially capable of interacting with reactive dyes due to their nucleotide binding site. In addition, PFK is known to exhibit two distinct binding sites where ATP is bound as a substrate and as regulatory ligand, respectively. Hence, it was of particular interest to investigate the affinity partition of PFK in dependence on its allosteric properties.

MATERIALS AND METHODS

Dextran M70 or T70 (mol wt. 70 000) was obtained from VEB Serumwerke (Bernburg, GDR) or from Pharmacia (Uppsala, Sweden). PEG 6000 (mol wt. 6000-7500) was from Serva (Heidelberg, FRG). The dyes were obtained from I.C.I., Organics Division (Blackley, Manchester, U.K.). The dye-PEG derivatives were prepared according to Johansson [4].

The reduction of the azolinkages of the dye-PEG derivatives by sodium borohydride and sodium dithionite, respectively, was performed as described for azodye-Sepharose according to Clonis [5]. The modified dye-PEG was purified by chloroform extraction and analyzed by thin-layer chromatography and spectroscopy.

Before use the enzymes commercially available or prepared according to [6] and [7], respectively, were exhaustively dialyzed against the respective buffer followed by a 2 h incubation in the presence of the appropriate effectors if desired.

The two-phase systems (4 g) were prepared by weighing from stock solutions of PEG (40 %, w/w), of dextran (20 %, w/w), and of stock buffer containing various additives as indicated in the figures. After adding 5 to 20 μl of the respective enzyme solution the samples were gently shaken for 15 s and kept at 0 °C. The systems were incubated for further 15 min (not in all cases), shaken again and centrifuged and at 5000 x g for 45 s at 0 °C. Enzyme concentrations were determined by measuring their activity, AP as described in [8], PFK according to [6], LDH according to [9] and G6PDH according to [10].

RESULTS AND DISCUSSION

The principle of affinity partitioning is summarized in Fig.1. A prerequisite of this technique is the unequal distribution of the affinity ligand which is achieved by its covalent coupling to one of the phase-forming polymers. In prevailing cases PEG is used because of the easiness of coupling and of sufficient separation of the reaction product and of the fact that in PEG/dextran-systems the protein bulk is mainly concentrated in the bottom phase.

In order to study the extent of ligand-protein interaction the composition of the two-phase system was adjusted in a manner that in the absence of ligand-PEG most of the protein is concentrated in the dextran-rich bottom phase. By replacement of a portion of PEG by dye-PEG, which is mainly located also in the upper phase, proteins showing affinity to the dye will be extracted into the ligand containing upper phase leaving the bulk protein in the opposite phase.

The efficiency of affinity partition is usually expressed by the $\triangle$ log K-value, i.e. the difference of the logarithms of the partition coefficients obtained in systems with and without dye-PEG. In many cases the dependence of the affinity partitioning effect on the concentration of the polymer-bound ligand gives rise to a saturation curve as presented in Fig. 2 from which the maximum of extraction ($\triangle \log K_{max}$) and the half-saturation point ($0.5 \times \triangle \log K_{max}$), usually calculated from double reciprocal plots, are obtained.

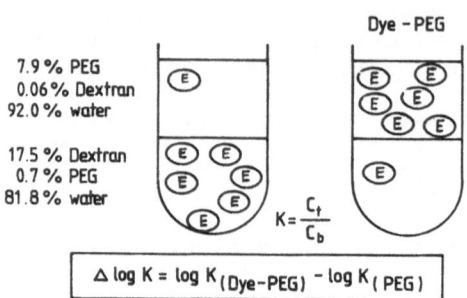

Figure 1. Principle of affinity partitioning

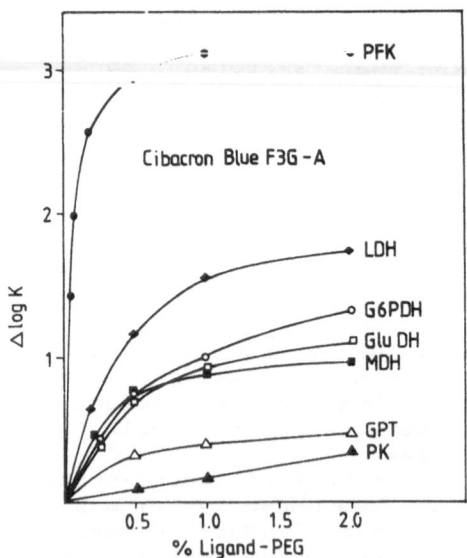

Figure 2. Dependence of affinity partitioning ($\triangle \log K$) on the concentration of Cibacron Blue F3G-A for diverse enzymes.
The system contained 7% dextran T500 and 5% PEG 6000 with different portions of dye-PEG, 50 mM sodium phosphate buffer, pH 7.0, 5 mM 2-mercaptoethanol, 0.5 mM EDTA. After cooling to 0 °C about 1.5 units of each enzyme were added to 4g of a system.

Among the great number of papers on affinity partitioning in only few contributions theoretical aspects of affinity partitioning are discussed. Following the approach of Flanagan and Barondes [2] which was experimentally verified in one case the difference in the logarithm of the partition coefficients of proteins $\Delta \log K_{max}$ should be equal to the number of binding sites per protein molecule times the logarithm of the partition coefficient of ligand-PEG assuming the same dissociation constants in both phases (Equation 1).

$$K_{E,max} = K_{e,0} \cdot K_L^n \qquad \text{\textcircled{1}}$$

$$K_E = K_{e,0} \left(\frac{1 + L_{tot,T} / K_{DT}}{1 + L_{tot,T} / K_{DB} \cdot K_L} \right)^n \qquad \text{\textcircled{2}}$$

$K_{E,max}$ = maximum partition coefficient of the protein,
K_E = actual partition coefficient of the protein,
$K_{e,o}$ = partition coefficient of the protein in the absence of ligand-PEG,
$L_{tot,T}$ = total ligand-PEG concentration in the upper phase,
K_L = partition coefficient of the ligand-PEG,
$K_{DT}; K_{DB}$ = dissociation constants of the protein-ligand complex in the upper and bottom phases, respectively

Equation 1, however, does not hold for the partition behaviour of the majority of the proteins analyzed. A more general equation was derived by Cordes and coworkers [11] (Equation 2) assuming that different dissociation constants have to be considered in the two phases. By calculation of the limits of this equation for very low and very high ligand concentrations, the following results were obtained:

$$\lim K_{E L_{tot,T} = \infty} = K_{e,0} \cdot \left(K_L \cdot K_{DB} / K_{DT} \right)^n$$

By determining the parameters $K_{e,0}$ and K_L independently the dissociation constant K_{DT} and K_{DB} as well as the number of binding sites, n, can be calculated by a fitting procedure of the experimental data.

There are several points of importance for interpreting saturation curves of affinity partitioning in terms of binding data. A number of enzymes binds affinity ligands at different binding sites. Therefore, Equation 2 was modified by Cordes [11] assuming 2, 4 and 6 binding sites, respectively. The calculated dissociation constants agree with the data obtained by other methods in most of the cases investigated. A further critical point of the approach seems to be the determination of the partition coefficient, K_L, of the free dye-PEG because stacking of dye molecules especially in the PEG containing phase can not be excluded.

In spite of these uncertainties in interpreting the saturation curves the term $\Delta \log K_{max}$ can be regarded as a measure of the strength of extraction power whereas the affinity of the

ligand to the protein correlates with the value of
0.5 x $\triangle \log K_{max}$. Both parameters are of interest for the em-
ployment of extractive procedures using liquid/liquid phases
for the purification of proteins. For instance, if the affinity partition effect, $\triangle \log K_{max}$ is about two and if the
volumes of both phases are approximately equal the conditions
for extraction can be adjusted that about 90 % of the target
protein which binds to the ligand is transferred into the
ligand containing phase whereas proteins without affinity to
the ligand are not affected. Some practical applications are
summarized in the last section of this paper.

Affinity partitioning is excellently suited for the recognition of small differences in the strength of protein-
ligand interaction. Thus, the method permits the screening of
protein-dye interaction in a simple manner by varying a number
of system parameters to obtain the following results:

1. Selection of a suitable dye-ligand for a certain protein
2. Selection of optimum conditions for association/disso-
   ciation of the protein-dye complex (i.e. metal-ion
   promoted protein-dye interaction)
3. Selection of corresponding effectors to compete in
   protein-dye interaction
4. Elucidation of the chemical basis of protein-dye
   interaction

As a rule, data obtained by this technique are valid also for
other affinity-ligand techniques like affinity chromatography
and affinity electrophoresis. The latter thesis was exem-
plarily demonstrated by Reuter et al. [12] for lactate dehydrogenase using a serie of triazine dyes coupled to PEG
(Table 1).

In Table 1 the $\triangle \log K_{max}$ of LDH are displayed for a set of
two-phase systems. Values about one and greater indicate
sufficient interaction with the dye. This holds for all of the
dyes tested with the exception of Procion Scarlet MX-G and
Procion Orange MX-G.

However, dyes listed in group II of Table 1 did not show
affinity to LDH if they have been coupled directly to the bead
cellulose. No significant difference in the $\triangle \log K$ was observed for Procion Green H-4G-PEG, Procion Yellow HE-3G and
all of the dyes of group I in affinity partition although an
opposite behaviour of these dyes have been observed in affinity chromatography. Since a steric hindrance by the matrix
was supposed to cause lacking affinity all the dyes of group
II were coupled via a spacer to the bead cellulose. Under
these conditions the dyes showed sufficient binding properties
to LDH [12].
In Fig. 3 the influence of competing effectors of enzyme-dye
interaction is demonstrated at constant concentration of the
dye-polymer for lactate dehydrogenase (LDH) and pyruvate kinase (PK). The binding of LDH to Cibacron Blue F3G-A was
diminished by increasing amounts of NAD^+ but seems to be not
completely abolished even at 2 mM of the effector.

TABLE 1

Interaction of heart muscle LDH with different triazine dyes
expressed by the affinity partitioning effect

Dye-PEG	$\Delta \log K$
Group I:	
Procion Red HE-3B	1.62
Procion Red HE-7B	1.54
Procion Yellow HE-4R	1.92
Procion Navy H-ER	1.82
Cibacron Blue F3G-A	1.33
Group II	
Procion Green H-4G	1.69
Procion Yellow HE-3G	1.86
Procion Scarlet MX-G	0.97
Procion Orange MX-G	0.94

Group I: Dyes directly coupled to bead cellulose bind LDH;
Group II: Dyes directly coupled to bead cellulose does not
bind LDH.

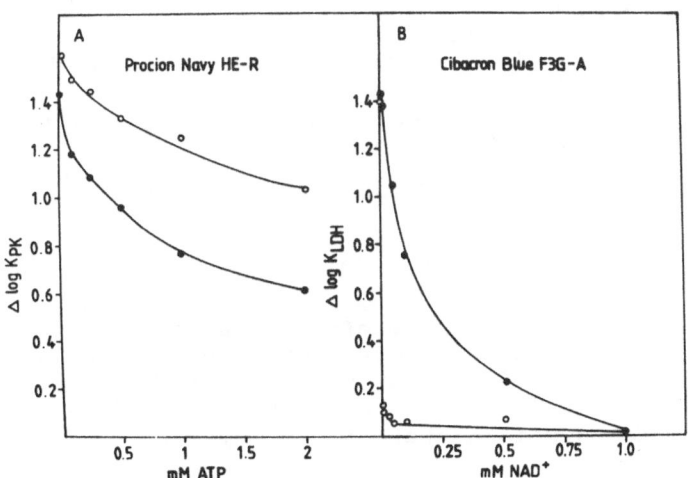

Figure 3. Dependence of affinity partitioning ($\Delta \log K$) of
pyruvate kinase and lactate dehydrogenase on the
concentration of competing effectors.
A) System conditions as described in Fig. 2 with
0.2% Procion Navy HE-R-PEG;
●——● ATP; o——o ATP + 2 mM $MgCl_2$
B) System conditions as described in Fig. 2 with
0.5% Cibacron Blue F3G-A-PEG;
●——● NAD^+; o——o NAD^+ + 2 mM sodium sulphite

However, addition of sodium sulphite reduced the affinity
partition effect completely at low concentration of NAD$^+$. This
component is known to form a ternary enzyme/NAD$^+$/sulphite
complex exhibiting an extremely low dissociation constant. In
the case of PK the affinity partitioning effect was diminished
partially by ATP. If Mg^{++} were supplemented which are known
to form a stable MgATP-complex and being the actual substrate
of the enzyme the competition did not increase. In contrast,
Mg^{++} seems to promote the protein-dye interaction as indicated by respective curve in Fig. 3.
   Fig. 4 shows the diversity of competing effect of NADP$^+$
with the dyes Procion Green H-4G, Procion Yellow HE-3G, and
Procion Blue MX-G at the binding sites of G6PDH although the
extent of the affinity partitioning effect of these three dyes
is similar.

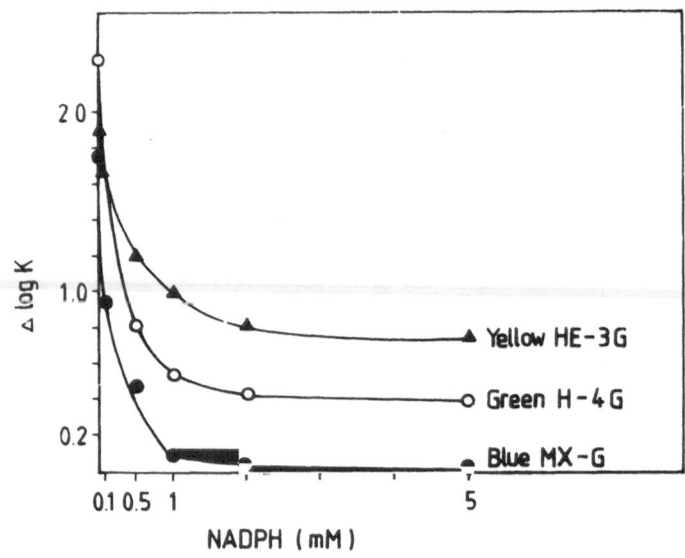

Figure 4. Dependence of the affinity partitioning ($\triangle$log K) of
          glucose 6-phosphate dehydrogenase on the concen-
          tration of NADP by varying the dye-PEG.
          System conditions as described in Fig. 2,
          the system contained 0.5% dye-PEG

Affinity partitioning has also been applied to the analysis of
the structural requirements of dye to the protein binding. As
an example, the interaction of alkaline phosphatase with
Procion Red HE-3B, H-3B, MX-8B and Procion Navy H-ER was
studied in more detail. These dyes show structural similarities as demonstrated in Fig. 5. Dyes like Procion Navy H-ER
exhibiting high affinity per se to AP provided the same
affinity partitioning effect before and after reduction of
their azo-groups with sodium borohydride. However, the affinity of Procion Red HE-3B to the enzyme increased signifi-

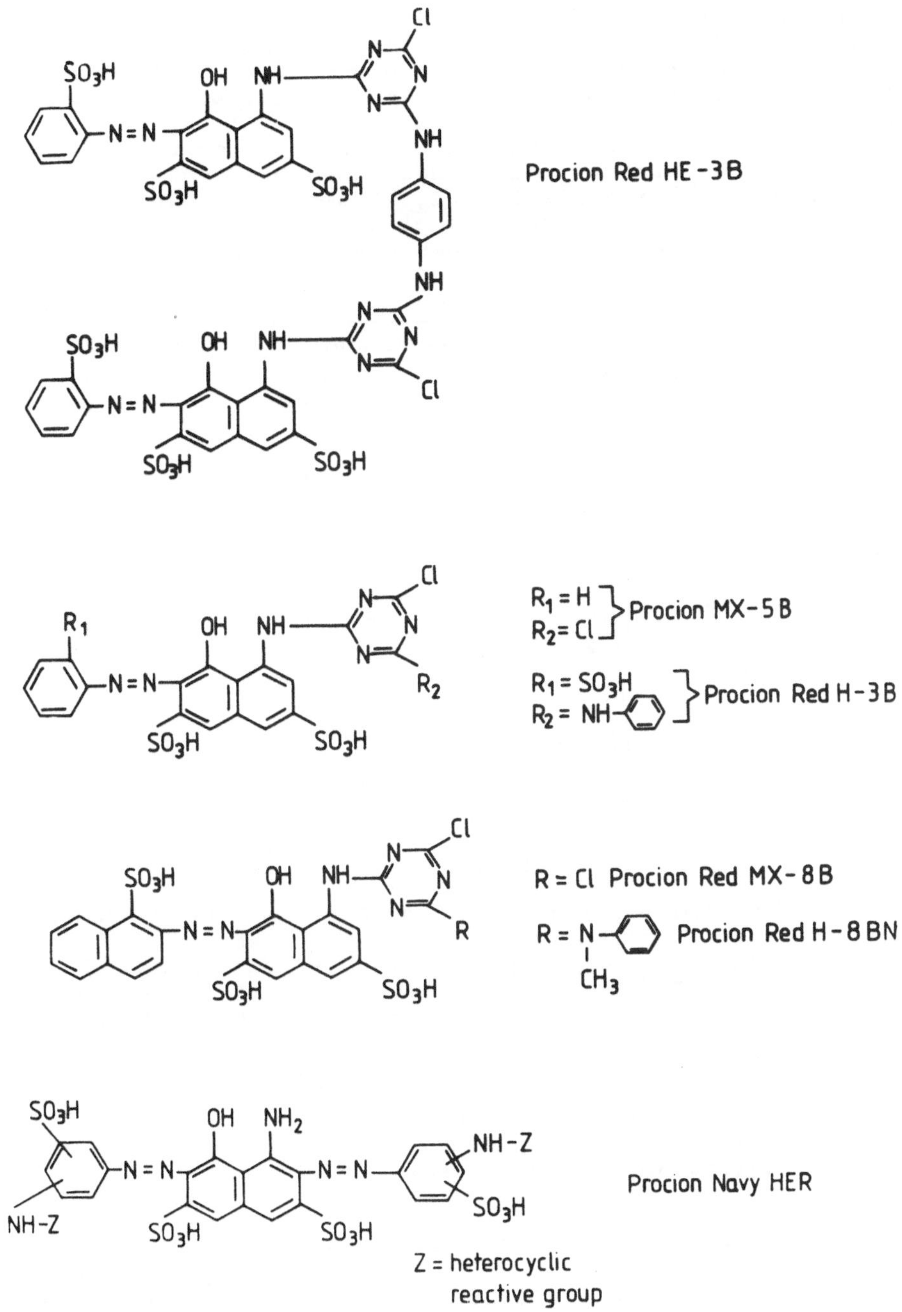

Figure 5. Formula of reactive azo-dyes

cantly when the azo-groups were either reduced or split off
with dithionite (Fig. 6). A similar but more pronounced behaviour was observed when Procion Red H-3B was modified under
the same conditions (Fig.6). The results suggest that the
binding of AP is mainly realized via the 1-amino-8-naphthol-
3,6-disulphonic acid residue of the dye molecule. Similar
results were obtained when Procion Red MX-8B-PEG was treated
with borohydride and dithionite, respectively. This conclusion
is confirmed by earlier studies in which the 1-amino-8-
naphthol-3,6-disulfonic acid was characterized as strong competitive effector in binding of AP to immobilized Procion Red
HE-3B [8].

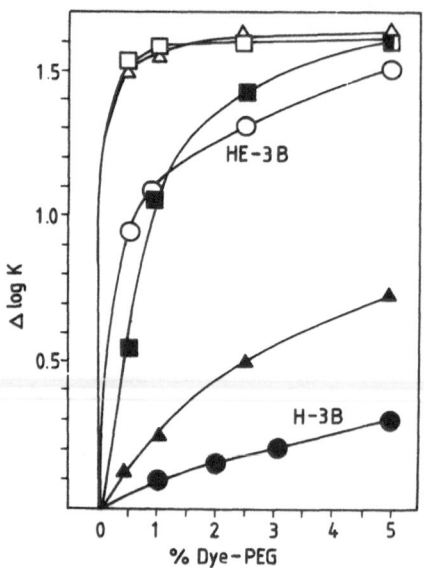

Figure 6. Affinity partitioning of alkaline phosphatase by
using modified dye-PEG derivatives.
The systems (4g) contained 9.75% dextran M70 and
6.5% PEG 6000 with different portions of dye-PEG,
10 mM Tris/HCl buffer, pH 7.5, 2 mM $MgCl_2$ and
10 units of the enzyme.
Procion Red HE-3B (open symbols),
Procion Red H-3B (closed symbols)
●;○: unmodified dyes
▲;△: $NaBH_4$ treatment
■;□: $Na_2S_2O_4$ treatment

Affinity partitioning of AP employing various aminonaphthol
dyes is interferred by the addition of anorganic phosphate.
Even 1 to 5 mM of this effector is capable of abolishing more
or less completely the affinity partitioning effect. However,
there is an opposite behaviour between the increase of the
affinity of the dye-ligand caused by modification and of the
specificity of the dye-enzyme interaction, expressed as the
effectiveness of phosphate to transfer AP into the lower phase

(Fig. 7). Whereas at 2 mM phosphate the △ log K became sero in the case of the unmodified dye Procion HE-3B the residual affinity of the modified dyes in presence of phosphate seems to be significant indicating that the split group might have a discriminating function in the selectivity of binding.

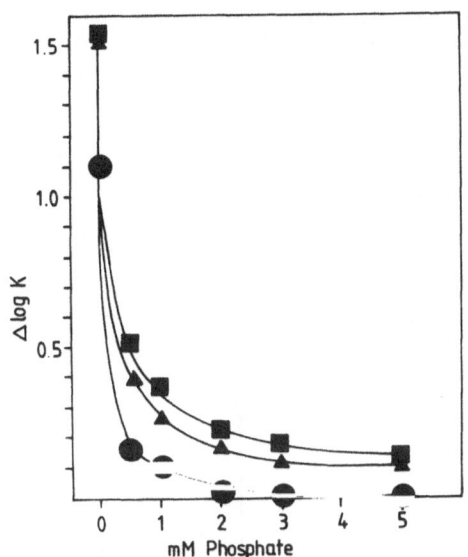

Figure 7. Dependence of affinity partition effect (△log K) of alkaline phosphatase on the phosphate concentration by using modified Procion Red HE-3B-PEG. System conditions as described in Fig. 6. with 1% dye-PEG.
●———● unmodified dye, ▲———▲ NaBH$_4$ treatment, ■ ---■ Na$_2$S$_2$O$_4$ treatment

Phosphofructokinase (PFK) from yeast was found to interact with Cibacron Blue F3G-A and other triazine dyes [13]. Applying affinity partitioning a 1000-fold increase in the partition coefficient of the enzyme was observed if a dye-PEG of high affinity was introduced into the system [14]. This enormous response generated by the dye-enzyme complex formation is a good prerequisite for recognizing also small changes in dye-protein interaction caused by, e.g., conformational changes of the enzyme. Due to the allosteric properties of yeast PFK, fructose 6-phophate (Fru 6-P) is capable of shifting the R - T equilibrium in favour of the R-state thus relieving the inhibitory effect of ATP. If certain triazine dyes mimic the natural nucleotide and bind closely to the ATP-binding site(s) any alteration of the allosteric properties of the enzyme accompanied by a change of the ATP inhibition should be paralleled by a change of the dye-enzyme interaction and should also be reflected in the affinity partition behaviour.

In order to examine this assumption the enzyme was preincubated with Fru 6-P before adding to the system containing 0.05
% Procion Red HE-7B as affinity ligand (Fig. 8). After attai-
ning equilibrium between both phases (this proceeds within a
few seconds) the time dependence of the partition coefficient
was observed. Since only 5 to 10 μl of enzyme solution were
added to a total of 4 g the substrate was diluted to micro-
molar level and can be roughly neglected. The preincubated
enzyme showed a significant lower partition coefficient immediately after adding to the system which continuously
increased and attained a saturation value within 30 min
(Fig. 8A).

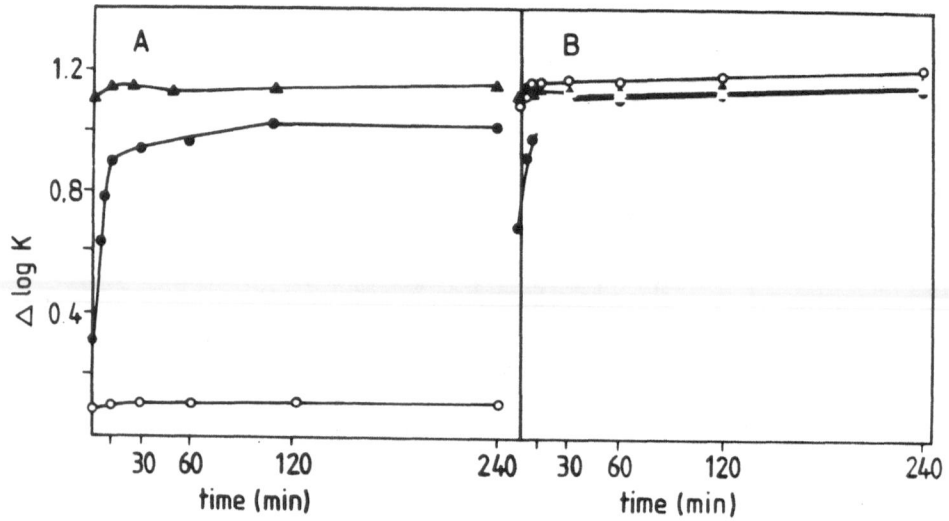

Figure 8. Time dependence of the affinity partitioning of
phosphofructokinase in presence of Procion Red
HE-7B-PEG.
The system contained 5% PEG 6000 in which 1/2000 of
the amount was replaced by dye-PEG, 7% dextran M70,
50 mM sodium phosphate buffer, pH 7.0, 0.5 mM EDTA,
5 mM 2-mercaptoethanol. About 1 unit of enzyme was
added to the system. The procedure of cross-linking
was performed according to [21].
A) native PFK, o——o 1 mM Fru 6-P in the system,
●——● no Fru 6-P in the system, preincubation
of the PFK at 1 mM Fru 6-P for 1 h, ▲——▲ system
free of Fru 6-P, no preincubation
B) ▲——▲ cross-linking in the absence of Fru 6-P,
no preincubation, ●——● cross-linking in the
absence of Fru 6-P, preincubation at 1 mM
Fru 6-P, o——o cross-linking at 10 mM Fru 6-P,
no preincubation

This behaviour was found for diverse dye-PEG derivatives exhibiting high affinity to the enzyme. PFK remained in the bottom phase if the system contained 1 mM Fru-6-P. No change in the partition coefficient was also observed when the enzyme was not preincubated with the substrate.

There are different possibilities to explain the effect of Fru 6-P on the partitioning of PFK, (i) a steric hindrance of the interaction between dye and enzyme caused by bound Fru 6-P, (ii) conformational changes by substrate binding which lowered the affinity of PFK to the ligand, and (iii) changes in the oligomeric state of the enzyme. In order to discriminate these models the enzyme was covalently modified by cross-linking using dimethylsuberimidate in the presence and the absence of Fru 6-P leading possibly to a more rigid protein structure. The curves in Fig. 8B were obtained with a cross-linked enzyme showing oligomeric species up to the tetramer in SDS-polyacrylamide gel electrophoresis and exhibiting full activity. Remarkable, cross-linking in the presence of Fru 6-P had no influence on the partition behaviour of PFK in comparison with the control cross-linked without Fru 6-P. However, the cross-linked PFK exhibit an only moderate time-dependence of partitioning when the enzyme was preincubated with Fru 6-P before adding to the system. The preliminary results with cross-linked PFK are not consistent enough to explain the alteration of the partition of PFK by preincubation with Fruc 6-P on the basis of a simple conformational change of the enzyme.

Dye-ligand affinity methods applying the principle of biorecognition have been extensively developed over the last decade in response to greater demands placed on protein purification technology. The number of papers published has been rapidly increased and today more than 100 proteins have been purified. Recently, the technique of affinity partitioning becomes more attractive also for preparative purposes. There are several advantages of this and in generally of other aqueous two-phase extraction technology:

1. Rapidity of separation particulary by using centrifugal separators as demonstrated
2. High protein content per unit of volume
3. Easy to scale up
4. Simple technical device

In Table 2 some enzymes purified by using affinity partitioning are listed.

In summary, affinity partitioning of proteins will proceed as a useful approach for protein purification, particularly in large scale. Since the relatively high costs of the dextran can be lowered by using crude dextran or alternatively by replacement of dextran by hydrolyzed starch [20] and the ligand containing phase is used several times without regeneration the economic advantages as demonstrated by formate dehydrogenase [18] might be generalized. Furthermore, the improvement of the binding properties of biomimetic dyes by employing the potential of computer-aided ligand design will increase the significance of any affinity ligand separation technique for downstream processing in biotechnology.

TABLE 2
Purification of enzymes by means of affinity partitioning

Protein	Dye-ligand	Purification-factor (n-fold)	Reference
Phosphofructokinase (yeast)	Cibacron Blue F3G-A	10	15
Glucose 6-phosphate dehydrogenase (yeast)	Procion Yellow HE-3G	25	16
	Procion Olive MX-3B	40	16
Lactate dehydrogenase (muscle)	Procion Yellow HE-3G	30	17
	Cibacron Blue F3G-A	25	17
Formate dehydrogenase (candida boidinii)	Procion Red HE-3B	6	18
Alkaline phosphatase (calf intestine)	Procion Red HE-3B	13	19

REFERENCES

1. Albertsson, P.-A., Partition of Cell Particles and Macromolecules. Almqvist & Wiksell, Stockholm, New York, 1960.

2. Flanagan, S.D. and Barondes, H., Affinity partitioning. J. Biol. Chem., 1975, 250, 1484 -89.

3. Johansson, G., Affinity partitioning of enzymes. In Methods in Enzymology, ed., W.B. Jakoby, Academic Press, New York 1984, Vol. 104, pp. 356 - 64.

4. Johansson, G. and Joelsson, M., Preparation of Cibacron Blue F3G-A-Poly(ethylene glycol) in large scale for use in affinity partitioning. Biotechnol. Bioeng., 1985, 27, 621 -25.

5. Clonis, Y.D., Affinity chromatography on immobilized triazine dyes. Post-immobilization chemical modification of triazine dyes. J. Chromatogr., 1982, 236, 69 -80.

6. Hofmann, E. and Kopperschläger, G., Phosphofructokinase from yeast. In:Methods in Enzymology, eds., S.P. Colowick and N.O. Kaplan, Academic Press, New York, 1982, Vol. 90, pp. 49 - 60.

7. Kirchberger, J. and Kopperschläger G., Preparation of homogeneous alkaline phosphatase from calf intestine by dye-ligand chromatography. Prep. Biochem., 1982, 12, 29 - 47.

8. Kirchberger, J., Seidel, H. and Kopperschläger, G., Inter-action of Procion Red HE-3B and other reactive dyes with alkaline phosphatase: A study by means of kinetic, difference spectroscopic and chromatographic methods. Biomed. Biochim. Acta, 1987, 46, 653 - 63.

9. Bergmeyer, H.U., ed., Methoden der Enzymatischen Analyse, Akademie Verlag, Berlin, 1970, p. 527.

10. Biochemica Information I, Boehringer Mannheim, 1973, p.99.

11. Cordes, A., Flossdorf, J. and Kula, M.R., Affinity partitioning: Development of mathematical model describing behaviour of biomolecules in aqueous two-phase systems. Biotechnol. Bioeng., 1987, 30, 514 - 520.

12. Naumann, M., Reuter, R., Metz, P.and Kopperschläger, G., Affinity chromatography of bovine heart lactate dehydrogenase using dye-ligands linked directly or spacer-mediated to bead cellulose. J. Chromatogr. in press.

13. Kopperschläger, G., Böhme, H.-J. and Hofmann, E., Cibacron Blue F3G-A and realated dyes as ligands in affinity chromatography. In: Advances in Biochemical Bioengineering, ed., A. Fiechter, Springer Verlag, Heidelberg, New York, 1982, Vol. 25, pp. 101 - 38.

14. Johansson, G., Kopperschläger, G. and Albertsson, P.-A., Affinity partition of phosphofructokinase from beaker's yeast using polymer-bound Cibacron Blue F3G-A. Eur. J. Biochem., 1983, 131, 589 - 94.

15. Kopperschläger, G. and Johansson, G., Affinity partitioning with polymer-bound Cibacron Blue F3G-A for rapid large-scale purification of phosphofructokinase from beaker's yeast. Anal. Biochem., 1982, 124, 117 - 24.

16. Johansson, G, and Joelsson, M., Partial purification of D-glucose 6-phosphate dehydrogenase from beaker's yeast by affinity partitioning using polymer-bound triazine dyes. Enzyme Microb. Technol., 1985, 7, 629 - 35.

17. Johansson, G. and Joelsson, M., Liquid-liquid extraction of lactate dehydrogenase from muscle using polymer-bound triazine dyes. Appl. Biochem. Biotechnol., 1986, 13, 15 - 27.

18. Cordes, A. and Kula, M.-R., Process design for large-scale purification of formate dehydrogenase from candida boidinii by affinity partition. J. Chromatogr., 1986, 376, 375 - 84.

19. Kirchberger, J., unpublished

20. Tjerneld, F., Berner, S., Cajarville, A. and Johansson, G., New aqueous two-phase system based on hydroxylpropyl starch useful in enzyme purification. Enzyme Microb. Technol., 1986, 8, 417 - 23.

21. Hajdu, J., Bartha, F. and Friedrich, P., Cross-linking with bifunctional reagents as a means for studying the symmetry of oligomeric proteins. Eur. J. Biochem., 1976, 68, 373 - 83.

AFFINITY PARTITION AND SOLUBILISATION OF ENZYMES BY USE OF POLYMER-BOUND TRIAZINE DYES

GÖTE JOHANSSON
Department of Biochemistry,
Chemical Center, University of Lund,
P.O.Box 124, S-221 00 Lund, Sweden

ABSTRACT

Textile dyes have been used in polymer-bound state for selective extraction of enzymes within aqueous two-phase systems composed of two polymers. The effectivity of this extraction depends on several factors including type, molecular weight and concentration of bulk polymers as well as the ligand-carrying polymer. Also the degree of substitution (mole ligand/mole polymer) is a parameter which now has been studied. A high degree of substitution does not only give rise to affinity partition but also allows steering of the partition of the ligand-polymer itself. Therefore, the same ligand-polymer can be used for extraction of enzymes both to the upper and to lower phase. Studies have been carried out with Procion yellow HE-3G bound to poly(ethylene glycol), dextran, hydroxypropyl starch, Ficoll, ethylhydroxyethyl cellulose and polyvinyl alcohol. Dye-poly(ethylene glycol) has also been used for selective solubilisation of enzymes at such concentration of poly(ethylene glycol) which causes precipitation of proteins.

INTRODUCTION

Aqueous liquid-liquid two-phase systems can be obtained by dissolving two polymers in water (1,2). The polymers are concentrated in opposite phases. By attachment, to one of the polymers, a dye ligand can (to large extent) be restricted to the phase rich in this polymer. By increasing the concentrations of phase-forming polymers the partition of the ligand-polymer will be more and more extreme. The most commonly used systems are composed of dextran and poly(ethylene glycol) (PEG). In such systems PEG has been used as ligand carrier with the dye Procion yellow HE-3G as ligand to extract enzymes e.g. glucose-6-phosphate dehydrogenase and lactate dehydrogenase from protein mixtures (3,4). The most effective purification is obtained when the systems contain relatively high concentration of polymers since the bulk proteins then partition strongly towards the lower (dextran-rich) phase while the target enzyme is extracted into the upper (PEG-rich) phase containing the dye ligand.

The effectivity of affinity partition also depends on a number of other factors, e.g. type and concentration of ligand, temperature, pH value, type of salt present and its concentration, molecular weight of the polymers and the presence of free (natural) ligand (5-10). Several other factors can be of interest to study in order to obtain high-resolution systems based on dye affinity ligands. In this work the effects of density of ligand (on the ligand-carrying polymer) and the use of a third polymer as ligand carrier are discussed.

MATERIALS AND METHODS

The polymers used are polyethylene glycol (PEG), polyvinyl alcohol (PVA), Ficoll (Fi), dextran (Dx), hydroxypropyl starch (HPS) and ethylhydroxyethyl cellulose (EHEC). They are all commercial available and have been used without further purification. Systems have been prepared from concentrated stock solutions of the polymers in water (normally 20-50 %), buffer, sample and water. The concentration of polymers are given in per cent (w/w) of the total system.

The partition of included enzyme or protein is described by the partition coefficient, K, defined as the ratio between the concentration of partitioned material in upper and lower phase, respectively. Protein has been determined according to Bradford (11). Enzymes were determined by photometric methods at 340 nm (12-14).

The coupling of dye ligands to the polymers were carried out in alkaline water solution and the products were purified by precipitation, chromatography or liquid-liquid extraction as has been described in detail elsewhere (15,16).

RESULTS

Dye density

The effect of the density of dye on a carrier polymer molecule has been studied (15) by using dextran 70 (Mw = 70,000) as ligand carrier. Dextran derivatives with the degree of substitution, n (= average number of dye molecules per dextran molecule), equal to 1.3, 2.3, 5.3 and 8.3 were prepared. The partition coefficient, K_{L-Dx}, of the dye-dextran (in a Dx-PEG system) is strongly dependent on the kind of salts present in the two-phase system, Fig. 1. Systems containing potassium chloride show the same K_{L-Dx} (= 0.07) independent of the n value. With phosphate, on the other hand, K_{L-Dx} increases from 0.23 to 27 when n is changed from 1.3 to 8.3. The high-substituted dextran can consequently be moved from one phase to the other by changing the salt content of the two-phase system.

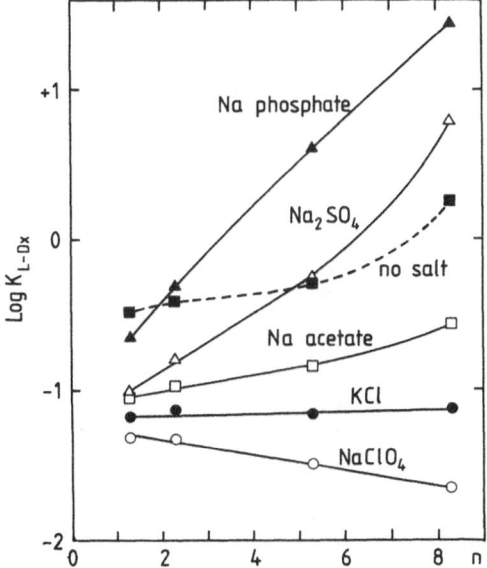

Figure 1: Influence of salts on the partition of Procion yellow HE-3G dextran 70 in a system containing 4.5 % (w/w) PEG 8000, 8 % (w/w) dextran 70, 50 μM dextran-bound dye and salt (100 mM except for phosphate, 50 mM). In all cases pH was adjusted to 7.9. Arrow indicates the partition of unsubstituted dextran. Temperature, 22 °C. Reproduced from Johansson and Joelsson (15).

Independent of the localization of the (dextran-bound) ligand the ligand-binding enzyme is extracted into the ligand-containing phase. The change in the logarithmic partition coefficient, $\Delta \log K_E$, caused by the ligand, is once to twice the logarithmic partition coefficient, $\log K_{L-Dx}$, of the ligand-dextran, Table 1.

Table 1

Log K_{L-Dx}, of Procion yellow HE-3G dextran 70, $\Delta \log K_E$ for 3-phosphoglycerate kinase (PGK), glucose-6-phosphate dehydrogenase (G6PDH), both from yeast and lactate dehydrogenase (LDH) from muscle, and a$_{app}$ (= $\Delta \log K_E / \log K_{L-Dx}$). System composition: 8.0 % (w/w) dextran 70, 4.5 % (w/w) PEG 8000, 30 μM dextran-bound dye, and 10 mM sodium phosphate buffer, pH 7.5, and enzyme (3, 7, and 12 U ml^{-1} for PGK, G6PDH, and LDH, respectively). Temperature, 22 $^{\circ}$C. Data from Johansson and Joelsson (15).

n	log K_{L-Dx}	PGK			G6PDH			LDH		
		K_E	$\Delta \log K_E$	a$_{app}$	K_E	$\Delta \log K_E$	a$_{app}$	K_E	$\Delta \log K_E$	a$_{app}$
0	–	0.48	0	–	0.28	0	–	0.14	0	–
1.3	-0.47	0.12	-0.59	1.3	0.04	-0.85	1.8	0.04	-0.57	1.2
2.3	-0.12	0.37	-0.11	0.9	0.15	-0.25	2.1	0.15	0.01	–
5.3	0.84	2.7	0.75	0.9	2.8	1.00	1.2	5.0	1.54	1.8
8.3	1.62	13	1.43	0.9	9.3	1.53	0.9	39	2.43	1.5

Model for affinity partition

A model for describing the affinity partition has been developed by Flanagan and Barondes based on thermodynamic reasoning (17). When the ligand-polymer (L-pol) is present in excess the model gives Eq. (1):

$$K_E = K_0 \, (K_{L-pol} \, D_B/D_T)^a \qquad (1)$$

where K_0 is the partition coefficient of enzyme in absence of ligand. D are the dissociation constants in top and bottom phase, respectively, and a is the number of binding sites for the ligand on the enzyme molecule. In logarithmic form the expression takes the following form, Eq. (2):

$$\Delta \log K_E = a \ (\log K_{L-pol} + \log (D_B/D_T))$$ (2)

If the dissociation constants are the same in the two phases, which is not necessarily true (18), the relation could be simplified to give Eq. (3):

$$\Delta \log K_E = a \log K_{L-pol}$$ (3)

For 3-phosphoglycerate kinase, the experimental values give, via Eq. (3), an apparent $\underline{a}$ value (a_{app}) close to one, Table 1. This enzyme is assumed to have one binding-site for the dye. Glucose-6-phosphate dehydrogenase (with two assumed binding-sites) shows a_{app} close to two when the ligand is concentrated to the lower phase, but the value is around one with ligand in the upper phase. This can be evaluated as an increased dissociation of the enzyme-ligand complex in the top phase. Lactate dehydrogenase, with 4 binding sites, has a_{app} values less then two.

If the difference between the dissociation constants in the two phases is not too extreme it is resonable to assume that one to two dextran molecule bind per enzyme molecule. This is illustrated in Fig. 2b where a ligand-dextran molecule and an enzyme molecule are drawn in scale.

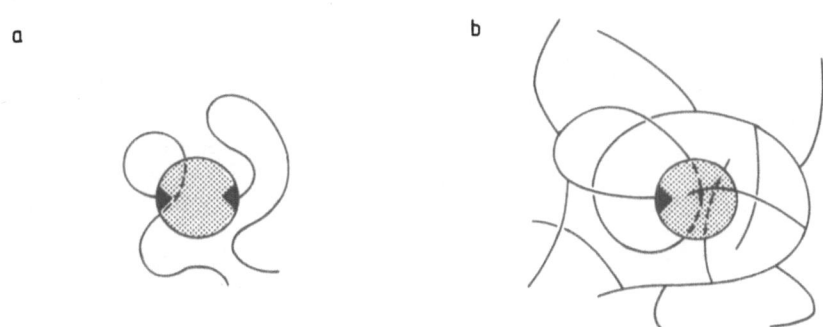

a   b

Figure 2: Enzyme molecule, Mw = 100,000, surrounded a) by two PEG molecules, Mw = 8000, and b) by one dextran molecule, Mw = 70,000.

Effect of a third polymer as ligand carrier

Affinity partition in aqueous two polymer two-phase systems can also be achieved by using a third polymer (present in low con-

Table 2

Partition of ligand-polymers (L-pol) in various two-phase systems and their effect on the partition on lactate dehydrogenase (LDH). All systems contained 25 mM sodium phosphate buffer, pH 7.5, and 7 U LDH per ml. Ligand concentration was 42 µM, except for EHEC (10 µM). Temperature, 22 $^{\circ}$C. Reproduced from Johansson and Joelsson (16).

Two-phase system	Ligand carrier	log K_{LDH}	log K_{L-pol}	a_{app}
7 % Dx 500 –	None	−0.75	–	–
5 % PEG 8000	PEG 8000	1.48	1.50	1.5
	PVA 14000	–	*	–
	EHEC	1.31	1.63	1.3
	Fi 70	1.36	0.90	2.3
	HPS	−0.16	0.32	1.8
	Dx 70, n=8.3	1.56	1.60	1.4
15 % HPS –	None	−1.16	–	–
5.8 % PEG 8000	PEG 8000	1.16	1.46	1.6
	PVA 14000	–	*	–
	EHEC	1.15	>2.3	<1.0
	Fi 70	0.25	0.48	2.9
	HPS	−2.52	−1.37	1.0
	Dx 70, n=8.3	0.54	1.26	1.3
4.2 % Dx 500	None	−1.11	–	–
6 % PVA 14000	PEG 8000	−0.59	0.96	0.5
	PVA 14000	−0.30	1.74	0.5
	EHEC	–	*	–
	Fi 70	−1.53	−0.30	1.4
	HPS	−0.12	0.64	1.5
	Dx 70, n=8.3	1.36	2.03	1.2
4 % Dx 500	None	−0.01	–	–
11.5 % Fi 400	PEG 8000	−0.35	−0.28	1.2
	PVA 14000	–	**	–
	EHEC	–	*	–
	Fi 70	−0.71	−0.30	2.3
	HPS	−0.80	−0.47	1.7
	Dx 70, n=8.3	−1.21	−0.86	1.4

* Ligand-polymer precipitated.
** Three phases were formed.

centration) as ligand carrier (16). Partition of lactate dehydrogenase as well as of ligand-polymer in such systems are
presented in Table 2. The a_{app} values have been calculated
according to eq. (3). In a few cases, e.g. when Ficoll is
ligand carrier, the a_{app} values are larger than two. These
favourable values may be caused of either by the arrangement of
more than two polymer molecules around the enzyme molecule, or
by different degrees of dissociation of the complex (in the two
phases) in such way that the affinity extraction is favoured.

Affinity solubilisation

When proteins are precipitated with aid of PEG some enzymes can
be kept in solution by including dye-PEG derivatives (19). This
process has been named Affinity constrained precipitation. The
affinity-solvated enzymes can be purified at least three times,
Table 3. This way of using dye-PEG can also be seen as an
affinity partition where the system is made up of PEG and bulk
proteins.

Table 3

Affinity solubilisation of enzymes in yeast extract using
12.5 % PEG 8000 at pH 4.35 and Procion yellow HE-3G PEG,
0.21 %. Temperature, 0 °C. From Johansson and Joelsson (19).

Enzyme	U/mg protein		Recovery (%)
	Original extract	Supernatant liquid	
Glucose-6-phosphate dehydrogenase	0.48	1.66	93
3-Phosphoglycerate kinase	5.55	16.8	86

DISCUSSION

Dyes can be useful ligands for extraction of enzymes by using
two-phase techniques. The results above show that the
interaction between ligand and enzyme may depend both on the
bulk polymers and the polymer carrying the ligand. The most

commonly used ligand carrier, PEG, may be far from the best choice.

As visualized in Fig. 2, the binding of one to two flexible macromolecules is enough to wrap the enzyme molecule in a polymer "atmosphere" and greatly restrict the surface of the protein from interaction with the surrounding. The arrangment of the polymer chain around the enzyme molecule is a function of the positive and negative interactions between the two. PEG, which is known to exclude many proteins, may form a shell with a certain distance from the protein molecule (except for the point of attachment). Between this PEG shell and the enzyme there is a buffer zone of water. Other polymers with less negative interactions, like dextran, may come in closer contact with the protein and therefore facilitate the dye-enzyme interaction.

The partition behaviour of the dye dextran derivatives, Fig. 1, is a result of the chemical structure of dye. The dye molecule contains both aromatic rings, which have positive interactions with PEG, and several negatively charged groups. The partition of charged polymers (1,2) has been shown to depend on the salt present in the two-phase system. When sodium acetate, Fig. 1, is used the partition of polyelectrolytes is normally independent of their charge. The increase in K_{L-Dx} would consequently be due to the PEG-aromatic group interactions. Phosphate, on the other hand, increases the partition coefficient of the ligand-dextran and the change in log K_{L-Dx} is proportional to the number of charged groups on the dextran molecule. By using potassium chloride, which normally weakly extract polyelectrolytes towards the dextran-rich phase, the effect from the aromatic groups can be levelled out and the partition is then independent of the degree of substitution. These effects allow us to direct the partition of a dye-polymer (preferentially with n>5) by using various salts. This increases the flexibility of using dye-affinity partition in aqueous two-phase systems for enzyme extraction.

CONCLUSIONS

The great number of interactions between the components of aqueous polymer two-phase systems containing polymer-bound affinity ligands makes affinity partitioning to a complex system. Detailed studies and understanding of the interplay between dye, polymers, proteins, and electrolytes will give rise to new and more effective extraction systems.

REFERENCES

1. Albertsson, P.-A., Partition of Cell Particles and Macro molecules, Wiley, New York, 1986, 3rd edition.

2. Walter, H., Brooks, D.E. and Fisher, D. E. (eds.), Partitioning in Aqueous Two-Phase Systems. Theory, Methods, Uses, and Applications to Biotechnology, Academic Press, Orlando, 1985.

3. Johansson, G. and Joelsson, M., Liquid-liquid extraction of lactate dehydrogenase from muscle using polymer-bound triazine dyes. Appl. Biochem. Biotechnol., 1986, 13, 15-27.

4. Tjerneld, F., Johansson, G. and Joelsson, M., Affinity liquid-liquid extraction of lactate dehydrogenase on a large scale. Biotechnol. Bioeng. 1987, 30, 809-16.

5. Kroner, H.K., Cordes, A., Schelper, A., Morr, M., Buckmann, A.F. and Kula, M.-R., Affinity partition studied with glucose-6-phosphate dehydrogenase in aqueous two-phase systems in response to triazine dyes. In Chromatography and Related Techniques, eds., T.C.J. Gribnau, J. Visser and R.J.F. Nivard, Elsevier, Amsterdam, 1982, pp. 491-501.

6. Johansson, G., Kopperschläger, G. and Albertsson, P.-A., Affinity partitioning of phosphofructokinase from baker's yeast using polymer-bound Cibacron blue F3-A. Eur. J. Biochem., 1983, 131, 589-94.

7. Johansson, G. and Andersson, M., Parameters determining affinity partitioning of yeast enzymes using polymer-bound triazine dye ligands. J. Chromatogr. 1984, 303, 39-51.

8. Johansson, G., Aqueous two-phase systems in protein purification. J. Biotechnol., 1985, 3, 11-8.

9. Kopperschläger, G., Lorenz, G. and Usbeck, E., Application of affinity partitioning in an aqueous two-phase system to the investigation of triazine dye-enzyme interactions. J. Chromatogr., 1983, 259, 97-105.

10. Joelsson, M. and Johansson, G., Sequential liquid-liquid extraction of some enzymes from porcine muscle using polymer-bound triazine dyes, Enzyme Microb. Technol., 1987, **9**, 233-7.

11. Bradford, M.M., A rapid and sensitive method for the quantitation of microgram quantities of protein utilizing the principle of dye binding. Anal. Biochem., 1976, **72**, 248-54.

12. Scopes, R.K., 3-Phosphoglycerate kinase of skeletal muscle. Methods Enzymol. 1975, **42**, 127-34.

13. Noltmann, E.A., Gubler, C.J. and Kuby, S.A., Glucose 6-phosphate dehydrogenase (Zwischenferment). I. Isolation of crystalline enzyme from yeast. J. Biol. Chem., 1961, **236**, 1225-30.

14. Bergmeyer, H.U., Methoden der enzymatischen Analyse, Verlag Chemie, Weinheim/Bergstr., 1970, 2nd ed., Vol. 1, p. 441.

15. Johansson, G. and Joelsson, M., Affinity partitioning of enzymes using dextran-bound Procion yellow HE-3G. Influence of dye-ligand density. J. Chromatogr., 1987, **393**, 195-208.

16. Johansson, G. and Joelsson, M., Effect of polymer structure on affinity partitioning of lactate dehydrogenase in polymer-water two-phase systems. J. Chromatogr., 1987, **411**, 161-6.

17. Flanagan, S.D. and Barondes, S.H., Affinity partitioning - a method for purification of proteins using specific polymer-ligands in aqueous polymer two-phase systems. J. Biol. Chem., 1975, **250**, 1484-9.

18. Cordes, A., Flossdorf, J. and Kula, M.-R., Affinity partitioning: development of mathematical model describing behaviour of biomolecules in aqueous two-phase systems. Biotechnol. Bioeng. 1987, **30**, 514-20.

19. Johansson, G. and Joelsson, M., Specifically increased solubility of enzymes in polyethyleneglycol solutions using polymer-bound triazine dyes. Anal. Biochem., 1986, **158**, 104-10.

THE COMPARATIVE STUDY OF THE INTERACTION OF YEAST AND HORSE LIVER ALCOHOL DEHYDROGENASES WITH REACTIVE DYES BY AFFINITY PARTITIONING IN TWO-PHASE SYSTEMS

J.-H.J.PESLIAKAS, V.V.ZUTAUTAS, A.A.GLEMZA
ESP "Fermentas", All-Union Research Institute of
Applied Enzymology, Vilnius, Lithuanian SSR,USSR

ABSTRACT

The affinity partitioning of two alcohol dehydrogenases - from yeast and horse liver is studied in two-phase systems containing dyes - polyethylene glycol and dextran. It was shown that both alcohol dehydrogenases are effectively extracted in the top phases containing $Cu(II)$ complexes of Light resistant yellow 2KT and Red-violet 2KT - PEG. $\Delta \log K_{max}$ of yeast alcohol dehydrogenase is 1.94 and 2.94 for Red-violet 2KT and Light resistant yellow 2KT respectively. $\Delta \log K_{max}$ of horse liver alcohol dehydrogenase was 1.81 and 1.64 respectively. The dependence of the enzymes $\Delta \log K$ on concentration of NaCl, NAD and chelating agents was also studied. It was shown that the presence of the latter in two-phase system reduces $\Delta \log K_{max}$ of yeast alcohol dehydrogenase by 80-90% and $\Delta \log K_{max}$ of horse liver alcohol dehydrogenase by 64 -84%.

INTRODUCTION

The interaction of yeast alcohol dehydrogenase (ADH) with a number of reactive dyes was studied in buffer and using immobilization by differential spectroscopy, circular dichroism and chromatography (1-5). It was demonstrated for the first time that the interaction of yeast ADH with some dyes-Light resistant yellow 2KT, Red-violet 2KT and Claret CT depends on the presence of metal ions (1,3,4). It was also shown that the specific binding of ADH by dyes (Light resistant yellow 2KT-Cu(II)) is due to the formation of the mixed complex of the dye and ADH with Cu^{2+} ions. This high selectivity complex is formed by the coordination between dye's donor groups and the imidazole donor nitrogen from the histidine residue of yeast ADH (4). This allows to suppose that the interaction of horse liver ADH with dyes also depends on the presence of metal ions. To clarify this assumption the affinity partitioning of yeast and horse liver ADH's was studied in two-phase systems consisting of polyethylene glycol (PEG)-dyes/dextran.

MATERIALS
Yeast alcohol dehydrogenase (ADH) (E,C 1.1.1.1.) (Boehringer, GFR or Calbiochem, Switzerland). Horse liver ADH (Reanal, Hungary). NAD and Tris (Serva, GFR). Polyethelene glycol (M_r 60 00). PEG 6000, Fluka, Switzerland) or (Serva, GFR). Dextran (M_r 60000 $\pm$ 10000) of Krasnojarsk Factory of Clinical Preparations. Reactive dyes: Light resistant yellow 2KT, Red-violet 2KT, Red-brown 2KT, Orange 5K, Bright-yellow 53, Bright-red 6C, Red-brown 2K, Bright-yellow 2KT, Claret 4CT - of native production; used without additional purification. Cibacron Blue MX-R, Procion Yellow M-4R, Procion Yellow MX-4R (Serva, GFR), Procion Green H-4G, Procion Red HE-3B (Sigma, USA), Procion Red HE-7B, Procion Red MX-2B, Procion Blue SP-3P, Procion Yellow H-3R (ICI, Great Britain), also supplied by prof. G.Kopperschlager of Karl Marx University (GDR). Salts "highly pure".

METHODS
The activity of alcohol dehydrogenase was determined according to the described procedure (6) slightly modified in that the concentration of NAD in ADH activity determination was $1.5 \cdot 10^{-2}$ M. The two-phase system was prepared from aqueos solutions of PEG 6000 (20%) and dextran 60000 $\pm$10000 (20%) and buffer of 0.I Tris-maleic acid-NaOH (pH 6.5), 50 mM $MgCl_2$, 0.5 mM EDTA, 0.05 mM β-mercaptoethanol and the required amount of ADH activity units. PEG-dye conjugate was synthesized

by the described method (7). Cu^{2+} ions were removed from the
conjugates PEG – Red-violet 2KT and PEG – Red-brown 2KT by washing these conjugates, adsorbed on anion exchanger, with 0.1
M of EDTA. For determining the partition coefficients of enzymes 2-10 acitivity units of yeast ADH and 2-7 activity units
of dialyzed horse liver ADH were added. The mixture was shaken
for 15 sec and then centrifuged at 2000 for 5 minutes at 4°C.
For assaying the ADH activity appropriate samples from each
phase were taken. The partition coefficient K is defined as
the ratio between the ADH activities in top phase and lower
phase. The recovery of ADH activities in both phases was greater than 90%. The alteration of ADH partition coefficients
(logK) was determined by adding the definite amount of PEG –
dye conjugate into the top phase and ΔlogK was defined as the
difference between the logarithmic partition coefficients of
enzymes in the presence (K) and in the absence of dye (K_o)
(ΔlogK = logK – logK_o).

RESULTS AND DISCUSSION

The effectiveness of enzyme partition between the PEG – dye
phase and dextran phase was evaluated by the maximum increase
of logarithmic partition coefficients of the enzymes ($logK_{max}$)
for several PEG – dye system. The alteration of ΔlogK of yeast
ADH as depending on PEG – dye conjugate concentration and the
values of $\Delta logK_{max}$ of enzyme is represented in Fig. 1 and
Table 1, respectively. As seen from the results in Fig. 1
and Table 1 the extraction of yeast ADH in the top PEG– dye
containing-phase is effective for Cibacron Blue F3GA ($\Delta logK_{max} \equiv$
1.20), Procion Blue SP-3R (1.28), Red-violet 2KT (1.94) and
Light resistant yellow 2KT (2.94). The partition coefficient
of Light resistant yellow 2KT changed for almost 3 orders.
The cited alterations of $\Delta logK_{max}$ value reflecting a change
of yeast ADH affinity to dyes correlate with the stability
of ADH – dye complexes that had been determined earlier(2)
by difference spectroscopy. We have studied the conditions
of ADH – dye complexes disruption: in the presence of NaCl,
specific agent NAD, and for PEG – dye – Cu(II) conjugates-pre-

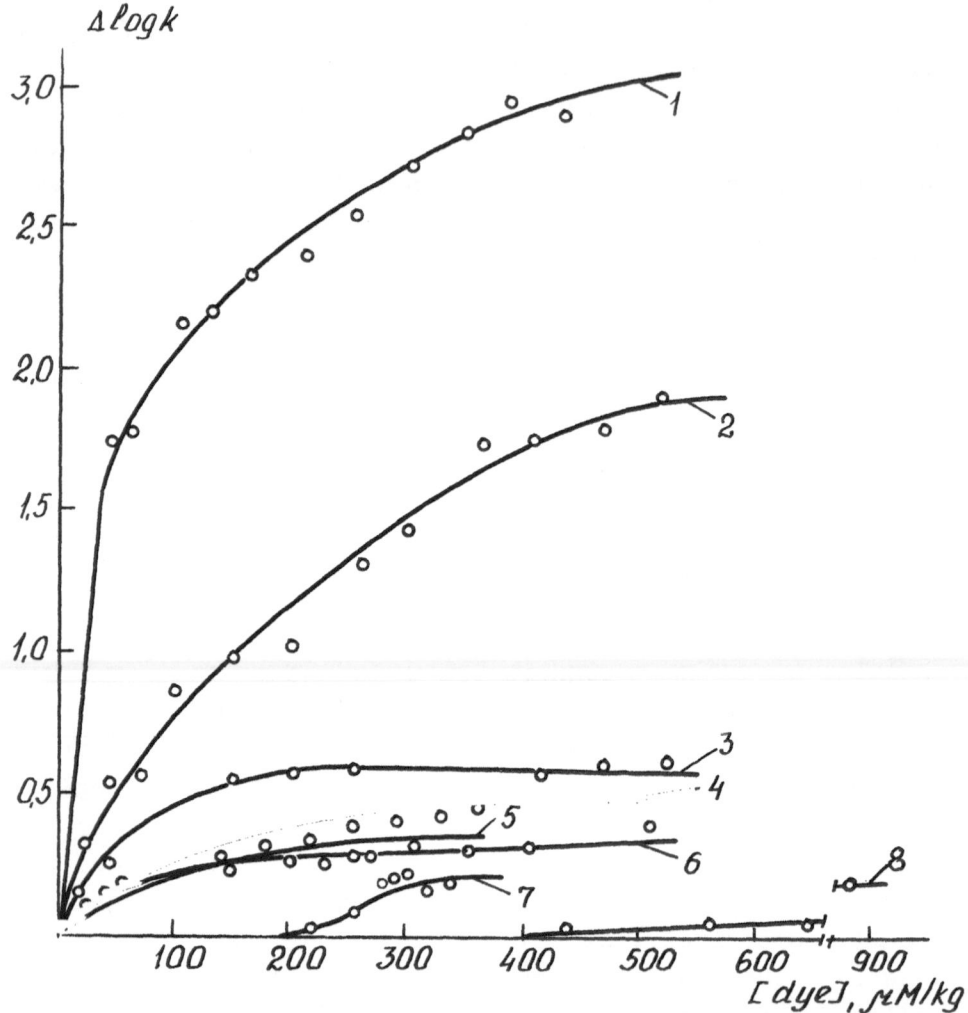

Figure 1.Alteration of the partition coefficient of alcohol
dehydrogenase from yeast depending on concentrations
of various dye-polyethylene glycol conjugates.The system (4g) contains 6.5% PEG 6000 with different portions of dye-PEG and 10% (w/w) dextran 60000±10000,
10mM Tris-maleic acid-NaOH buffer (pH 6.5),5mM $MgCl_2$
0.5mM EDTA and 0.05mM 2-mercaptoethanol.Dyes:(1)Light
resistant yellow 2KT,(2)Red-violet 2KT,(3)Bright-red
6C,(4)Red-brown 2KT,(5)Bright-yellow 53,(6)Red-brown
2K,(7)Claret 4CT,(8)Bright yellow 2KT.

TABLE 1
Maximum alteration of the logarithmic partition coefficient (Δlog K_{max}) of yeast ADH depending on dye-
PEG conjugates. Phase composition and affinity partition conditions as in Fig. 1

Dye - PEG	Δlog K_{max}
1. Bright yellow 2KT	0.20
2. Procion Red MX-2B	0.27
3. Claret 4CT	0.29
4. Procion Red HE-3B	0.33
5. Bright yellow 53	0.35
6. Procion Yellow MX-4R	0.39
7. Procion Red HE-7B	0.39
8. Orange 5K	0.43
9. Procion Green H-4G	0.48
10. Procion Blue MX-R	0.55
11. Red-brown 2KT	0.55
12. Red-brown 2K	0.54
13. Bright-red 6C	0.54
14. Procion Yellow H-3R	0.88
15. Cibacron Blue F3G-A	1.20
16. Procion Blue SP-3R	1.28
17. Red-violet 2KT	1.94
18. Light resistant yellow 2KT	2.94

Position 3, 17, 18 - Cu(II) complexes of cited dyes

ence of chelating agents.Analysis of the dependence of yeast
ADH ΔlogK alteration on NaCl (Fig. 2, Table 2) showed that
ADH interacts with dyes Bright-red 6C, Bright yellow 53, Cibacron blue F3GA, Procion blue MX-R, Procion Yellow MX-4R due
to the dominating electrostatic binding forces as the presence of 0.2 M NaCl in the two phase system decreases ΔlogK$_{max}$
more than 60%. In case of dyes Red-violet 2KT, Light resistant yellow 2KT, Bright yellow 53, Procion Blue MX-R (Fig. 2,
Table 2) further increase of NaCl concentration (from 0.2 M
till 0.75 M) causes the increase of ΔlogK of yeast ADH, i.e.
the extraxtion of enzyme into the top phase containing dye -

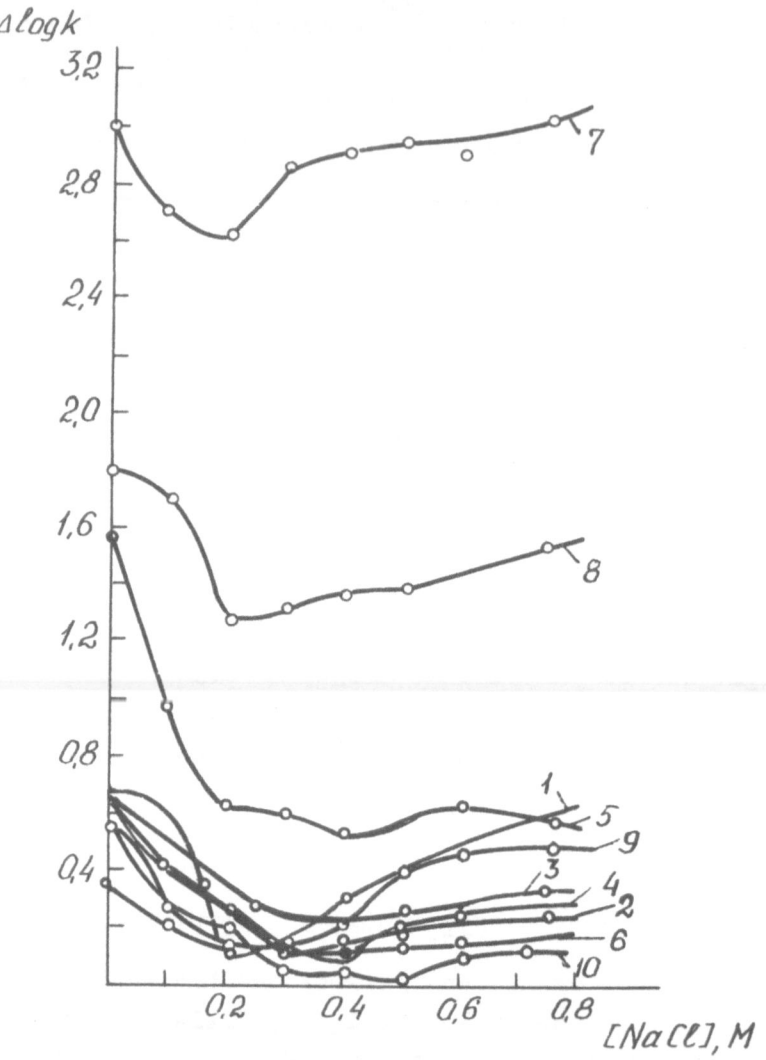

Figure 2. Effect of sodium chloride on partition of yeast ADH.
Phase composition and affinity partition conditions
as in Fig.1. Dyes and dye-PEG concentration in two-
phase system: (1)Procion Blue MX-R (270uM/kg),(2)Pr-
ocion Yellow MX-4R (400),(3)Procion Green H-4G(320),
(4)Cibacron Blue F3GA (185),(5)Procion Blue SP-3R
(3% of total PEG).(6)Procion Red HE-7B (3%),(7)Lig-
ht-resistant yellow 2KT(350uM/kg),(8)Red-violet 2KT
(420),(9) Bright-red 6C (265).

TABLE 2
Dependence of partition (ΔlogK) of yeast ADH on the
concentration of sodium chloride. Phase composition
and partition conditions as in Fig. 2

Dye - PEG	NaCl concentration reducing ΔlogK$_{max}$ by 50%	Percent of ΔlogK$_{max}$ reduction at	
		0.2M NaCl	0.6M NaCl
1. Bright-yellow 53	0.09	76	16
2. Bright-red 6C	0.10	71	80
3. Red-violet 2KT	0.15	28	19
4. Light-resistant yellow 2KT	0.14	13	–
5. Cibacron Blue F3GA	0.07	75	77
6. Procion Blue SP-3R	0.14	59	62
7. Procion Blue MX-R	0.16	77	26
8. Procion Yellow MX-4R	0.15	64	66
9. Procion Gree H-4G	0.18	55	56
10. Procion Red HE-7B	0.30	29	65

PEG. This might be due to the fact that higher NaCl concentrations (greater then 0.3 M) while quenching the part of electrostatic forces of binding increase the part of hydrophobic binding forces. The alteration of yeast ADH partition coefficient in the presence of NAD for some dye-PEG conjugates is presented in Fig. 3, Table 3. As could be seen 5 mM of NAD disrupt the yeast ADH-dye complexes effectively - ΔlogK decreases by 61-100% for dyes from Bright-yellow 53 to Procion Red HE-3B (except for Light-resistant yellow 2KT). In case of dye Cu(II) complexes - Light-resistant yellow 2KT and Red-violet 2KT (Fig. 2,3; Table 2,3) the presence of NaCl (up to 0.2 M) or NAD (up to 5 mM) reduces ΔlogK only partially - 13-28% (0.2 M NaCl) or 17-65% (5 mM NAD). The effective disruption of ADH-dye complex (80-90% decrease of ΔlogK) is achieved only with the addition of 3 mM chelating agents - imidazole, adenine or EDTA (Fig. 4) into the two-phase system,

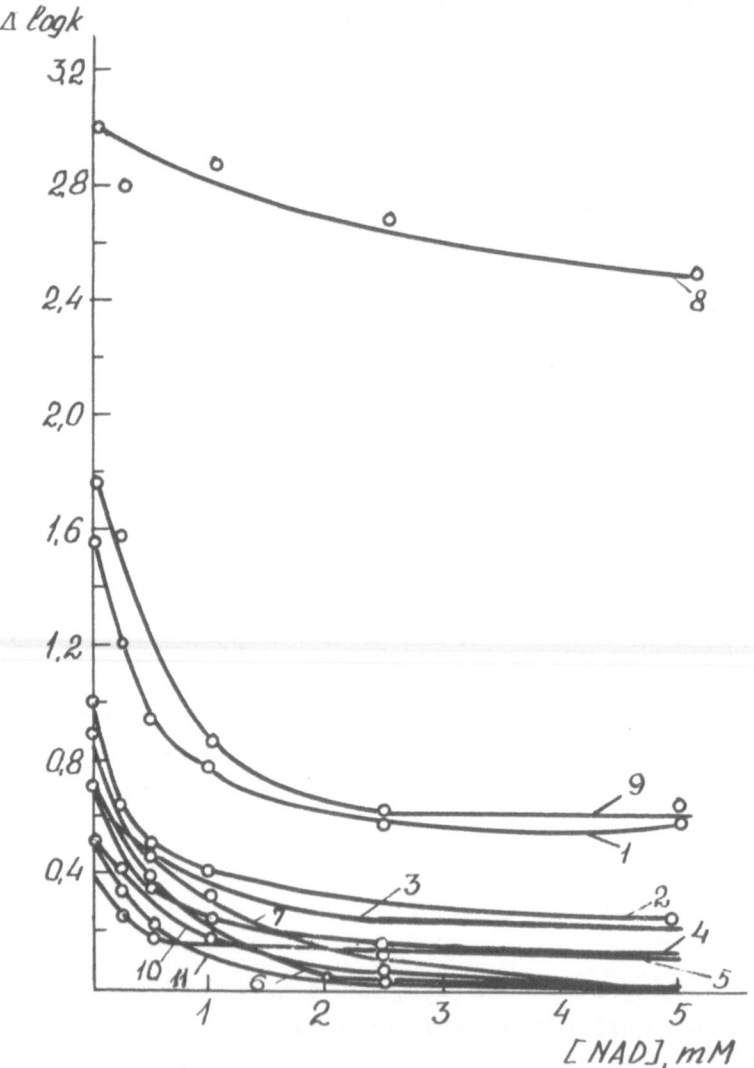

Figure 3. Effect of NAD on partition of yeast ADH. Phase composition and affinity partition conditions as in Fig.
Dye-PEG concentration in two-phase systems as in Fig
2. Dyes:(1)Procion Blue SP-3R,(2)Procion Yellow H-3R
(3% of total PEG),(3)Procion Yellow MX-4R,(4)Procion
Red HE-3B,(5)Procion Green H-4G,(6)Procion Blue MX-R
(7)Cibacron Blue F3GA,(8)Light-resistant yellow 2KT,
(9)Red-violet 2KT,(10)Bright-yellow 53,(11)Brigt-red

TABLE 3
Dependence of partition ($\Delta\log K$) of yeast ADH on the
concentration of NAD. Phase composition and partition
conditions as in **Fig.3**

Dye - PEG	Concentration of NAD reducing $\Delta\log K_{max}$ by 50%	Percent of $\Delta\log K_{max}$ reduction at		
		0.5mM NAD	1mM NAD	5mM NAD
1.Bright-yellow 53	0.80	24	66	100
2.Bright-red 6C	0.45	53	80	99
3.Red-violet 2KT	1.00	24	50	65
4.Light-resistant yellow 2KT	-	3	5	17
5.Cibacron Blue F3GA	0.40	53	68	98
6.Procion Blue SP-3R	1.00	39	50	61
7.Procion Blue MX-R	0.65	44	67	95
8.Procion Yellow MX-4R	0.35	47	59	66
9. Procion Green H-4G	0.50	48	66	80
10.Procion Red HE-3B	0.30	62	66	68

consequently ADH is extracted effectively into the top phase. To identify the role of metal ions on the horse liver -
ADH - dye interaction PEG - dye conjugates were used: Light-
resistant yellow 2KT, Red-brown 2KT, Red-violet 2KT and their
Cu(II) complexes , claret 4CT- Cu(II) also Bright-yellow 53,
Procion Blue SP-3R, Orange 5K. Affinity partitioning of horse liver ADH was investigated under the same conditions as
yeast ADH. The dependence of alteration of partition coefficient of horse liver ADH ($\Delta\log K$) on the concentration of PEG-
dye conjugate and $\Delta\log K_{max}$ value of enzyme is presented in
Fig. 5, Table 4. Tha data presented in Fig. 5 and Table 4 demonstrate that from all dyes which do not contain metal ions:
Bright-yellow 53, Procion Blue SP-3R and Orange 5K only in

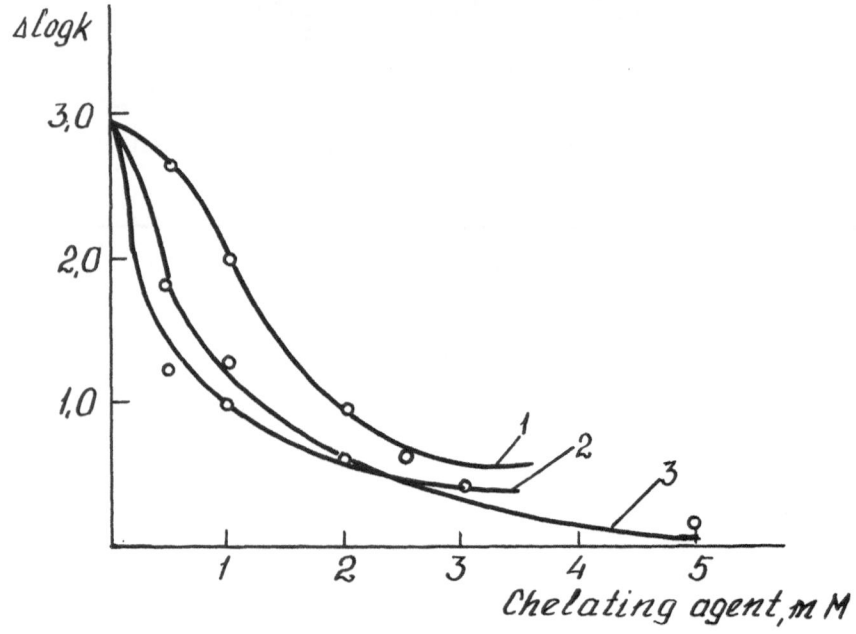

Figure 4. Effect of chelating agents on partition of yeast
ADH in two-phase system containing PEG-Light-resis-
tant yellow 2KT-Cu(II) '(380uM/kg) and 0.75M NaCl.
Phase composition and affinity partition conditions
as in Fig. 1. Agents: (1)Imidazole, (2)Adenine,(3)
EDTA

the latter case $\Delta \log K_{max}$ of enzyme is remarkable (0.70).
The dyes containing coordinated Cu^{2+} ions - Light-resistant
yellow 2KT, Red-brown 2KT and Red-violet 2KT - in contrast to
their Cu^{2+} noncontaining- PEG conjugates alter significantly
the partition coefficient of horse liver ADH; $\Delta \log K_{max}$ is
1.64 - 1.81. This allows to suppose that the horse liver ADH-
dye interaction is also dependent on the presence of metal
ions as well as the yeast ADH interaction. This assumption
maight be proved by the four-fold increase of the $\Delta \log K_{max}$
value due to the presence of Cu^{2+} ions which in turn increases the strength of horse liver ADH binding to dyes. The extraction of horse liver ADH into top phase containing PEG -
Light-resistant yellow 2KT-Cu(II) or PEG - Red-violet 2KT-

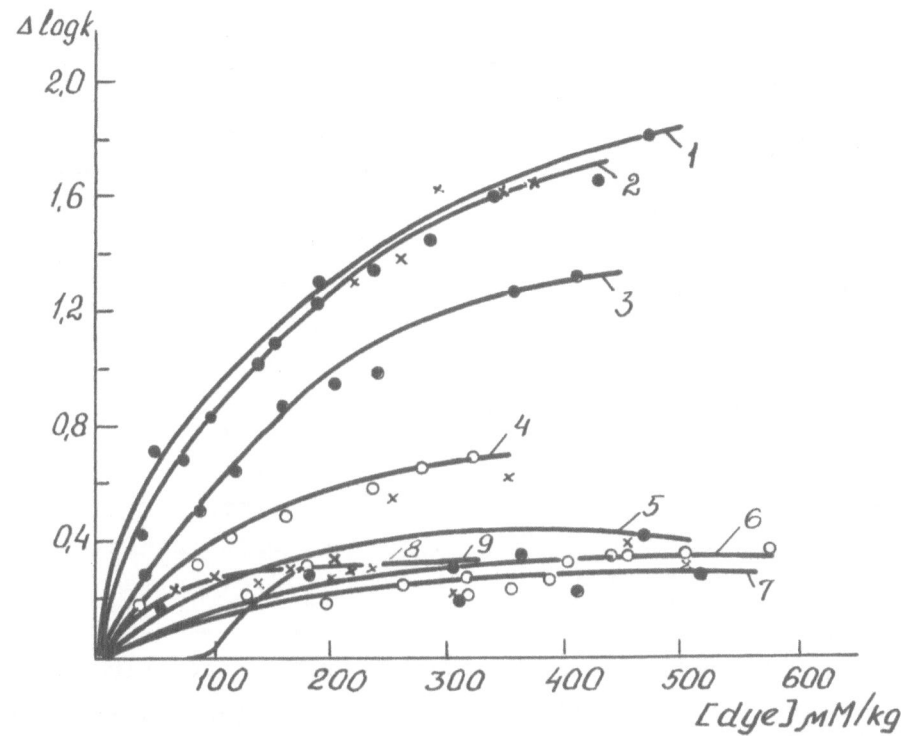

Figure 5. Alteration of the partition coefficient of alcohol
dehydrogenase from horse liver depending on concentrations of various dye-polyethylene glycol conjugates. The system (4g) contains 6,5% (w/w) PEG 6000
with different portions of dye-PEG and10% (w/w) dextran 60000 ± 10000, 10 mM Tris-maleic acid-NaOH
buffer (pH 6.5), 5 mM $MgCl_2$, 0.5 mM EDTA and 0,05
mM 2-mercaptoethanol. Dyes:1)Red-violet 2KT - Cu(II),
(2)Red-brown 2KT - Cu(II), (3)Light-resistant yellow 2KT - Cu(II), (4)Orange 5K, (5)Light-resistant
yellow 2KT, (6)Red-brown 2KT,(7)Red-violet 2KT,(8)
Bright-yellow 53, (9)Claret 4CT.

NaCl reduces the value of $\Delta logK_{max}$ of horse liver ADH by 42-
55% and NAD (5 mM) by 57-39% in systems PEG - Light-resistant
yellow 2KT - Cu(II) and PEG - Red-violet 2KT - Cu(II). Thus
the comparative study of yeast and horse liver ADH interact-

TABLE 4

Maximal alteration of the logarithmic partition coefficient ($\Delta \log K_{max}$) of horse liver ADH depending on dye-
PEG conjugates. Phase composition and affinity partition
conditions as in Fig. 5

Dye – PEG	$\Delta \log K_{max}$
1. Bright–yellow 53	0.31
2. Procion Blue SP-3R	0.60
3. Orange 5K	0.70
4. Claret 4CT-Cu(II)	0.36
5. Light-resistant yellow 2KT	0.38
6. Light-resistant yellow 2KT-Cu(II)	1.64
7. Red-brown 2KT	0.39
8. Red-brown 2KT-Cu(II)	1.65
9. Red-violet 2KT	0.40
10. Red-violet 2KT-Cu(II)	1.81

-Cu(II) is almost of the same effectiveness, the values of
$\Delta \log K_{max}$ is almost identical - 1.64 - 1.84. In case of ye
ast ADH (Table 1) the value of $\Delta \log K_{max}$ of enzyme in the
presence of Cu^{2+} ions differed significantly for the complexes with Red-brown 2KT, Red-violet 2KT and Light-resistant
yellow 2KT: from 0.55 (Red-brown 2KT) to 2.94 (Light-resistant yellow). From this we may conclude that in respect of th
investigated dyes yeast ADH possesses more restricted select
ivity as compared to the selectivity of horse liver ADH. The
se differences could be inferred from the fact that horse l
ver ADH possesses larger substrate specificity that yeast AD
(8). The study of the disruption of horse liver ADH complexe
with conjugates PEG - Light-resistant yellow 2KT - Cu(II),
PEG - Red-violet 2KT - Cu(II) in the presence of NaCl (up t
1.5 M), NAD (up to 5 mM) and chelating agents - imidazole (u
to 5 mM) has shown that, as well as in the case of yeast ADH
the value of $\Delta \log K_{max}$ is reduced significantly by imidazole
5 mM of immidazole reduces $\Delta \log K_{max}$ of enzyme by 64 - 84%.

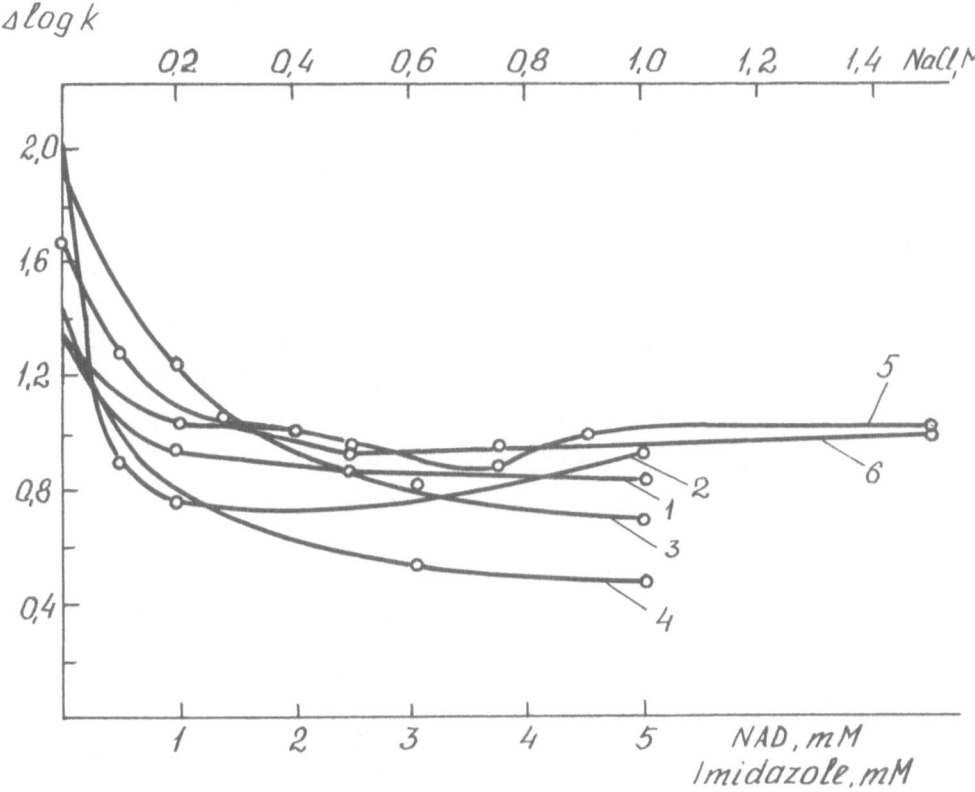

Figure 6. Effect of sodium chloride, NAD and imidazole on partition of horse liver ADH in two-phase systems containing PEG- Light-resistant yellow 2KT - Cu(II)
330 uM/kg (1,4,6,) and PEG - Red-violet 2KT- Cu(II)
330 uM/kg (2,3,5). Phase composition and affinity
partition conditions as in Fig. 5. Agents: (1,2)-
NAD, (3,4) - Imidazole, (5,6) - NaCl.

ion with reactive dyes by affinity partitioning in two-phase
system made it possible to show for the first time that the
interaction of horse liver ADH with dyes also depends on the
presence of metal ions.This increases the strength of binding
of dye by enzyme and the possibility of disruption the enzyme-
dye complexes by chelating agents. The dyes analysed , Light-
resistant yellow, Red-violet 2KT as their Cu^{2+} complexes were

effective specific ligands for yeast ADH and horse liver ADH.
Since the values of $\Delta \log K_{max}$ of both ADH's are high (1.64 -
2.94) in the presence of the mentioned dyes, it is consider-
ed(9) that the latters may be applied in their PEG - conjugates for large-scale enzyme purification by affinity partitioning.

REFERENCES

1. Flaksaite S.S., Sudzhuviene O.F., Pesliakas J.-H.J.,
   Glemzha A.A. Role of metal ions in specific binding of
   yeast alcohol dehydrogenase by free dyes and dye-adsorbents. Bioorg. Khimiya, 1984, v.10, №1, p.25-31.
2. Flaksaite S.S.,Sudzhuviene O.F., Pesliakas J.-H.J., Glemzha A.A. Group specific adsorbents with immobilized dyes
   for purification of the yeast alcohol dehydrogenase. Pri-
   kladnaya Biokhim. Mikrobiol., 1985, v.21, iss1, p.25-34.
3. Flaksaite S.S., Sudzhuviene O.F., Pesliakas J.-H.J.,
   Glemzha A.A. Role of metal ions to specific binding of
   yeast alcohol dehydrogenase by free and immobilized dyes.
   Inorgan. Chim. Acta, 1983, v.79, p.157-158.
4. Flaksaite S.S., Sudzhuviene O.F., Pesliakas J.-H.J.,
   Glemzha A.A. The interaction between dyes-Cu(II) complexes with the yeast alcohol dehydrogenase. The possibility of the involvment of the histidine residue of the enzyme in the formation of a mixed Cu(II)-complex with
   light-resistant yellow 2KT. Biorg. Khimiya, 1987,v.13,
   №3, p.293-299.
5. Flaksaite S.S., Sudzhuviene O.F., Pesliakas J.-H.J.,
   Glemzha A.A., Studies on the interaction of NAD(H)-de-
   pendent dehydrogenases with reactive dyes and their transient metal ion complexes. Biokhimiya, 1987, v.52, iss1,
   p.73-81.
6. Kotshetov G.A. Praktitszeskoe rukovodstvo po enzimologyi.
   M. Vischaya Schkola, 1971, p.127-130.
7. Johansson G., Joelsson M. Preparation of Cibacron Blue
   F3GA (polyethylene glycol) in large-scale for use in affinity partitioning. Biotecnol. Bioengn., 1985,v.XXVII,

p.621-625.

8. Jonathan J. The specificity of dehydrogenases. <u>Experientia Suppl.</u>, 1980, N°36, p.85-125.

9. Kopperschlager G., Lorenz G., Usbeck E. Application of affinity partitioning in an aqueous two-phase system to the investigation of triazine dye enzyme interactions. <u>J.Chromatogr.</u>, 1983, v.259, p.97-105.

AFFINITY ULTRAFILTRATION AND AFFINITY PRECIPITATION USING A WATER-SOLUBLE COMPLEX OF POLY(VINYL ALCOHOL) AND CIBACRON BLUE F3GA

R. R. FISHER, B. MACHIELS, K. C. KYRIACOU, AND J.E. MORRIS
Department of Chemical Engineering BF-10
University of Washington
Seattle, Washington 98195   USA

ABSTRACT

Two applications using anthraquinone dyes for protein separation and purification are presented. In both cases, Cibacron Blue F3GA is attached to water-soluble poly(vinyl alcohol) to facilitate a homogeneous binding to the protein. To effect precipitation, the poly-dye's solubility can be altered either via binding to a multimeric protein, or via independent control of the polymer solubility. In the case of ultrafiltration, the poly-dye provides sufficient size to prevent the target protein from entering the UF membrane. These early results indicate that PVA is a good polymer for UF, but that the precipitation poly-dye is not correctly formulated.

INTRODUCTION

In this work, the dye Cibacron Blue F3GA (also know as Basilen Blue E3G) is used as a ligand for dehydrogenases to impart increased specificity to ultrafiltration and precipitation. Our overall interest is to develop and characterize a number of protein recovery methods that offer inherent engineering advantages to scale-independent operation. Presented here is preliminary work on a macroligand, combining dye and soluble polymer, which gives insight into affinity ultrafiltration and affinity precipitation. We outline results of synthesis of the ligand complex and preliminary results of affinity ultrafiltration and affinity precipitation experiments.

Affinity ultrafiltration refers to a hybridization of biospecific binding with ultrafiltration and has been described previously [1-7]. Affinity precipitation can be achieved by two general methods. In the *first* , a bifunctional ligand is contacted with a protein possessing multiple sites for ligand binding [8,9]. As a result, a network of proteins connected by ligand bridges grows until insoluble and precipitates. In the *second* method of affinity precipitation, the ligand is covalently attached to a soluble polymer, the solubility of which is controllable independently [10] of the ligand binding. Contacting this polymer-ligand compound with the protein solution first allows affinity-specific adsorption, after which the complex can be precipitated, usually by changes in pH [10], ionic strength, or temperature [11]. The recovered precipitate can be resolubilized and the protein eluted by further solution changes. The ability to independently precipitate the polymer-ligand complex can be used again after elution of the target protein as a method to recover and recycle the precipitant.

Polydye as a macroligand

The common and important feature between the affinity UF and affinity precipitation presented here is the macroligand used for both methods. The macroligand is Cibacron Blue F3GA (CB) as substituted onto poly(vinyl alcohol) (PVA). The reasons for the choice of PVA as a ligand support are:
  •PVA is a simple polymer,uncharged itself so as to not interact electrostatically
      with proteins.
  •PVA possesses hydroxyl groups, providing sites of attachment to the reactive
      triazine chlorine of the dye.
  •The homogeneous binding inherent to using a soluble macroligand reduces the
      steric factors typically associated with a surface that may interfere with binding.
      The PVA, in particular, being hydrophilic, is not expected to bind with the UF
      membrane or other surfaces.
  •For affinity precipitation, PVA offers the potential for precipitation either by the
      bridging mechanism or through polymer solubility control. The solubility of PVA
      is well studied and is known to be a function of ionic strength and temperature.
      Also, of particular interest to us is the formation of strong, large precipitates,
      important for downstream solid-liquid separation after formation of the solid
      phase. High molecular weight polymers have been shown to give these
      benefits previously [12].

DYE LINKAGE TO THE POLYMER

The synthesis and purification of the macroligand was studied to find the parameters which control the degree of substitution of the dye onto the polymer. The CB (Aldrich) contains some inorganic salt and was purified by precipitation [13] before use. The PVA (Scientific Polymer Products, Inc.) is 88% hydrolyzed (12% acetyl groups), with a weight averaged molecular weight of 125,000, unless otherwise noted.

The method for dye-polymer linkage relies on a solid phase reaction of the dye and polymer. The procedure is as follows: a CB solution (0.04g/ml) was mixed with a 5% PVA solution, which was then added to an NaCl solution without stirring, causing gelation of the polymer and entrapment of the dye. NaOH was introduced and the mixture was heated to 80°C to cause the linkage reaction. This polymer was then recovered by methanol washing, aqueous dissolution, and organic reprecipitation. After redissolution, this polymer solution was chromatographed on a Sephadex G25 GPC column, showing one leading blue fraction. Determination of degrees of substitution were based on a mass balance of the free dye, measured spectrophotometrically at 600nm. The effect of dye, salt, and hydroxide concentrations on substitution are presented in Figures 1-3.

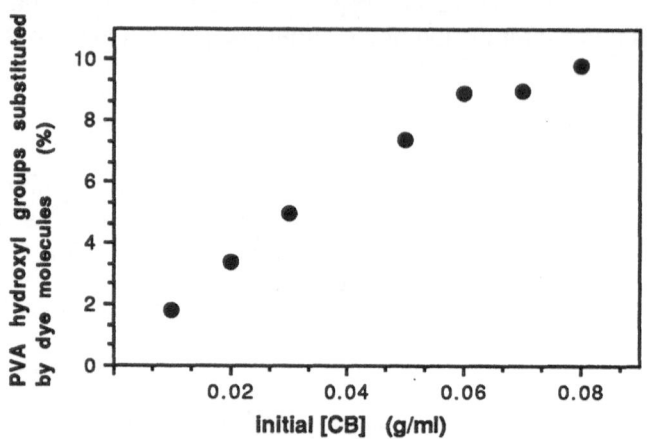

Figure 1. Effect of initial CB concentration on extent of hydroxyl substitution using the gel-phase reaction scheme.

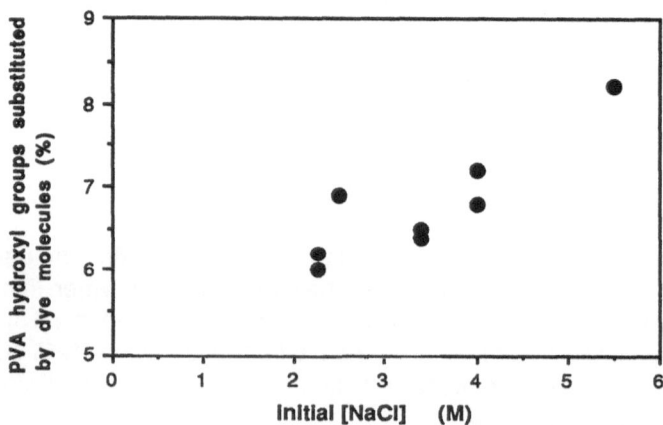

Figure 2. Effect of initial NaCl concentration on extent of hydroxyl substitution using the gel-phase reaction scheme.

The effect of CB concentration shown in Figure 1 is expected. As shown in Figure 2, the NaCl is considered to improve physical adsorption of the dye onto the polymer, enhancing the coupling. From Figure 3, the NaOH concentration at 0.5M is clearly sufficient for attachment, and has no improved effect at higher levels.

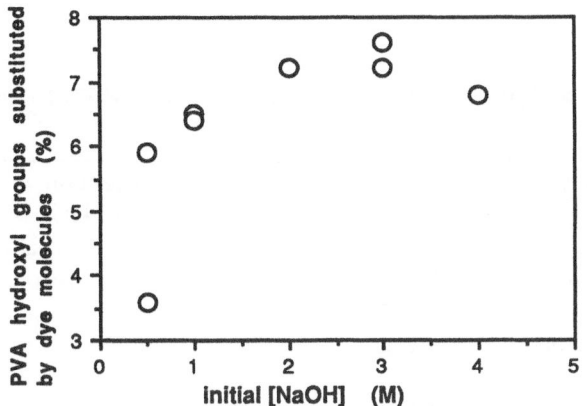

Figure 3. Effect of initial NaOH concentration on extent of hydroxyl substitution using the gel-phase reaction scheme.

AFFINITY ULTRAFILTRATION

Ultrafiltration was performed on Nucleopore S43-70 70ml stirred cell using 43mm polysulfone membranes (Amicon, PM-10, nominal MW cutoff of 10,000; XM-300, nominal MW cutoff of 300,000). Pressure was provided by N_2, stirring was at 720rpm. The polydye used is 6.5% substituted, MW 125,000 PVA, as above. The protein was horse liver alcohol dehydrogenase (HLAD), a dimer of MW 80kD.

UF results are given in Table 1. Buffers were selected to give good protein-dye binding; rejection coefficients (R) have been calculated based on permeate concentrations.

These results allow the following observations: the polymer, dye, and protein separately each behaved as expected in ultrafiltration experiments; the attachment of the dye to the polymer greatly increases the rejection of the polymer, due in part to the increase in total MW (from 125,000 to 270,000), but also due to the intermolecular repulsion of the charged dye, giving an extended conformation to the polydye in solution; the polydye successfully increased the rejection of the protein, and the rejection was a function of the buffer, which affects the binding of the dye to the protein.

AFFINITY PRECIPITATION

Affinity precipitation was attempted using the PVA-Cibacron Blue macroligand to precipitate bovine heart lactate dehydrogenase (LDH), a tetrameric protein with MW

140kD. Polydye was dissolved in 0.05M phosphate buffer, pH 7.5, and mixed with protein (subunit final concentration of 24μM) and 0.8M sodium pyruvate or 0.08M sodium oxalate to give final dye:enzyme subunit ratios of approximately 1.6:1 or 4.8:1. No precipitate formation was evident after 24h at 4°C.

Table 1. Conditions and results for ultrafiltration runs using Amicon membranes, as described.

solute & wt% in feed	solvent	membrane	ΔP (psig)	retention or R	flux (ml min^{-1}cm^{-2})
PVA 1.%	water	PM-10	25	~40-80%	-.----
polydye 0.04%	water	XM-300	50	99%	-.----
CB[a] 0.06%	water	XM-300	10	negligible	-.----
HLAD ~0.3%	0.02M NaAc[b] pH 5.5	XM-300	10	R=21%	
polydye 0.04%	"	XM-300	10	R=86%	0.08
polydye 0.02%[c] HLAD 0.04%[c]	"	XM-300	10	R=63%[d]	-.----
HLAD 0.04%	0.010M tris pH 7.5	XM-300	10	R=52%	0.066
polydye 0.04% HLAD 0.04%	"	XM-300	10	R=97%[d]	0.050

[a] ethanol precipitated to remove salt [13]
[b] sodium acetate
[c] corresponding to a dye unit:enzyme subunit molar ratio of 10:1
[d] rejection coefficient of HLAD reported

To verify that dye-protein binding was occurring, the binding of the enzyme to the dye was quantified by observing inhibition of enzymatic activity. Figure 4 show that polydye inhibits the reaction, implicating protein-dye binding. The difference spectrum of the polydye compared to free dye, shown in Figure 5, is virtually identical to that for the dye in a dioxane solvent of greater than 97% [14], indicating that the dye in the polydye is encountering a very hydrophobic environment, possibly other dye molecules in close proximity. Such a hydrophobically interactir dye is not expected to be freely extending from the polymer, thus decreasing its ability to bridge and induce precipitation.

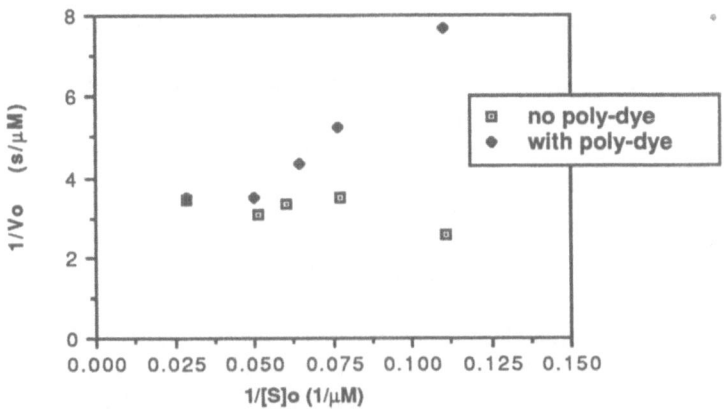

Figure 4. Lineweaver-Burk plot of LDH enzymatic activity as function of macro–ligand (initially 125,000 MW , 88% hydrolyzed) with 1% of the hydroxyls substituted with CB. LDH (8.8nM subunits) in 0.05M phosphate buffer, pH 7.5, with 0.17mM sodium pyruvate was mixed with polydye (as above) and NADH (denoted S) to give a dye:enzyme subunit ratio of ~2000:1. Reaction velocities were measured spectrophotometrically as NADH depletion. The control contained no polydye.

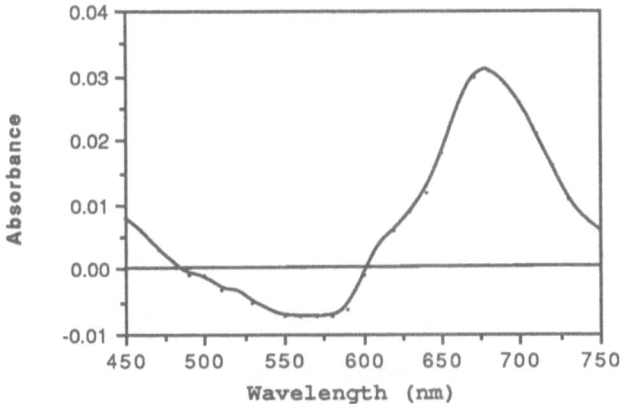

Figure 5. Difference spectrum of macroligand compared to free CB. Both solutions unbuffered in deionized water. Initially 100% hydrolyzed PVA, MW 115,000, 1% substituted with dye.

Although the solubility behavior of PVA is known and can be controlled with solution ionic strength or temperature, the attachment of the dye (substitution of 6.5% of the hydroxyls) greatly alters solubility behavior and makes prediction of precipitation conditions difficult. In addition, the macroligand used here requires approximately 0.3M NaCl to cause precipitation. At these high ionic strengths, CB-protein binding is weakened. This was observed when a solution of 0.47µM LDH was mixed with macroligand (to give a CB concentration of 100µM) and the ionic strength increased to 0.5M. Almost complete precipitation of the macroligand resulted in an only 5% decrease of LDH activity in the supernatant.

CONCLUSIONS

The synthesis and application of the polydye Cibacron Blue F3GA coupled onto poly(vinyl alcohol) has been shown. Affinity ultrafiltration has been demonstrated, and the effect of the dye to expand or stiffen the polymer will be further studied. The polydye's application to affinity precipitation has been hampered by the strong effect that the dye and/or protein have on the polymer solubility. These effects are the subject of current research.

This material is based upon work supported in part by the National Science Foundation under Grant CBT-8708819.

REFERENCES

1. C.R. Lowe, D.A.P. Small, and A. Atkinson, Int. J. Biochem. 13, 33 (1981).
2. A.S. Michaels, Ultrafiltration, in E.S. Perrry (ed.), Advances in Separation and Purification, John Wiley, New York (1968).
3. K.B. Male, J.H.T. Luong, and A.L. Nguyen, Enzyme Microb. Technol. 9, 374 (1987).
4. B. Mattiasson and M. Ramstorp, Ann. N.Y. Acad. Sci. 413, 307 (1983).
5. T.B. Choe, P. Masse, and A. Verdier, Biotechnol. Lett. 8, 163 (1986).
6. E. Pungor, Jr., N.B. Afeyan, N.F. Gordon, and C.L. Cooney, Bio/Technology 5, 604 (1987).
7. J. Bonnerjea, J. Jackson, M. Hoare, P. Dunnill, Enzyme Microb. Technol. 10, 357 (1988).
8. P.-O. Larsson, S. Flygare, and K. Mosbach, Meth. Enzymology 104, 364 (1984).
9. M. Hayet and M.A. Vijayalakshmi, J. Chromatogr. 376, 157 (1986).
10. M. Schneider, C. Guillot, and B. Lamy, Ann. N.Y. Acad. Sci. 369, 257 (1981).
11. A.S. Hoffman, K.A. Auditore-Hargreaves, C.A. Cole, R.L. Houghton, N. Monji, R.C. Nowinski, J.B. Plastino, and J.H. Priest, "Two novel applications of polymer chemistry and physics for immunoassays and bioseparations," in Proc. of International Meeting on Polymers in Medicine and Surgery V (The Netherlands, September 10-12, 1986).
12. R.R. Fisher and C.E. Glatz, "Polyelectrolyte precipitation of proteins. II. Effects of the reactor conditions," Biotechnol. Bioeng. (to appear).
13. R.S. Beissner and F.B. Rudolph, Arch. Biochem. Biophys. 189, 76 (1978).
14. S. Subramanian, Arch. Biochem. Biophys. 216, 116 (1982).

AFFINITY PRECIPITATION OF (NAD DEPENDENT DEHYDROGENASE)
LACTATE DEHYDROGENASE USING CIBACRON BLUE DIMERS.

B. RIAHI and M.A. VIJAYALAKSHMI
Laboratoire de Technologie des Séparations,
Univ. Technol. Compiègne, B.P. 649, 60206 Compiègne, France.

SUMMARY

The selective precipitation of lactate dehydrogenase, by using two structural variants of the Cibacron Blue dimers is demonstrated. While both the
dimers are showing good selectivity, the dimer having a hexamethylene group
as the bridge is more effective, both in terms of purification fold and
yield. Thus, this dimer with hexamethylene bridge is able to precipitate
selectively the lactate dehydrogenase from a crude extract of rabbit muscle
with 110 fold purification, and 4750 % yield, while a second dimer having
a diaryl group gives a 90 fold purification with 2500 % yield from the same
extract. The purified enzyme shows a single band in the SDS-PAGE analysis.

INTRODUCTION

Affinity precipitation of enzymes, a hybrid technique derived from the principles of affinity chromatography and immunoprecipitation, was first introduced by Mosbach (1). He used bis-NAD as the biospecific ligand dimer for
the selective precipitation of NAD dependant enzymes. On the other hand,
Cibacron Blue F3GA is widely used as the NAD mimicking affinity ligand for
the purification of co-factor dependant enzymes such as lactate dehydrogenase (EC 1.1.1.27) (LDH). We introduced the use of Cibacron Blue dimers for
the selective precipitation of LDH in 1986 (2). In this paper, the influence of the nature of the bridging group between the two monomer units in the

affinity precipitation is studied.

MATERIALS AND METHODS

The dimer bis Cibacron Blue F3GA with a hexamethylene bridge hereafter named as Dimer HD was synthesized according to the previously published procedure (2). The second dimer of Cibacron Blue F3GA with no spacer but direct
diaryl function (Dimer V) was a gift from Outumuro of Vilmax Foundation,
Argentina. The structures of these dimers are shown in Fig. 1. The pure LDH
preparations tested were either from Sigma or from Boehringer, France,
and pyruvate was from Sigma. All the other chemicals were of reagent grade.

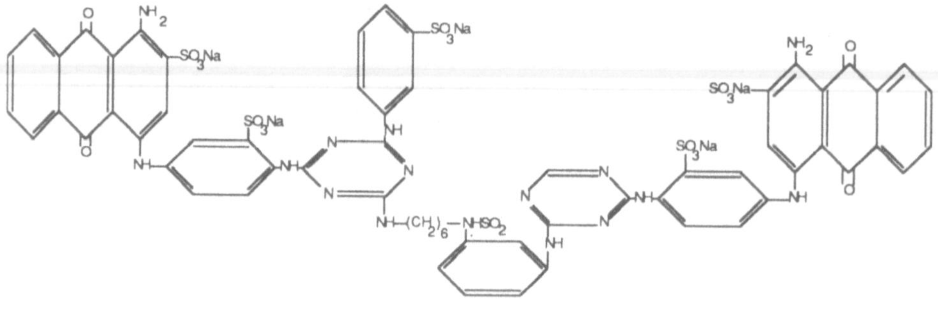

Figure 1a. Structure of the Dimer HD.

Preparation of the crude rabbit muscle extract : The crude extract
was prepared according to the method described by (3). To 50g of fresh rabbit muscle, 60ml of 50mM Phosphate buffer pH 7.5, containing 1mM mercapro-
ethanol were added. After homogeneization with a waring blender, the sus-

pension was centrifuged at 10000 g for 15 mn and the turbid supernatent was
recentrifuged at 20000 g for 30 mn. The clear supernatent was dialysed overnight against 50mM Phosphate buffer pH 7.5 at 4°C, and recentrifuged.

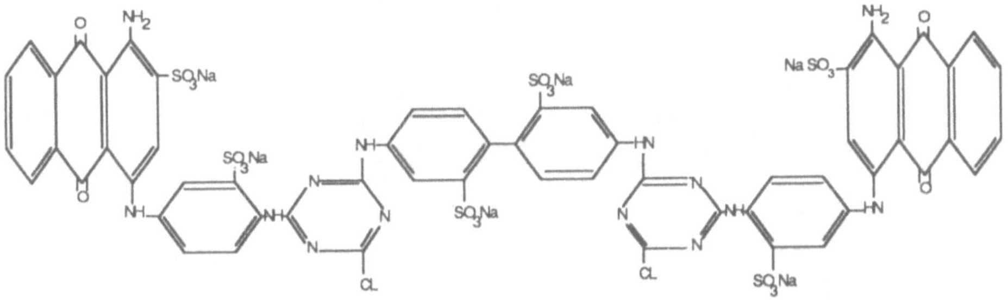

Figure 1b.   Structure of the Dimer V.

Precipitation procedure : A typical precipitation procedure was as
following. Two ml of either a solution of pure LDH at a concentration of
1.2 mg/ml in 50mM Phosphate buffer pH 7.5 or the crude extract containing
16.4mg of protein/ml were taken in a test tube and 1 ml of the bis dye solution corresponding to 35 nmole in the case of pure LDH and 60 nmole in the
case of crude extract along with 0.5 ml of sodium pyruvate (0.8 M) were added and after mild agitation overnight at 4°C, the mixture was centrifuged
at 5000 g for 15 mn, in order to separate the filtrate from the precipitate. Then, the precipitate was washed with the starting buffer, before dia-
filtering with 1mM NAD solution containing pyruvate to dissociate the enzyme dimer complex. The dye free retentat containing the enzyme was analysed for the protein content and enzyme activity. The purity was checked by
SDS-PAGE.

RESULTS

Table 1 shows the comparison of the precipitation of LDH, by the two dimers of the CBF3GA namely Dimer HD and Dimer V. The enzyme preparations used underwent extremely rapid denaturation. The low yield as compared to our previous results (2) could be partially attributed to this fact.

TABLE 1
Affinity precipitation of pure LDH by the two dimers

		Quantity of protein (mg)	Activity (U.I.)	Specific activity	% yield activity	% yield protein
	Starti g solution	1.2	151920	126600	100	100
DIMER HD	Supernatant	0.45	20256	45013	13	37
	Precipitate	0.7	81024	115 749	53	58
DIMER V	Supernatant	0.68	32410	47661	21	56
	Precipitate	0.505	37980	75208	25	25

Figure 2a, b show the precipitation yield as a function of the dimers concentrations. From these curves, we can see that the Dimer HD and Dimer V have 54.4 mg/µmole and 49.0 mg/µmole precipitation capacities, respectively for the LDH. Mosbach (4) has reported that a quantitative precipitation yield of LDH could be obtained at a NAD eq/LDH subunit ratio of 1.3. Our results show a similar ratio of dye monomer/LDH subunit, of 1.32 in the case of Dimer HD and 1.47 in the case of Dimer V. This indicates that the mode of precipitation in both cases are similar and the dye dimers mimick the NAD dimer.

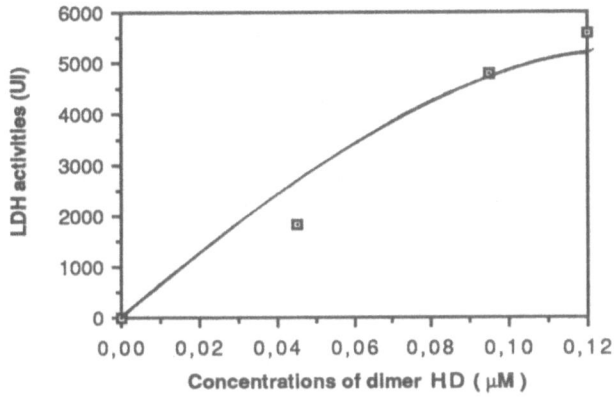

Figure 2a.   Precipitation of LDH as the function of Dimer HD concentration.

Affinity precipitation of LDH from a crude rabbit muscle extract

The table 2 shows the affinity precipitation of the crude rabbit muscle extract, using the two dimers. The Dimer HD gives an enzyme yield in terms of
LDH activity of 4750 % and 107 fold purification. The Dimer V, with the direct diaryl linking gave only 2750 % yield and 90 fold purification. The
supernatents in both cases contained 57 and 60 % of the initial proteins
respectively, without any significant LDH activity.

Fig. 3 shows the SDS-PAGE of the crude extract, the supernatents (S_1,
S_2) and the purified LDH (P_1, P_2) and the pure commercial LDH (L_1, L_2). We
can see that the LDH purified by affinity precipitation using the two dimers was as pure as the commercially available pure LDH preparations.

The improved efficiency of the Dimer HD compared to the Dimer V can
be attributed to the chemical nature of bridge. Mosbach has reported an

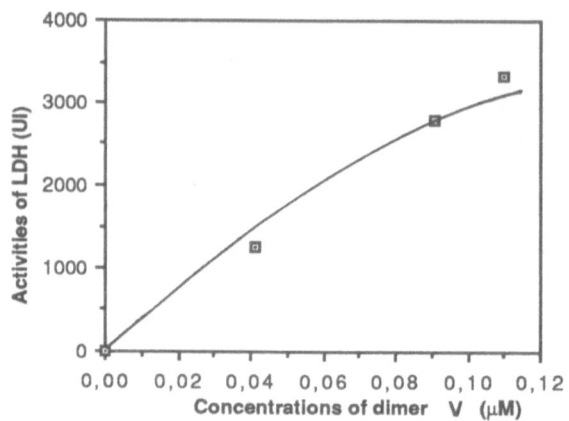

Figure 2b. Precipitation of LDH as the function of Dimer V concentration.

TABLE 2
Affinity precipitation of LDH from a crude rabbit muscle
extract by the two dimers.

		Quantity of protein (mg)	Activity (U.I.)	Specific activity (U.I. mg)	% yield activity	% yield protein	Purifi cation factor
	Starting solution	16.4	121.5	7.4	100	100	1
DIMER HD	Supernatant	9.3	−	−	−	56.7	−
	Precipitate	7	5770.4	795.77	4750	42	107.39
DIMER V	Supernatant	11.4	−	−	−	69.5	−
	Precipitate	5	3342.4	668.4	2750	30	90

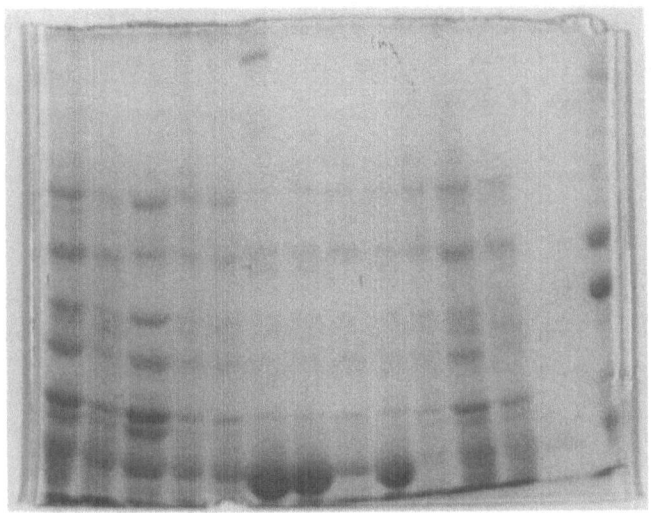

Ext. S_1 S_2 P_2 P_1 L_1 L_2 P_1 L_1 P_2 S_2 S_1 Std

Figure 3.   SDS-PAGE analysis of the crude extract as well as LDH purified from the crude extract.

optimum length of 17 Å between the monomer units. The Dimer HD contains a hexamethylene group as the bridge with one of the terminal amino benzene sulfonyl group free, while the Dimer V has no spacer between the two mono- mer units. This latter molecule has good symetry and both the terminal SO_3 functions are intact. This difference between the two structures can result in significant differences of rotational flexibility of the monomers. The Dimer V is planar while the Dimer HD can acquire an angle which will facili- tate the binding to two molecules of LDH, thereby provoking precipitation.

This affinity precipitation using the dimers of functional ligands seems to work, based principally, on the specific site oriented binding me- chanism rather than the other dye precipitation techniques cited by (5- 7) . The dye mediated precipitation using simultaneous pH effects as des- cribed by (5) can involve non-specific aggregation. It is worth noting in our affinity precipitation method, that the pH used is the optimum pH for the LDH activity and also that pyruvate is added to keep the active enzyme confirmation intact.

This preliminary work clearly demonstrates the feasibility of biomimetic bifunctional reagents for the specific one step affinity precipitation of the oligomeric proteins. Further work using different spacers between the two monomers are in progress.

ACKNOWLEDGEMENTS

The kind gift of the Dimer V and collaboration of Drs Mazza and Outumuro of the Vilmax Company, Argentina are thankfully acknowledged.

REFERENCES

1. Larsson P.O.; Flygare S. and Mosbach K. Affinity precipitation of dehydrogenases, In "Methods in Enzymology" Ed. W.B. Jakoby Acad. Press, 1984, Vol. 104, pp. 364-369.

2. Hayet M. and Vijayalakshmi M.A. Affinity precipitation of proteins using bis-dyes. J. Chromatogr., 1986, 376, pp. 157-161.

3. Flygare S.; Griffin T.; Larsson P.O. and Mosbach K. Affinity precipitation of dehydrogenases. Anal. Biochem., 1983, 133, pp. 409-416.

4. Mosbach K. In "Affinity Chromatography and Biological Recognition". Ed. I. Chaiken, Acad. Press Inc., 1983, pp. 209-212.

5. Birkenmeyer G. In this volume.

6. Bertrand O.; Cochet S.; Kroviarski Y.; Truskolaski A. and Boivin P. Protein precipitation induced by textile dyes. Precipitation of human plasminogen in the presence of Procion Red HE-3B. J. Chromatogr., 1985, 346, pp. 111-124.

7. Lowe C.R., in this volume.

Chapter 5

Role of Added Ions in Dye–Protein Interactions

METAL ENHANCED INTERACTIONS OF PROTEINS WITH TRIAZINE DYES AND APPLICATIONS TO DOWNSTREAM PROCESSING

Peter Hughes, British Bio-technology Limited, Brook House, Watlington Road, Cowley, Oxford, OX4 5LY, U.K.

ABSTRACT

Protein binding to immobilised triazine dyes can be promoted by metal ions, particularly those of the first row transition series. For example, Zn^{2+} selectively enhances the adsorption of yeast hexokinase to immobilised Cibacron blue F3G-A, calf intestinal alkaline phosphatase to Procion Yellow H-A, tyrosinase to Procion blue HE-RD and carboxypeptidase G_2 to Procion red H-8BN.

Kinetic studies with carboxypeptidase G_2 and analogues of Procion red H-8BN indicate the metal ion increases the affinity of the dye for the enzyme near its active site through the formation of a highly specific ternary complex. These effects have been exploited for the purification of yeast hexokinase and carboxypeptidase G_2 using triazine dye - Sepharose matrices and also for the purification of lactate dehydrogenase (rabbit muscle) by precipitation with bis-Cibacron blue F3G-A.

INTRODUCTION

Immobilised triazine dyes have been used extensively as group specific ligands in protein purification technology for some time. However, the possibility of using multivalent metal ions to promote protein binding to these matrices has only been realised relatively recently.

Initially a number of enzymes, for example isocitrate dehydrogenase (1), carbamyl phosphate synthetase I (2) and aminoacyl tRNA synthetase (3), were shown to be dependent on relatively high concentrations of Mg^{2+} for binding. Subsequently, a more comprehensive range of effects have been established using low concentrations of divalent metal ions of the first row transition series to agarose based dye matrices (4) and also to silica based dye supports used in HPLC (5). This paper outlines these interactions with reference to a number of proteins, discusses the mechanism by which they operate and describes their application to protein purification by both chromatographic and precipitation techniques.

Metal Promoted Binding of Proteins to Triazine Dye Adsorbents
Metal ions enhance the binding of a number of proteins to immobilised triazine dyes at neutral pH (Table 1). In three cases, carboxypeptidase G_2, hexokinase and alkaline phosphatase, adsorption is promoted by either the metal co-factor required for activity or the same metal ion constitutive to the protein. However, the binding of tyrosinase, a copper containing enzyme is promoted by Zn^{2+} and ovalbumin binding is unique to Al^{3+}.

TABLE 1
THE EFFECT OF METAL IONS ON BINDING OF DIFFERENT
PROTEIN TO IMMOBILISED TRIAZINE DYES.

IMMOBILISED DYE	METAL		PROTEIN	% BOUND	
				Control	Metal
RED H-8BN	Zn^{2+}	(2mM)	CPG_2	15	100
YELLOW H-A	Zn^{2+}	(2mM)	Alk. Phos.	12	80
BLUE F-3GA	Mg^{2+}	(10mM)	Hexokinase	20	95
GREEN H-4G	Al^{3+}	(2mM)	Ovalbumin	3	97
BLUE HE-RD	Mg^{2+}	(10mM)	Hexokinase	20	71
	Zn^{2+}	(2mM)	Tyrosinase	31	77

Columns (1ml) were equilibrated at pH 7.3 in the presence and absence of metal ions as indicated. Protein samples were applied and eluted with 5 volumes of buffer. Bound material with 0.75m KCl in buffer.

CPG_2: Carboxypeptidase G_2; Alk. Phos.: Alkaline Phosphatase

Carboxypeptidase G_2: A more detailed investigation of these effects has revealed that binding of the zinc metalloenzyme carboxypeptidase G_2 to Procion red H-8BN-Sepharose is also facilitated by other divalent metal ions of the first row transition series. Optimum binding is observed with 2mM Zn^{2+}, whilst higher concentrations of Co^{2+}, Mn^{2+}, Ni^{2+} and Cu^{2+} are required to promote a similar effect. Other divalent ions e.g Ba^{2+}, Ca^{2+} and Mg^{2+}, the monovalent ions K^+, Na^+, Cs^+ and Ag^+ and the trivalent ions Al^{3+}, Fe^{3+} and Cr^{3+} have no effect. When columns are washed with a chelating agent e.g EDTA, prior to equilibration, metal independent binding of this enzyme is completely eliminated. This observation raises the possibility that many proteins which bind to dye matrices may do so by the unrealised presence of metal ions in the system.

Alkaline Phosphatase: Calf intestinal alkaline phosphatase is also a Zn^{2+} metalloenzyme and binding to Procion Yellow H-A is also enhanced by Zn^{2+}. In this case the effect is unique to this metal ion and in addition is buffer system dependent. In tris and phosphate based systems binding is severely inhibited whereas in Goods buffers e.g. HEPES and MOPS quantitative binding is observed at 4mM Zn^{2+}.

Hexokinase: Yeast hexokinase requires Mg^{2+} for activity in the form Mg-ATP. This metal ion, at a concentration of 10mM, promotes quantitative enzyme binding to both Cibacron blue F3G-A and to a lesser extent Procion Green H4-G. At lower concentrations, divalent transition metal ions also promote binding to these immobilised dyes. Again, enzyme binding is best promoted by Zn^{2+} whilst Ni^{2+}, Cu^{2+}, Co^{2+} and Mn^{2+} are less effective.

Tyrosinase: The metalloenzyme tyrosinase (mushroom) contains Cu^+ and Cu^{2+} (6) but neither of these ions exert an effect on enzyme binding to dye matrices. Using immobilised Procion blue HE-RD, metal enhanced binding occurs preferentially with Zn^{2+} and also Ni^{2+} and Co^{2+}.

Ovalbumin: In contrast to the examples already described which are either metalloproteins or enzymes requiring a metal co-factor for activity, ovalbumin is a non enzymic protein without any know metal requirement. Furthermore, ovalbumin does not normally bind to triazine dye matrices (7,8). Although this appears to be correct for most dyes tested in both the presence and absence of monovalent, divalent and most trivalent ions, Al^{3+} enhances protein binding to Cibacron blue - Sepharose and to a lesser extent immobilised Procion Orange Mx-G. In both these cases binding is Al^{3+} concentration dependent.

This result, together with the observations for tyrosinase, indicates that there is unlikely to be a relationship between the metallo-nature or co-factor requirement of a protein and the identity of the metal required to promote its binding to immobilised dyes.

Mechanism of Metal Promoted Binding

The interactions of metal ions with purified dyes have been monitored in free solution at neutral pH using the technique of difference spectroscopy. The dye-metal ion K_D values subsequently determined demonstrate that both Procion red H-8BN and Cibacron blue F3G-A display low affinity for metal ions with figures in the mM range. In addition, the order of metal ligation observed in each case is completely dissimilar to the orders observed for the chromatography of carboxypeptidase G_2 on immobilised red H-8BN and hexokinase on immobilised Cibacron blue F3G-A (Table 2). This suggests that specificity is not generated by binary complex formation between the metal and the dye alone. These results contradict the Irving-Williams series of metal complex stabilities (9). This order of metal ligations correlates increased complex stability with increased electrostatic attraction and decreasing ionic radius. These changes in expected order could be attributed to charge transfer and hydrophobic interactions associated with the dye. However, it is also worth noting that the Irving-Williams series is empirically determined using ligands containing N and O as electron donors. The possibility that S may function in this capacity due to the presence of a thiol containing protein may alter the sequence of stabilities with preference to Zn^{2+}. Since Zn^{2+} has a full d10 configuration it has greater polarising ability than Mn^{2+} and Ni^{2+} which possess incomplete d-subshells. Consequently, Zn^{2+}-S complexes are stabilised preferentially due to the latter's highly polarisable nature and availability of electrons of similar symmetry.

Difference spectral titrations performed with hexokinase and Cibacron blue F3G-A in the presence of divalent metal ions indicate that selectivity originates through the formation of a highly specific ternary complex involving protein, metal ion and dye. In the absence of metal ions at pH 7.5 and $22^{o}C$, the K_D value for the dye-enzyme complex is $250\mu M$. In the presence of 0.1mM Zn^{2+} under the same conditions, affinity is increased 2-3 fold with similar effects observed at higher concentration (1.5mM) for Cu^{2+}, Co^{2+}, Mn^{2+} and Ni^{2+}.

TABLE 2
OBSERVED ORDERS OF METAL LIGATION

Red H-8BN$_2$ Sepharose
M^{2+}
Carboxypeptidase G2 $Zn > Co > Mn > Ni > Cu$

Red H-8BN/M^{2+} $Co > Ca > Ni > Zn > Mn$

Cibacron Blue Sepharose
M^{2+}
Hexokinase $Zn > Cu > Ni > Co > Mn > Ca$

Cibacron Blue/M^{2+} $Ni > Ca > Mg > Mn \ K > Ca > Zn$

Irving Williams Series $Zn < Cu > Ni > Co > Fe > Mn$

Inactivation studies with hexokinase and Cibacron blue F3G-A in the presence of metal ions under the same conditions, suggest these interactions occur near the active site of the protein (10). In the absence of metal ions, inactivation by the dye is slow with 90% of the original activity remaining after 30 minutes. In the presence of 1mM Zn^{2+}, the rate of inactivation is increased with approximately 25% of the original activity remaining over the same time period. Pre-incubation of the enzyme with the substrates glucose and ATP, provide protection from metal mediated inactivation indicating the interaction is biospecific. Total elimination of inactivation observed with 0.5mM ATP cannot be attributed to metal chelation by the nucleotide since progressive protection is still observed when the free metal ion concentration is maintained constant.

Procion red H-8BN however is unable to inactivate carboxypeptidase G$_2$ either at neutral pH and room temperature, or under more reactive conditions at pH 8.5 and 35°C. This is paralleled by the inability of the dye to bind to the enzyme monitored spectrophotometrically on a mole/mole basis following the removal of free dye and reconstitution of the putative complex in 8M urea (Table 3) (11).

TABLE 3
DISSOCIATION CONSTANTS AND BINDING RATIOS OF PROCION RED H-8BN AND ITS ANALOGUES FOR CARBOXYPEPTIDASE G_2

Dye	Zn^{2+}	Dye bound to enzyme subunit		Residual activity	K_D (μM)
		mol/mol		%	
Procion red H-8BN	+	0.024		99	41
Procion red MX-2B	+	0.87	<	10	237
	-	0.96	<	10	780
Procion red MX-5B	+	1.15	<	10	159
	-	1.23	<	10	149
Procion red MX-8B	+	0.98		10.0	26
	-	1.3	<	10.0	300
Procion red MX-8B + methotrexate	+	0		100	-
Procion red MX-8B + Procion red H-8BN	+	0.13		98	-

Binding ratios were determined by quantitative enzyme inactivation at pH 8.0 and 30°C. The reaction mixture of 1ml contained Tricine -NAOH pH 8.0 (100μmol); enzyme (10μmol); dye (200μmol). Free dye was removed by gel filtration and ethanol precipitation. Precipitated complexes were reconstituted in 8M urea prior to spectrophotometric determination. Dissociation constants were determined from double reciprocal plots of Kobs versus dye concentration.

At pH 8.0 and 30°C however, more reactive dichlorotriazinyl analogues of red H-8BN inactivate the enzyme to within 10% of its original activity over a period of 25 minutes. Irreversible covalent modification of the enzyme by these dyes results in essentially equimolar binding of dye per enzyme sub-unit in each case regardless of the presence of Zn^{2+}. When the enzyme substrate methotrexate is pre-incubated with the enzyme prior to the addition of red MX-8B, total protection from inactivation occurs and is paralleled by the inability of the dye to bind to the enzyme. A similar protective effect is also noted when the enzyme is pre-incubated with 2mM Procion red H-8BN, but only in the presence of Zn^{2+}.

These observations indicate that Procion red H-8BN, its dichlorotriazinyl analogues and methotrexate are mutually competitive and bind at the active region of the enzyme. Comparison of the K_D values obtained from the inactivation data (Table 3) demonstrate that differences in affinity occur with different dye structure. These values demonstrate the importance of the azo bond in the interaction as reduction of native red H-8BN with borohydride (11) reduces the metal mediated affinity of the dye for the enzyme 18-fold. With the exception of red MX-5B, Zn^{2+} increases the affinity in each case. Since this is the only analogue without an ortho-sulphonic acid substituent on the neighbouring benzene ring, this residue also appears to be an important structural feature. These functions are potential electron donors and may well chelate the metal ion subsequently facilitating adsorption of the protein via its active site to form a specific ternary complex. As Zn^{2+} normally forms tetracoordinate complexes, it is proposed that metal promoted binding involves metal ion coordination with the dye via its azo, sulphonic acid and hydroxyl moieties and an appropriate ligand on the protein. (Figure 1).

FIGURE 1

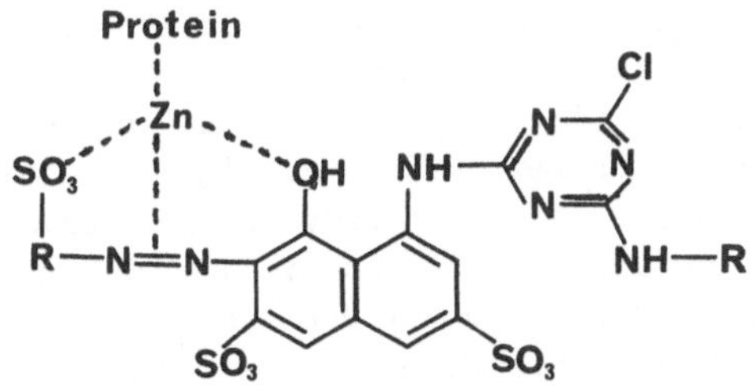

Figure 1. Proposed structure of the enzyme -metal ion-dye
ternary complex

Application to Protein Purification

Enzyme elution can be facilitated using chelating agents e.g EDTA following metal promoted binding of proteins to immobilised dyes (4). This is unsatisfactory for carboxypeptidase G_2 elution from red H-8BN because of the large volumes of eluant required. By washing the column with EDTA at pH 5.8 and then raising the pH to 7.3, homogeneous enzyme is eluted in 0.2 column volumes with 70% recovery. The 26-fold purification obtained is a considerable improvement over a previously used method involving an immobilised dye (13).

The Mg^{2+} promoted adsorption of hexokinase has been exploited for purification from a crude yeast extract (14). After binding in the presence of 10mM Mg^{2+}, a proportion of activity is recovered simply by omitting the metal ion from the irrigation buffer. However, 87% recovery and a 7-fold purification is obtained using 10mM Mg^{2+} and 20mM ATP in combination.

Purification has also been achieved using metal ions to promote protein precipitation with bis-dyes in free solution. For example, bis-Cibacron blue F3G-A forms a precipitate with rabbit muscle lactate dehydrogenase at pH 7.5 in the presence of 1mM Co^{2+} (15). Under these conditions approximately 90% of the enzyme activity is precipitated from the crude supernatant. The enzyme is rapidly resolubilised from the isolated precipitate through the addition of 1mM EDTA. This provides 86% recovery with an approximate 2-fold enhancement of specific activity.

CONCLUSIONS

With the exception of the Co^{2+} dependent precipitation of lactate dehydrogenase, which is inhibited in the presence of NaCl, transition metal promoted adsorption of proteins to triazine dyes appears to be bio-specific. This process is likely to be similar in principle to metal chelate chromatography, though influenced by the more complex range of interactions which characteristically occur between proteins and triazine dyes. These include charge transfer, electrostatic and hydrophobic effects. Nevertheless, this phenomenon has an application to protein purification particularly, although not exclusively, for metalloproteins.

REFERENCES

1. Nagoaka, T., Hatchimora, A., Takeda, A. and Samejima, T. (1977). J Biochem. 81, 71-78

2. Mori, M. and Cohen, P.P., (1978) Fed. Proc. 37, 1341

3. Moe, J. and Piszkicwicz, D. (1976) FEBS Lett, 72, 147-150

4. Hughes, P., Lowe, C.R. and Sherwood, R.F. (1982) Biochem. Biophys. Acta. 700, 90-100

5. Small, D.A.P., Atkinson, A. and Lowe, C.R. (1981) J. Chromatogr. 215, 175-190

6. Duckworth, H.W. and Coleman, J.E. (1970) J Biol. Chem. 245, 1613-1625

7. Easterday, R. and Easterday, I. (1974) Adv. Exp. Med. Biol. 42, 123-133

8. Stellwagen, E. (1977) Acc. Chem. Res. 10, 92-98

9. Irving, H. and Williams, R.J.P. (1948) Nature (London) 162, 746-747

10. Hughes, P., Sherwood, R.F. and Lowe, C.R. (1982) Biochem. J. 205, 453-456

11. Hughes, P., Sherwood, R.F. and Lowe, C.R. (1984) Eur. J. Biochem. 144, 135-142

12. Sherwood, R.F., Melton, R.G., Alwan, S.M. and Hughes, P. (1985) Eur. J. Biochem. 148, 447-453

13. Baird, J.K., Sherwood, R.F., Carr, R.J.G. and Atkinson, A. (1976) FEBS Lett 70, 61-66

14. Clonis, Y.D., Goldfinch, M.J. and Lowe, C.R. (1981) 197, 203-211

15. Lowe, C.R. and Pearson, J.G. Some applications of bio-mimetic dyes. In Affinity Chromatography and Biorecognition. Ed. I.W. Chaiken, Academic Press Inc (London). 1983 pp 421-432

METAL MEDIATED MODULATION OF
ENZYME KINETICS BY STRUCTURALLY
RELATED DYES

F. CADELIS and M.A. VIJAYALAKSHMI
Laboratoire de Technologie des Séparations;
Université de Technologie de Compiègne, B.P. 649, 60206 Compiègne France

ABSTRACT

The synergesic effects of metals on the enzymes inhibition by dyes has been shown by the pioneering works of HUGHES et al (1982). We have shown for the first time, the competitive nature of certain metal complexes vis-a-vis the dehydrogenase(s) dye reversible association or the enzyme inhibition by the dyes (1987). In this work, the effect of a few bivalent metals on the kinetics modulation of Lactate Dehydrogenase in the presence of three structurally related Procion dyes are reported. The data confirm our previous observations that Cu^{2+} and Zn^{2+} have completely different and opposite effects on the inhibition pattern. Cu^{2+} acts always as an enhancer of the inhibition, while Zn^{2+} has a mixed pattern. The modulation of the enzyme inhibition by the dye in the presence of Zn^{2+} is a concentration (Zn^{2+}) and time dependant phenomenon, suggesting a "dynamism in the space" of the confirmational changes of the enzyme. Moreover, the pattern of modulation by Zn^{2+} depends only partially on the bulkiness of the terminal groups of the dye. The Zn^{2+} effects are qualitatively same in all the three dyes while there is significant differences in terms of Zn^{2+} concentration and the zig-zag patterns of the kinetics. Based on these observations, the spatial confirmational orientation of the functional groups as well as minor localized pH changes at or near the substrate/ inhibitor binding sites are proposed as the mechanism of the metal mediated kinetic modulations of the enzyme kinetics in the presence of dye inhibitors.

INTRODUCTION

The triazine dyes are widely used as ligands in "affinity" based purification systems (1, 2). The binding of the dye ligand to the proteins involves a wide range of interactions, including hydrophobic, ionic and charge transfer effects. The more specific interactions observed between the dyes and the nucleotide cofactor dependant enzymes such as dehydrogenases, certainly involve the same interactions, with appropriate spatial geometry, hence, the nucleotide mimicking properties of these dyes.

It is obvious that any modifications in these interactions by additives forming ternary complexes e.g. metal ions should reflect in a modification of the association/dissociation of the dye-protein complexes. Hughes et al. (3) in fact has shown the improved binding of the proteins to the dye immobilized columns in the presence of transition metal ions. S. Rajgopal and M.A. Vijayalakshmi have reported the antagonist effects of certain metal ions in the dye-protein complex formation (4). While the potential of practical applications of these reports is obvious, very little is known in terms of the effects of the chemical nature of the dye and metal, in modifying the kinetics of the enzymes. In an effort to understand the mechanism of metal mediated kinetic modulation of the nucleotide cofactor dependant enzyme(s) we undertook to study the effect of transition metal ions, particularly Zn^{2+}, Cu^{2+} and Mg^{2+} on the binding of a few structural variants of triazine dyes to a few adenosine nucleotide dependant dehydrogenases.

METHODS

Materials

Luciferase from Photinus pyralis (EC 1.13.12.7) was from Boehringer Mannheim, France; lactate dehydrogenase (EC 1.1.1.27) (LDH); ATP, NAD^+, NADH and luciferin were from Sigma. Sepharose 4B was from Pharmacia (Sweden) and the dyes CBF3GA and CBB-II were from Ciba-Geigy, Basle, Suiss. Zinc sulfate ($Zn\ SO_4$, $7H_2O$) and copper sulfate ($CuSO_4$, $5H_2O$) were from Prolabo, France. All other chemicals were of reagent grade.

Elution of luciferase from the dye columns using ATP or the metal complexes;

Disposable columns (Bio-Rad Laboratories, France, 8.30) were containing
1.0 ml bed volume of dye-immobilized Sepharose 4B, (1 0) which was equilibrated with 10 volumes 0.01 M MOPS buffer, containing 1 mM DTT, pH 7.4. A
sample of the dialysed extract (1.68 mg protein) was loaded onto the column
and washed with 10 volumes of equilibrating buffer containing ATP, or the
corresponding metal complex as indicated in Table 1. Fractions of a 1 ml
were collected at 35 ml h^{-1} and dialysed against Tris-acetate (0.02 M), pH
7.8, for 3-4 h before checking for luciferase activity and protein concentration at 278 nm.

Elution of LDH from CBF3GA-Sepharose 4B dye columns using the metal compounds:
Cibacron blue F3GA was coupled to Sepharose 4B by the procedure of (10).
All chromatographic elutions were performed at room temperature (25°C).
Disposable columns (Biorad Laboratories, France, 8 × 30 mm) containing 1,8 m
bed volume CBF3GA-Sepharose 4B was equilibrated with 10 volumes of 0,1 M
Phosphate buffer pH 7.5. About 0,03 mg protein was loaded onto the column
and sequentially washed using twice the column volume of equilibrating buffer to remove unbound protein. The bound protein was then eluted from the
column using twice the column volume of buffer containing either 1 mM NAD^+
and 0,1 mM Pyruvate or different biomimetic complex solutions, as indicated
in Table 2. Fractions of 1 ml were collected at 35 ml/hr and was tested for
LDH activity.

Enzymes assay

Luciferase

Luciferase assay was carried out by a nucleotimeter (Interbio, France). The
reaction mixture containing 100 µl of 10^5 pg ml^{-1} ATP in 0.01 M MOPS (pH
7.4.) with 10 mM $MgSO_4$, 100 µl luciferin (0.168 µg ml^{-1}) and 50 µl of luciferase sample was introduced into the nucleotimeter in a special cuvette
and the maximum intensity of light recorded in millivolts. The unit of activity is the maximum intensity of light recorded in millivolts per picogram ATP per milligram of protein at 562 nanometers.

Lactate dehydrogenase

The enzyme lactate dehydrogenase was assayed at 25°C as described by (5).
One unit of enzyme activity is defined as the amount that catalyses the

conversion of 1 µmol of substrate to product per minute at 25°C.

Inhibition studies
Dye-metal mediated kinetic modulations
The inactivation studies are carried out by incubating the enzyme with the corresponding dye and the metal, at concentrations as described in each case in the corresponding buffer ∿ 20°C and aliquots (50 µl luciferase and 20 µl for LDH) were taken and assayed for the activity as described above. Controls were performed with the dye without added metal ion and the metal ion (2mM) without added dye.

RESULTS

EFFECT OF METAL IONS ON THE BINDING OF ENZYMES TO TRIAZINE DYE COLUMNS
Hughes et al (3) have shown that the transition metal ions such as Zn^{2+}, Cu^{2+}, Mg^{2+} and Al^{3+} promote the binding of a number of proteins to the immobilized dye columns and the bound proteins could be successfully eluted with chelating agents such as EDTA, substituted pyridines etc...
We have recently demonstrated that some of the bivalent metal ions, Mg^{2+}, Zn^{2+} and Co^{2+} in particular, combined with a chelating agent and a sugar could be excellent specific desorbing agents for the enzymes bound to the dye columns (6). The Tables 1 and 2 show the efficiency of the different metal-EDTA-sugar complexes, in eluting quantitatively the enzymes firefly luciferase and lactate dehydrogenase bound to a Cibacron Blue sepharose columns. Mg-EDTA-Ribose and Mn-EDTA-Ribose are the two complexes showing quantitative desorption of the bound enzymes from the dye columns, comparable to the conventional specific desorption using the nucleotide cofactor ATP for luciferase. The corresponding Zn^{2+} complex was slightly less effective.
In our studies, no enhancement of luciferase or LDH binding to CBF3GA column in the presence of Zn^{2+} or Co^{2+} was observed. However, it is to be noted that Co^{2+} or Ni^{2+} improved the retention (results not shown) of these proteins onto CBB-II or Procion Blue HB dyes immobilized columns. In any case, Zn^{2+} did not show any synergic effect on protein binding to the dye columns. Another triazine dye of the azo type -Procion Red HE-3B- is known to be a low affinity ligand for the LDH. In order to test whether the metal-chelate-sugar complexes are competitive only vis-a-vis the site specific immobilized

ligand, we ran chromatogrammes of LDH with Sepharose 4B linked with Procion Red HE-3B in similar conditions as in the previous case. The elution with 1 mM NAD was able to desorb only 37% of the bound enzyme, while 3 M NaCl gave a 80% desorption. However, when metal-chelate-sugar complexes are used either before passing 3 M NaCl or along with 3 M NaCl, a quantitative desorption was possible (Table 3).

TABLE 1
Elution of luciferase retained
on a Cibacron Blue (CBF3GA) Sepharose 4B
column, using different biomimetic complexes

Eluent % Activity recovery	1	2	3	4	5	6	7	8	9	10
ATP	0.5mM	–	–	–	–	–	–	–	–	–
EDTA	–	40mM	40mM	40mM	40mM	40mM	40mM	40mM	40mM	40mM
Mg^{2+}	20mM	20mM	20mM	20mM	20mM	20mM	–	–	–	–
Mn^{2+}	–	–	–	–	–	–	20mM	–	–	–
Zn^{2+}	–	–	–	–	–	–	–	20mM	–	–
Co^{2+}	–	–	–	–	–	–	–	–	20mM	–
Ni^{2+}	–	–	–	–	–	–	–	–	–	20mM
Ribose	–	–	5mM	–	–	–	5mM	5mM	5mM	5mM
Mannose	–	–	–	5mM	–	–	–	–	–	–
Glucose	–	–	–	–	5mM	–	–	–	–	–
Fructose	–	–	–	–	–	5mM	–	–	–	–
DTT	1mM	1mM	1mM	1mM	1mM	1mM	1mM	1mM	1mM	1mM
Activity recovered	180	20	120	67	0	65	180	75	0	0

The metal-EDTA-sugar complex alone was absolutely inefficient. It is noteworthy that the same metal-EDTA-ribose complex was extremely specific in dissociating the LDH bound to a Cibacron Blue F3GA Sepharose 4B column. Cibacron Blue F3GA is known to be a site specific ligand, while Procion-Red HE-3B is not a site specific one (1).

TABLE 2

Elution of LDH retained on a Cibacron Blue (CBF3GA)-
Sepharose 4B column, using different biomimetic complexes.

Eluent	1	2	3	4	5	6	7	8	9	10	11	12
NAD^+	1mM	-	-	-	-	-	-	-	-	-	-	-
Pyruvate	0.1M	0.1M	0.1M	-	-	0.1M	-	0.1M	0.1M	0.1M	0.1M	0.1M
EDTA	-	-	40mM	40mM	40mM	40mM	-	40mM	40mM	40mM	40mM	40mM
Mg^{++}	-	-	20mM	20mM	20mM	20mM	-	15mM	25mM	30mM	20mM	20mM
Ribose	-	-	5mM	5mM	-	-	5mM	5mM	5mM	5mM	10mM	20mM
% Activity recovery	95	83	172	154	52	78	0	173	160	171	145	157

TABLE 3
Desorption of LDH from a Procion Red HE-3B
Sepharose 4B column, with different eluents.

Eluent	NAD	3M NaCl	NAD + 3MNaCl	Metal (Mg) com.	Zn complex	Mg + 3MNaCl	Zn complex + 3MNaCl
% recovery	36.6	79.6	42.5	0	0	102.8	99.5

INFLUENCE OF METAL IONS IN THE DYE BINDING TO PROTEIN IN SOLUTION

In order, to better understand and exploit this interrelation, we undertook to study the enzyme kinetics, in solution, of a few nucleotide cofactor dependant enzymes in the presence of these biomimetic ligands namely a few structural variants of the dyes and the metal ions with or without chelator and sugar. Fig. 1a and 1b illustrate the structure of Cibacron Blue F3GA and Cibacron Brilliant Blue II, a meta para mixture and an isomer of the dye. The meta, para and ortho positions are determined with respect to the $SO_3^{\ominus}$ group.

Kinetics of firefly luciferase

Fig. 2a, 2b show the metal dependant inhibition pattern of an ATP dependant enzyme luciferase with the CBF3GA and CBB-II, respectively.

Low concentration of metals of the first row transition series, e.g. Zn^{2+}, Co^{2+}, Mn^{2+}, Ni^{2+}, Cu^{2+}, Fe^{2+} and to a lesser extent Mg^{2+} influence the interaction of luciferase with the triazine dyes CBF3GA and CBB-BR II respectively. Cu^{2+} promotes the dye binding to the enzyme in both cases and hence its inhibition in both cases, while Zn^{2+} is unfavurable to the protein-dye binding in the case of CBF3GA and not in the case of CBB-II. The order of protection of the enzyme from dye inhibition by the metal, differs from dye to dye. In the case of CBF3GA, the order of degree of protection is $Zn^2 > Co^{2+} > Ni^{2+} > Mg^{2+} > Fe^{2+}$, where as in the case of CBB-II the ortho isomer some protection is offered by Fe^{2+} and Ni^{2+}, with Zn^{2+} and Co^{2+} showing absolutely no protection effect and the order of protection, however less, is $Fe^{2+} > Ni^{2+} > Mg^{2+} > Mn^{2+} > Zn^{2+} > Co^{2+}$.

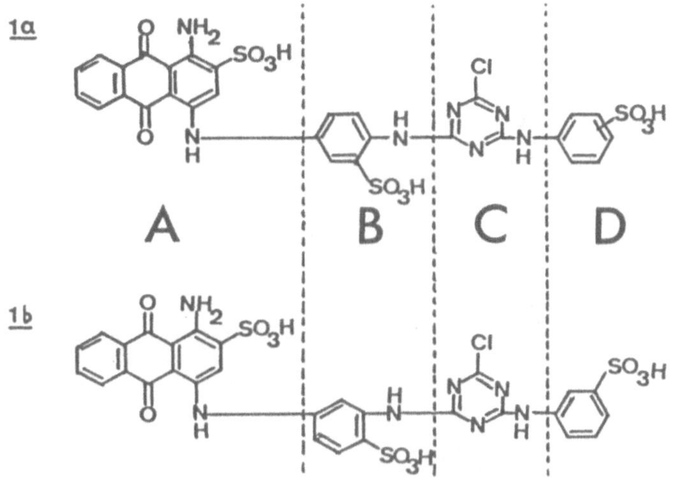

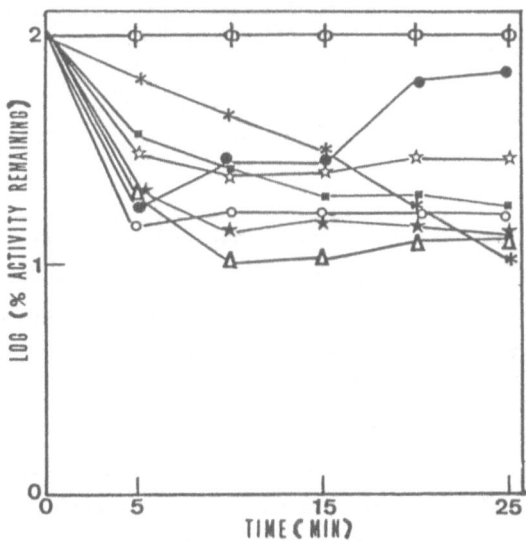

Figures 1a and 1b.   Structures of CBF3GA and CBB-II respectively.

Figure 2a.   Effect of metal ions on the inactivation of luciferase by Cibacron Blue F3GA. Luciferase (0.4 units) was inactivated by CB
F3GA (4.5 nmol) in the presence of 2 μmol of metal in 0.1 M Tricine, pH 7.8. A sample without the dye served as a control. See
text for further details, ϕ, Control; ● effect of Zn^{2+}; ✿, Co^{2+};
■, Ni^{2+}; ○, Mg^{2+}; ★, Mn^{2+}; ▲, Fe^{2+}; ✳, in absence of metal ion.

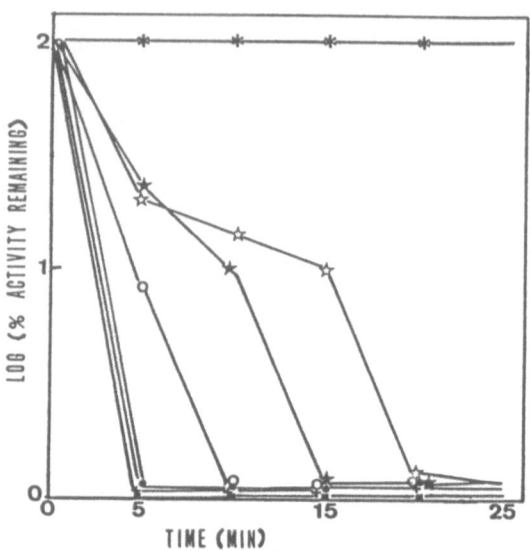

Figure 2b. Effect of metal ions on the inactivation of luciferase by Cibacron Brilliant Blue BR-II. Luciferase (0.4 units) was inactivated by CBB-II (0.005 μmol) in the presence of 2 μmol of metal
in 0.1 M Tricine, pH 7.8. A sample without the dye served as a
control; *, Control; ☆, effect of Fe^{2+}; ★, Ni^{2+}; o, Mg^{2+}; •, Mn^{2+};
+, Zn^{2+}; ■, in absence of metal ion. The curves indicating the
effect of Co^{2+} and Cu^{2+} are identical to that in the absence of
metal ions.

It is also very interesting to note that Zn^{2+} protects the enzyme from dye
inhibition only at low concentration (1 to 4 μmol) and at these low concentrations, even an enhanced enzyme activity is observed. However, at concentrations higher than 4 μmol, Zn^{2+} favors the enzyme inhibition by the same
dye (Fig. 3).

Figs 5a, b, c show the inhibition patterns of LDH with Procion Red HE-3B
and two structural variants of the same as shown in Figs 4a, b, c in the
presence of Mg^{2+}, Zn^{2+} and Cu^{2+} ions. It can be seen from these figures
that Zn^{2+} ions have a protective effect against the inhibition/inactivation
by both Procion Red HE-3B and the Dye 1. However, this protection against
inhibition seems to be time dependant phenomenon, showing an initial enhancement of the inhibition followed by protection and again an increase in
the inhibition. It is interesting to note that Zn^{2+} is without any protec-
ting effect in the presence of Dye 2, but a regular increase, however small,

in inactivation is observed. The effect of 2 mM Mg^{2+} does not seem to be very significant in all the cases, whereas Cu^{2+} always increases the inactivation almost immediately in all three cases. The kinetic patterns obtained with 2 mM of these metal ions, in the absence of any dyes show that, while Mg^{2+} does not affect the LDH activity, Zn^{2+} shows a mild inhibition and Cu^{2+} shows a strong inhibition. This inhibition by Cu^{2+} alone is, however, slightly less than that observed in the presence of added dyes.

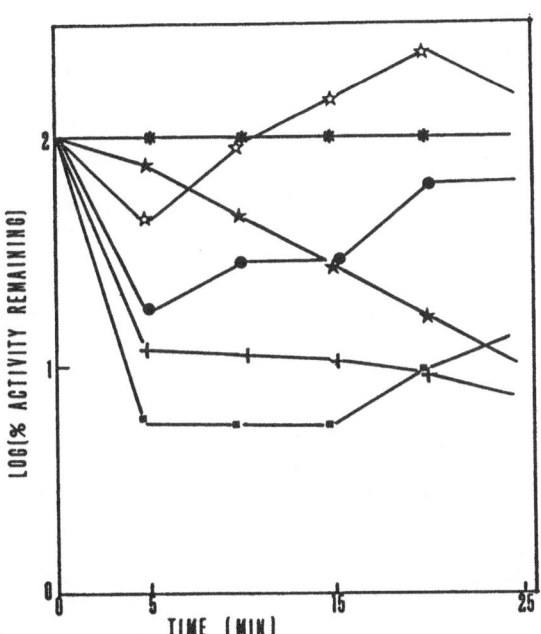

Figure 3. Effect of different concentrations of Zn^{2+} on the inactivation of luciferase by Cibacron Blue F3GA. ✳, Control; ✩, effect of 1 μmol Zn^{2+}; ●, 2 μmol Zn^{2+}; ★, in absence of Zn^{2+}; ■, 4 μmol Zn^{2+}; +, 6 μmol Zn^{2+}.

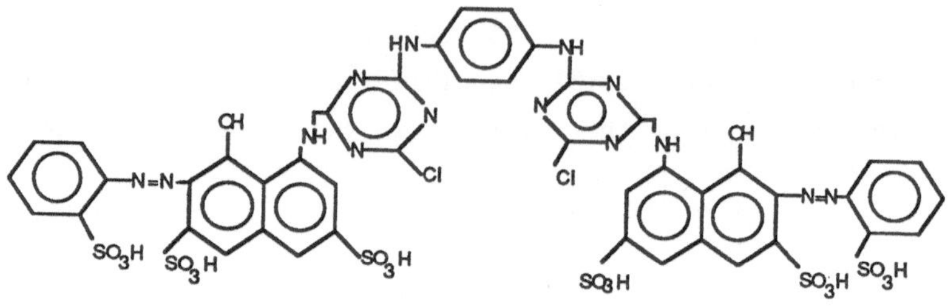

Figure 4a. Structure of Procion Red HE-3B

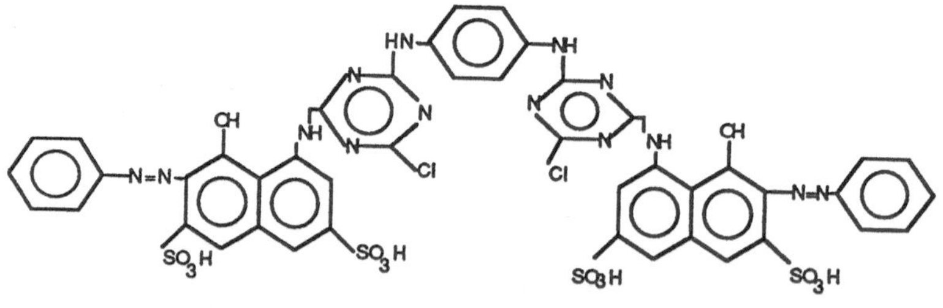

Figure 4b. Structure of Dye 1.

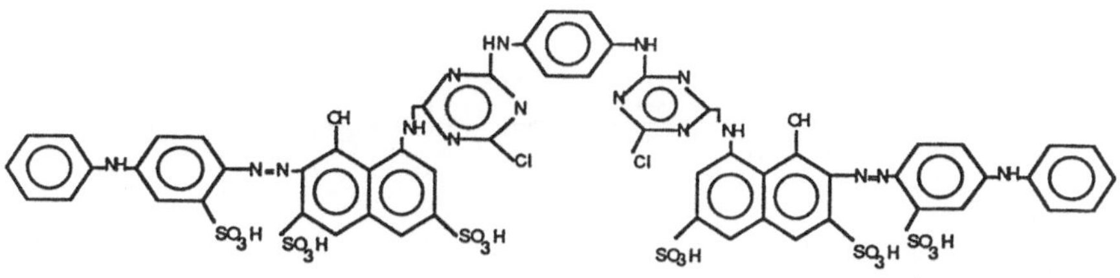

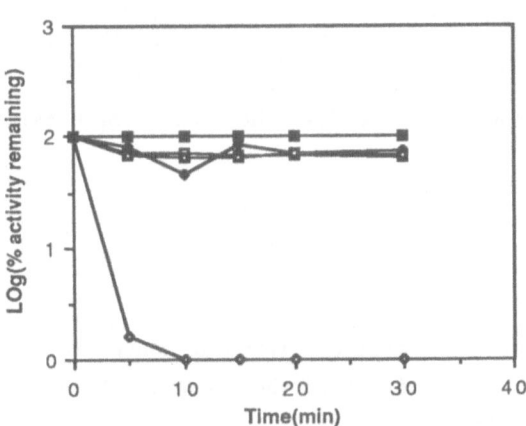

Figure 4c. Structure of Dye 2.

Figure 5a. Effect of metal ions on the inactivation of LDH by Procion Red
HE-3B. LDH (8.2 nM) was inactivated by Procion Red HE-3B
(3.125 μM) in the presence of 2 mM of metal in 0.05 M Phosphate
buffer, pH 7.5. A sample without the dye served as a control.
■ , Control; ◆ , Zn^{++} ; ▣ , Mg^{++} ; ● , Cu^{++} ; ▢ , in the absence
of metal ion.

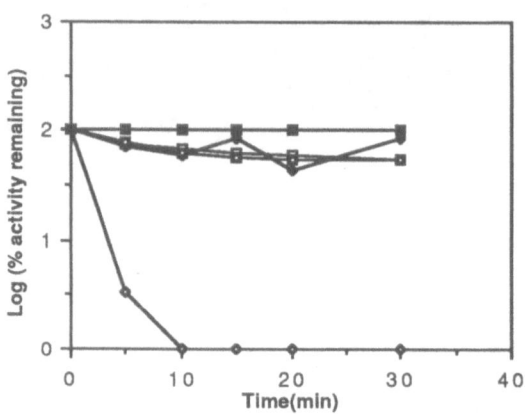

Figure 5b. Effect of metal ions on the inactivation of LDH by Dye 1. LDH
(8.2 nM) was inactivated by Dye 2 (50 µM) in the presence of
2 mM of metal in 0.05 M Phosphate buffer, pH 7.5. A sample
without the dye served as a control.-■- , Control; -◆- , Zn++ ; -■- ,
Mg++ ; -◆- , Cu++ ; -▣- , in the absence of metal ion.

Figs 6a, b, c show the effect of different concentrations of Zn^{2+} ions in
the modulation of LDH activity in the presence of these three structurally
related dyes, Procion Red HE-3B, Dye 1 and Dye 2 respectively. We can observe significant zig-zag patterns with 1 mM and 2 mM concentrations of
Zn^{2+} added in the presence of Procion Red HE-3B and the Dye 1. However, the
maximum and minimum LDH activities observed with 2 mM Zn^{2+} vary for these
two dyes. The Procion Red HE-3B showing the minimum at 10' and maximum at
15', whereas the Dye 1 shows the maximum at 15' and minimum at 20'. In
the case of Dye 2, all the concentrations tested show only an increase of
the enzyme inactivation.

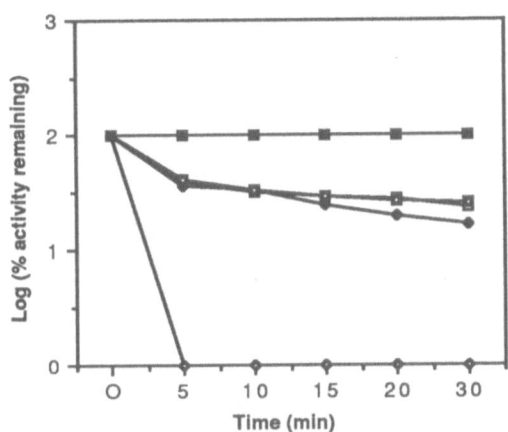

Figure 5c. Effect of metal ions on the inactivation of LDH by Dye 2. LDH
(8.2 nM) was inactivated by Dye 2 (3.125 µM) in the presence
of 2 mM of metal in 0.05 M Phosphate buffer, pH 7.5. A sample
without the dye served as a control.■ , Control; ◆ , Zn⁺⁺ ;■ ,
Mg⁺⁺ ; ◆ , Cu⁺⁺ ;□ , in the absence of metal ion.

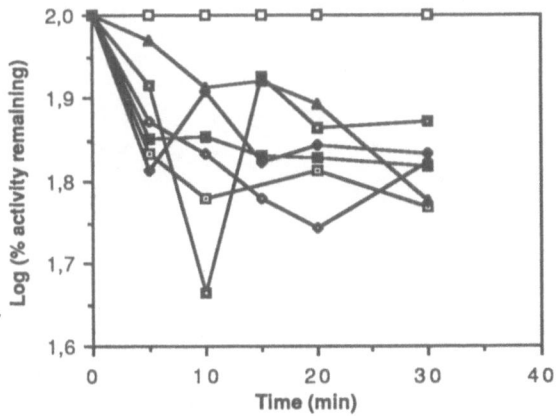

Figure 6a. Effect of different concentrations of Zn²⁺ on the inactivation
of LDH by Procion Red HE-3B (3.125 µM). □ , Control; ■ , effect
of 0.5mM Zn⁺⁺ ; ◆ , 1mM Zn⁺⁺ ; ■ , 2mM Zn⁺⁺ ; ◆ , 4mM Zn⁺⁺ ; ■ ,
in the absence of Zn⁺⁺ ; ▲ , in the presence of Zn⁺⁺ (2mM), wi-

thout dye.

230

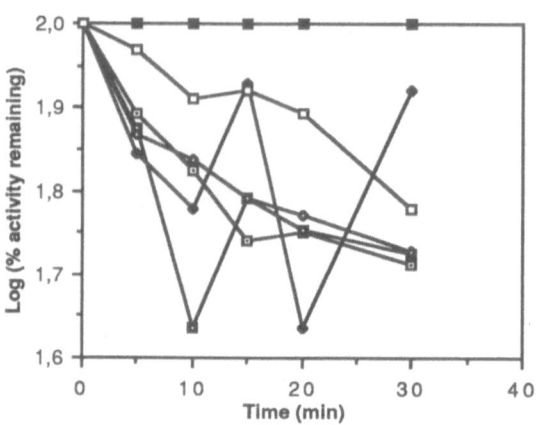

Figure 6b.   Effect of different concentrations of Zn^{2+} on the inactivation
of LDH by Dye 1 (50 µM). ■, Control; ▣, effect of 1mM Zn^{++};
◆ , 2mM Zn^{++}; ▣ , 4mM Zn^{++}; ◆ , in the absence of Zn^{++}; ▣ , in
the presence of Zn^{++} (2mM), without dye.

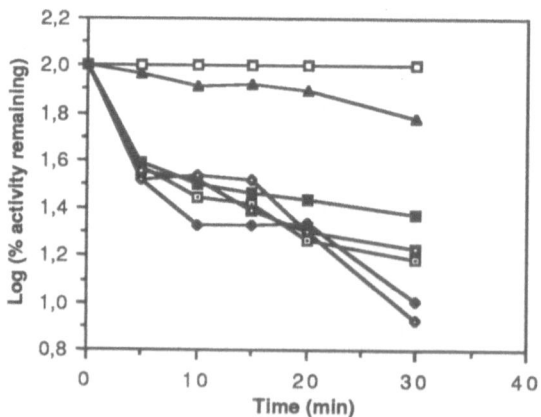

Figure 6c.   Effect of different concentrations of Zn^{2+} on the inactivation
of LDH by Dye 2 (3.125 µM). ▣ , Control; ▣ , effect of 0.5mM
Zn^{++};   , 1mM Zn^{++}; ▣ , 2mM Zn^{++}; ◆ , 4mM Zn^{++}; ■ , in the absence of Zn^{++}; ▲ , in the presence of Zn^{++} (2mM), without dye.

DISCUSSION

It has already been shown by (3) that different metal ions influence in different manners depending upon the nature of the immobilized dyes and the protein studied. The Zn^{2+} based binding improvement, is attributed to the favorable conformational changes occuring due to the interaction of Zn^{2+} with the functional groups of the dye chromophore (3). It is also known that the dyes do not assume one single unique conformation while binding to different proteins. Moreover, due to the full d10 electronic configuration of the Zn atom, no crystal fields occur and the stereo chemistry of the metal complex is decided entirely by electrostatic and covalent binding forces and by the ligand size. Based on these facts, any perturbation of the Zn^{2+}-dye complexation by other chelating agents such as EDTA will certainly have an influence on the ternary complex metal-dye-protein through the conformational changes induced. However, contrary to Hughes (3) successful elution of dye column bound proteins, using added EDTA in the buffer, we could achieve a quantitative elution of the bound protein, luciferase, lactate dehydrogenase, hexokinase etc..., only on the addition of 5mM ribose along with 15 to 30mM of the bivalent metal ions and 40mM of EDTA (Tables 1 and 2). From the results of competitive elutions of the nucleotide enzymes using metal complexes, as shown in Tables 1 and 2, it is clear that the addition of 5mM of sugar, particularly ribose adds to the efficiency of the biomimetic properties of the metal complex, in dissociating the dye ligand bound enzymes. When we compare the effect of Zn^{2+} in the immobilized dye system to that of the same dye in solution, the dissociation efficiency is rather different. While Zn^{2+} protects more efficiently than the Mg^{2+}, the enzyme inactivation of the cofactor dependant enzymes by the dyes (Figs. 3 and 5a, b, c), Mg^{2+} complex with added sugar is more efficient in eluting the dye column bound enzyme (Table 1). These data suggest that the immobilized and free dyes do not show the same mechanism towards the metal ion mediated dye-enzyme association dissociation. Though other sugars galactose, mannose and fructose are fairly efficient in improving the luciferase desorption, ribose seems to be the best both in the case of Mg^{2+} or Mn^{2+} as the metal ion. The results in Table 2, clearly show that, in the absence of ribose, only 50% of the active enzyme (LDH) could be recovered. The addition of the substrate (pyruvate) could only improve the yield to 80%. (eluents 5 and 6). On the other hand, the addition of 5mM ribose, with or without pyruvate, significantly improves the active enzyme recovery. It is also interes-

ting to note that ribose concentrations 5mM (eluents 11 and 12) show a
tendency to decrease the enzyme yield. This can be attributed to the increase in the bulkiness of the metal-EDTA-sugar complex due to extensive
hydrogen bonding resulting from the ribose in solution. The structure of
these complexes are not yet elucidated. Hence, it is too premature to ad-
vance any hypothesis regarding the ATP or NAD$^+$ structure resemblence of
this complex. Nevertheless, it is interesting to note that, these metal-ED-
TA-ribose complexes are specific only in dissociating the site specific competitive ligand, namely Cibacron Blue F3GA, bound LDH, and they are inefficient in desorbing the same enzyme bound to a Procion Red HE-3B column.
These phenomenological observations which seem valid in the dissociation
of many of the Adenosine nucleotide binding proteins, bound to cibacron
blue dye columns are interesting and useful in a better exploitation of the
interrelations between the two classes of the biomimetic ligands, namely
the triazine dyes and the metal-chelates.

The metal mediated enzyme kinetics in the presence of different triazi
dyes show that Cu^{2+} always behaves as an inhibition enhancer while Zn^{2+} always shows an antagonist effect, irrespective of the enzymes studied. Hughe
(3) has shown the enhanced inhibition of carboxy-peptidases G_2 and hexokinase in the presence of 2mM concentration of Zn^{2+}. However, if we consider
the metal/protein ratios, his studies use a ratio of 1:7 while our studies
use 650:1 and 40:1, for Procion Red HE-3B and Dye 1 respectively. The large
excess of Zn^{2+} ions used raise the question regarding the predominence of
Zn^{2+} effect, without the dye. But, the Zn^{2+} alone at the same concentration
ratio in the absence of dyes, behaves as a mild inhibitor. Hughes (7) pro-
poses a metal-enzyme-dye complex where one of the metal co-ordination is occupied by the amino acid in the enzyme. It is well known that most of the
nucleotide binding cavities in these enzymes have histidine (8), which can
interact with Zn^{2+} and Cu^{2+}. Bruce Martin (9) mentioned that the acidity of
bound water is appreciably enhanced on binding of imidazole rings to metal
ions. This could partially explain the metal ion effects on the enzyme kinetics, by modifying the potential and the localized pH at the binding site
of the enzyme. In fact, Hughes (3) in his work has clearly shown that the
Zn^{2+} ions can enhance the enzyme binding to the dye columns only at neutral
and alkaline pHs. However, the drastic differences observed between Cu^{2+} an
Zn^{2+} can be attributed mainly to their differences in their stereo selectivities with histidine or with histidine peptides. In fact, L.D. Petit and

R.J.W. Hefford (10) have shown significant differences both in the stereo selectivities of Zn^{2+} and Cu^{2+} histidine complexes as well as thermodynamic differences of these complexes. It is also worthwhile considering the dye structures acting as chelators, contributing to the orientation of the metal for the metal-histidine complexation to be favorable.

The results obtained with three structural variants of the Procion Red HE-3B do show differences in terms of metal mediated kinetics. Zn^{2+} ion acts as a protector, when used with the Procion Red HE-3B and the Dye 1, while it is without any significant effect with the Dye 2. The native Red HE-3B and the Dye 1 are different in their terminal sulfonic group and the Dye 2 has the terminal SO_3^- group, but has an addition amino benzene ring at both termini. It is hence, obvious that the terminal sulfonic group is not the determinant factor in the metal complexation. Perhaps these results show more a special case of metal-chelation where the dyes are the chelating ligands, quite different from the chelators reported by Porath (11).

CONCLUSION

Though the mechanism of competition between the dye ligand and the metal complex is not yet clear, the interrelation between these ligands, in terms of synergic as well as antagonist effects are established. Further studies, using crystallography are needed to elucidate this interesting phenomenon, which can throw more light into the enzyme oscillation due to extremely localized pH variations.

ACKNOWLEDGEMENTS

Stimulating discussions with Drs A. Friboulet and J.F. Hervagault of the Laboratoire de Technologie Enzymatique, University of Compiègne and Dr J. Kirscheberger from KMU, Leipzig GDR, are thankfully acknowledged. We are grateful to Dr Sunanda R. Narayan for her initial contribution to this work in our laboratory.

234

REFERENCES

1. Lowe, C.R.; Clonis Y.D.; Goldfinch M.J., Some preparative and analytical application of triazine dyes. In "Affinity Chromatography and Related Techniques", Ed. T.C.J. Gribnau et al, Elsevier Publishing Co, 1982, pp. 389-398.

2. Kopperschlager G.; Bohme H.J. and Hofmann E. In "Adv. Biochem. Engg., Ed. A. Fiechter, Springer Verlag, 1982, 25, 108.

3. Hughes P., Lowe C.R., Sherwood R.F. Metal ion promoted binding of proteins to immobilized triazine dye adsorbents. Biochem. Biophys. Acta, 1982, 700 90.

4. Rajgopal S., Vijayalakshmi M.A. Role of metal ions in triazine dye affinity chromatography, Enzyme Microb. Technol., 1984, 6, (12), 555.

5. Bergemeyer H.U., Bernt E., Hess B. UV assay of lactic dehydrogenase with pyruvate and NADH, 1974, 1 574.

6. Rajgopal S. and Vijayalakshmi M.A. New elution complex more specially for affinity chromatography and affinity precipitation, 1987, US Patent n° 4.666.604.

7. Hughes P. In this volume.

8. Lehninger A.L., Biochemistry Worth publishers Ltd, 1975, pp. 484-485.

9. Bruce Martin R. In "Metal ions in biological systems", Ed. H. Sigel, Marcel Dekker Publishers , 1979, Vol. 9, pp. 2-39.

10. Petit L.D. and Hefford R.J.W. In "Metal ions in biological systems", Ed. H. Sigel, Marcel Dekker Publishers, 1979, Vol. 9, pp. 173-212.

11. Porath J., Carlsson J., Olsson I. and Belfrage G. Metal chelate affinity chromatography, a new approach to protein fractionation, Nature, 1975, 258 (5536), pp. 598-599.

STUDY OF THE INTERACTION OF NAD(H)-DEPENDENT DEHYDROGENASES WITH REACTIVE DYES AND THEIR COMPLEXES WITH TRANSITION METAL IONS

S.S.FLAKSAITE, O.F.SUDZHIUVIENE, J.-H.J.PESLIAKAS
A.A.GLEMZHA
ESP "Fermentas", All-Union Research Institute of
Applied Enzymology, Vilnius, Lithuanian SSR, USSR

We studied the interaction of two NAD-dependent dehydrogenases - lactate dehydrogenase (LDH) from rabbit muscle and alcohol dehydrogenase (ADH) from yeast with some reactive dyes - light resistant yellow 2KT (LRY 2KT) (as Cu^{2+} complexe), claret ST (C ST) (Cu^{2+}), yellow 2KT (Y 2KT), scarlet 4ZT (S 4ZT) and orange 5K (O 5K). Interaction of above mentioned enzyme with dyes were studied by difference and circular dichroism (CD) spectroscopy, affinity partitioning and chemical modification of enzyme.

Data given in the table 1 show that ADH interacts with

TABLE 1
ADH and LDH interaction with dyes

Dyes	ADH				LDH			
	Diff. spectra extremum			K_d	Diff. spectra extremum			K_d
	max (nm)	min (nm)	isobestic point (nm)	(μM)	max (nm)	min (nm)	isobestic point (nm)	(μM)
LRY 2KT-Cu^{2+}	490	420	470	4,75	500	430	480	35,0
Y 2KT	450	390	430	450	490	400	480	20,0
C ST-Cu^{2+}	550	520	530	24,0	-	-	-	-
S 4ZT	530	500	510	468	-	-	-	-
O 5K	526	470	-	n.d.	530	460	490	0,60

LRY 2KT-Cu^{2+}, C ST-Cu^{2+} and S 4ZT, and this let us to determine dissociation constants (K_d) of the dye-ADH complexes. Linearity of the curves of ADH titration by dyes in the Scatchard coordinate assume the formation of the stechiometric 1: complexes of ADH subunit:dye. Dye O 5K interacts with ADH mor complicated. Introduction of NAD to O 5K-ADH complexe caused not the "quenching" of the difference absorbance maxima as ir the case of LRY 2KT-Cu(II) (fig.1) but the change of differential absorbance maxima position. Dissociation constants (K_d) of the complexes of ADH with dyes are different. As show in table 1 ADH interacts strongly only with dyes containing Cu^{2+} ions in opposite to analogues of the same dyes containir no metal ions (dyes Y 2KT and S 4ZT).

LDH from rabbit muscle interacts with all of the studied dyes but there is significant difference in its binding. In all cases, in opposite to ADH, LDH produce complexes with dye and their difference absorption maxima are effectively "quenched" by NAD. In opposite to ADH, LDH interacts with LRY 2KT and Y 2KT with similar strenght, K_d of the complexes are the same. The strongest binding of LDH is realised in the case of O 5K, K_d is lowest (0,6 μM), so we can conclude that later dy binds to enzyme specifically, (fig.2).

CD spectroscopy, as it can be seen from data given in table 2, showed the different interaction of ADH with metal-

TABLE 2
Extremum in CD spectra at ADH and LDH interaction
with dyes

Dyes	CD extremum (nm) (maxima·position)	
	ADH	LDH
LRY 2KT-Cu^{2+}	480 (+)	460 (+)
Y 2KT	–	
O 5K	480 (–)	480 (+)
Cibacron blue F3GA	600 (–)	600 (+)

containing and non-containing metal dyes.

In the case of LDH, as it is seen in table 2, in all the

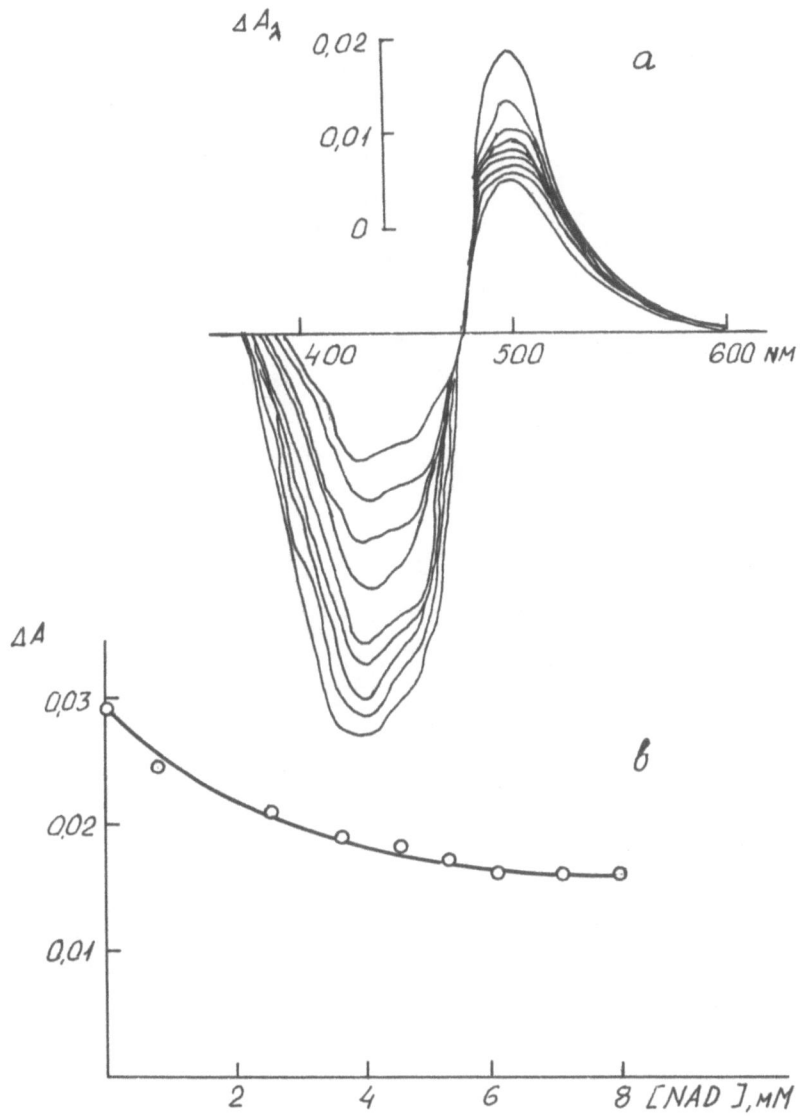

Figure 1. Difference spectra of ADH-LRY 2KT-Cu(II) complex
in the presence of NAD (a) and dependence of differential absorbance at 430 nm on concentration
of NAD (b). Initial concentration of dye 17 μM,
ADH 10 μM (per subunit), NAD 0,16 M, 10 mM Tris-
HCl buffer (pH 6,5)

cases irrespective of the presence of metal ions (0 5K, Cibacron Blue F3GA) dye interaction with enzymes induce CD
spectra with characteristic positive maximum at long wave-

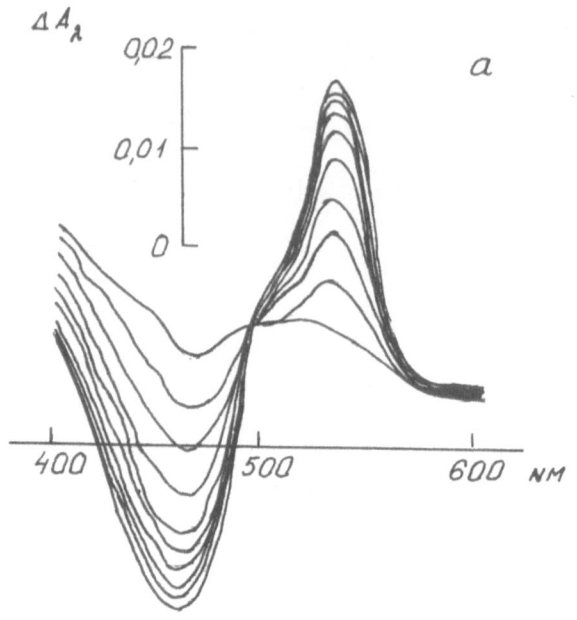

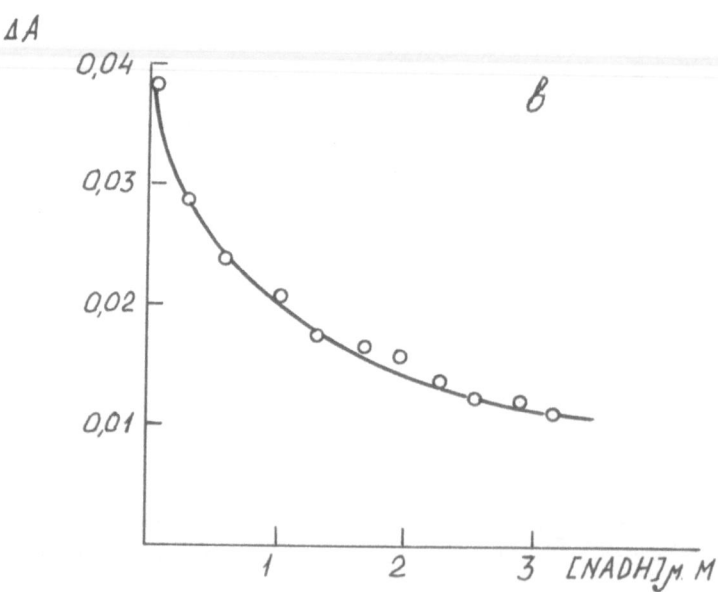

Figure 2. Differential spectrophotometric titration of **LDH** by
O 5K in 10 **mM** Tris-HCl buffer (pH 7,0) (a). Depen-
dence of differential absorbance of LDH-O 5K comp-
lex at 530 nm on the concentration of NADH (b).
Initial concentration of LDH 23,1 μM, dye 1,94 μM,
NADH 5,39 **mM**

lengths. This is evidence that dyes interact with enzyme probably in identical conformational states. In oder to study the role of metal, when forming ADH-dye complexes, Cu^{2+} ions has been excluded (from LRY 2KT-Cu(II)) and substituted by Zn^{2+}, Ni^{2+}, Mn^{2+}. Also dissociation constants of the complexes dye-metal ion-ADH have been measured, additional metal ions have been introduced into LRY 2KT and their interaction with ADH has been analized. As it can be seen from table 3, after

TABLE 3

Dissociation constants (K_d) of dye-metal ion complexes and dissociation constants (K_d) of dye-ADH complexes

Dyes	Constants	Metal ions			
		–	Cu^{2+}	Zn^{2+}	Mn^{2+}
LRY 2KT	K_d	–	0,084	0,090	0,60
LRY 2KT-ADH	K_d	320	4,75	12,5	–
Y 2KT	K_d	–	0,84	1,40	–
Y 2KT-ADH	K_d	–	–	–	–

removal of Cu^{2+} from LRY 2KT-Cu(II), K_d of the dye-ADH complex increased approximately 100 times. Inverse introduction of Cu^{2+} either its substitute by Zn^{2+} reduced K_d to 5-10 μM. Other studied metals did not possess such properties. Thus, it is quite evident that metal ions participate in the formation of dye-ADH complex. To approve this, ADH-dye complexes destruction adding special reagents have been studied. Figure 3 shows that CD spectra induced by ADH interaction with LRY 2KT-Cu(II) are effectively reduced in the presence of chelating agents 8-oxyquinoline-5-sulfonic acid (Oxinsa), 1,3-diamino-2-propanol-N,N,N',N'-tetraacetic acid (PDTA). The latters are capable to remove enzyme from dye-Cu(II) complex. At the same time AMP, NAD are less effective, as well as sodium chloride. On the other hand ADH complexes with O 5K, noncontaining ion metals are destroyed by NaCl, AMP and NAD more effectively incomparison with chelating agents. From our point of view, it is evident, that ADH with dyes, containing bound metal interact through the metal, forming bond between them

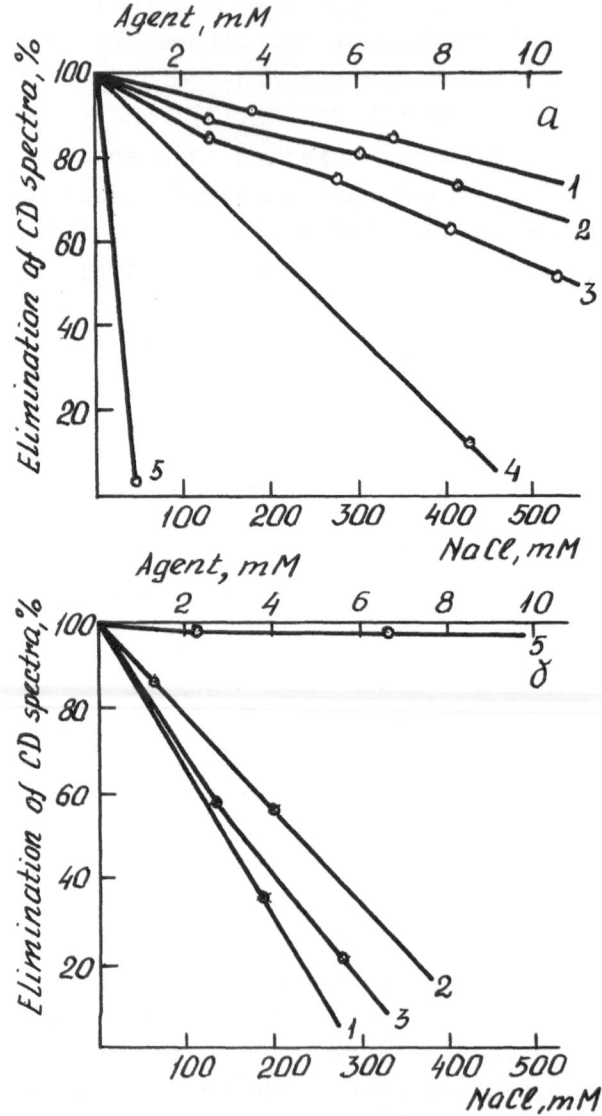

Figure 3. "Quenching" of CD spectra maxima of **ADH-LRY** 2KT-
Cu(II) complex (a) and ADH-O 5K complex (b) in
the presence of NaCl (1), **AMP** (2), ÑAD (3),
PDTA (4), Oxinsa (5)

and a certain side-chain radical of some amino acid available
in the coenzyme-binding part.

ADH with dyes, non-containing metal ions interact considerably weaker. In this case, however, the participation of

other parts of enzyme can be observed.

LDH active centre structure is considerably different from ADH, so LDH is more effective in the formation of metal ion-non-containing dye complexes. Since the structure of these enzymes has been studied sufficiently properly (and on the basis of some data of complex stability), we suggested that histidine of the ADH active centre can be introduced in the formation of ADH-LRY 2KT-Cu(II) complex. In opposite to ADH, histidine with the same function is not observed in the NAD-binding part of LDH.

Study to approve this assumption has been made. ADH has been modified by diethylpyrocarbonate - the specific modificator of imidazole ring of histidine residue. Changes in optical density at 237 nm showed the quantity of modified histidine residues.

Difference spectra of modified ADH is illustrated in figure 4. As a result of ADH modification, within 2 hours, less than one residue of histidine on one subunit has been blocked, preserving 20% of initial activity, (figure 5). It should be mentioned, that modification did not cause irreversible changes in the structure of enzyme. After the action of hydroxylamine for modified ADH, enzyme activity has been completely restored.

On the other hand, as it is presented in figure 6, modification has been of great importance in ADH-dye complex formation. Minimum and maximum can be observed in CD spectra at interaction of native enzyme with dye. Modified ADH with dye produce quite different CD spectra. It can be approved by nature changes of interaction. Since diethylpyrocarbonate chemically modifies imidazole ring of histidine, it has been of great importance that the latter can participate in the complex formation with dye.

Treatment of modified ADH by hydroxylamine reactivates enzyme activity, it is capable to restore the form of the induced CD spectra.

Thus, the illustrated experimental data gives evidence of participation of Cu^{2+} ions and imidazole ring of histidine of ADH in enzyme-dye complex formation.

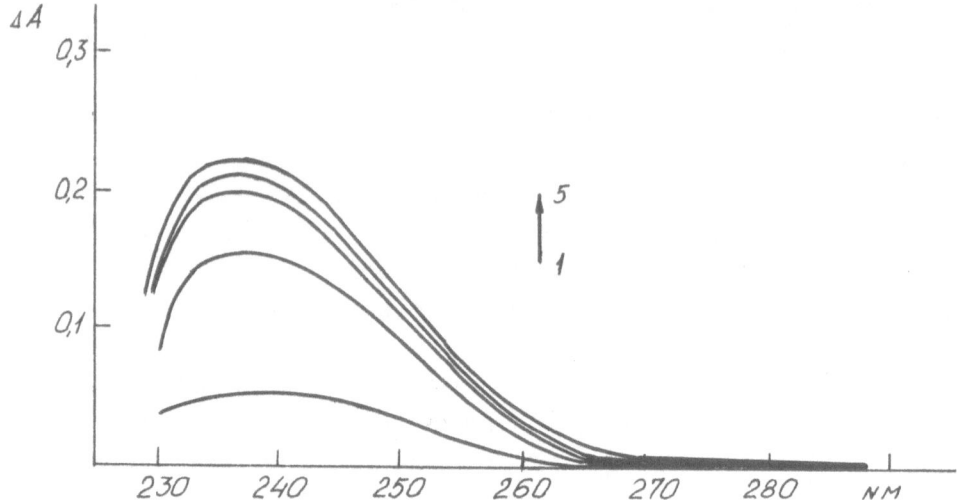

Figure 4. Difference spectra of the modified ADH (23 μM
calculated per one subunit). ADH was modified by
0,18 mM diethylpyrocarbonate in 0,1 M potassium
buffer (pH 6,5). Incubation period: 2 min (1),
12 min (2), 24 min (3), 37 min (4), 40 min (5)

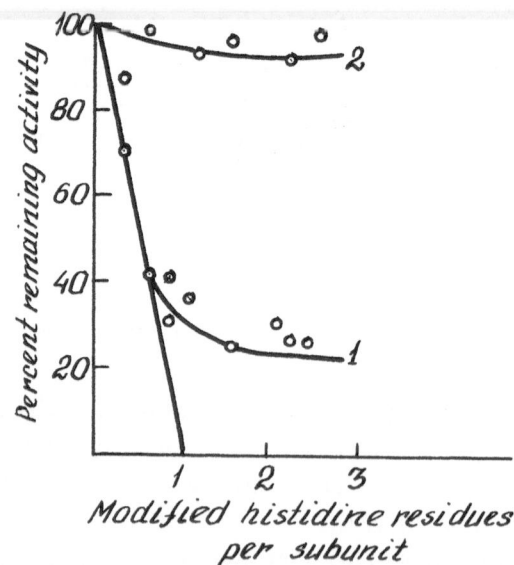

Figure 5. Dependence of the ADH residual activity on the
quantity of the modified histidine residues per
subunit of an enzyme. ADH (28 μM on subunit) was
inhibited by 0,18 mM diethylpyrocarbonate in 0,1 M
potassium phosphate buffer (pH 6,5) (1). Reacti-
vation of modified enzyme by 0,1 M hydroxylamine
at pH 7,5 (2)

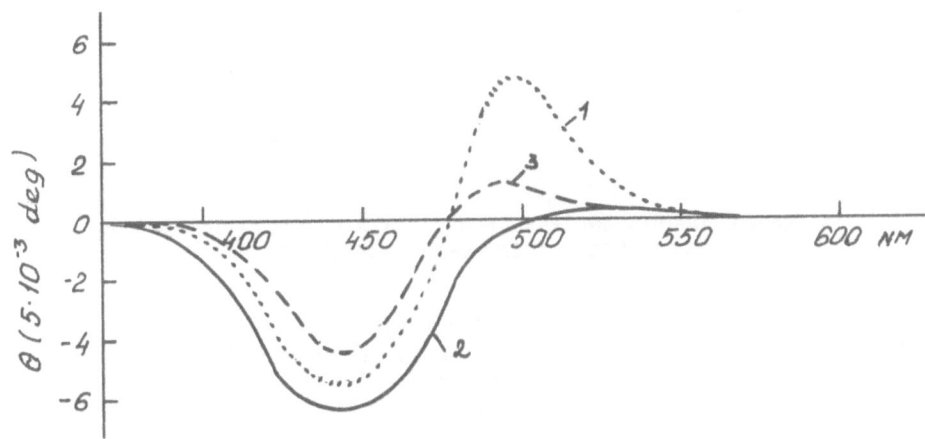

Figure 6. CD spectra of LRY 2KT-Cu(II) with native ADH (1),
ADH modified in the period of 25 min by 0,18 mM
diethylpyrocarbonate (2) and ADH modified in the
period of 40 min by 0,18 mM diethylpyrocarbonate
and reactivated during 35 min by 0,1 M hydroxylamine at pH 7,5 (3). Dye concentration is 100 μM,
ADH 28 μM (per subunit)

IMMUNOGLOBULIN AND PREALBUMIN INTERACTIONS WITH REMAZOL YELLOW GGL : PREALBUMIN REQUIRES AN ANION

PETER G H BYFIELD
Endocrinology Research Group,
Clinical Research Centre,
Harrow HA1 3UJ, U.K.

ABSTRACT

Immobilised Remazol yellow GGL will extract prealbumin from serum but only if a sufficient concentration of a large inorganic anion is present. Short chain mono- and dicarboxylate ions are also effective. The dye will also interact with immunoglobulins and this property may be exploited in a separation method for polyclonal antibodies in serum from human and other species. In autoimmune human sera oligoclonal antibodies against the same epitope may be resolved.

PREALBUMIN

A large number of dyes were screened as ligands for the affinity chromatography of the major thyroxine-binding proteins found in serum [viz. thyroxine-binding globulin (TBG) and prealbumin]. Some of the dyes were selected, because their structures had a distant resemblance to that of thyroxine, in the hope that there might be interactions at the natural ligand-binding site of the protein as is seen for Cibacron blue F3GA and NAD-dependent enzymes [1]. Only one dye, Remazol yellow GGL, interacted and only with prealbumin when serum samples were passed through columns of the dye coupled on to Sepharose-4B (yellow gel). This interaction was weak and the prealbumin was not adsorbed but merely delayed in its passage. Interestingly, the dye was one of those with a superficial resemblance to thyroxine (Fig. 1) and so further studies were undertaken.

One ml volumes of serum with or without selected additives, were applied to 1 ml columns of yellow gel equilibrated in appropriate buffer solutions and washed through with 2 ml of the same solution. The

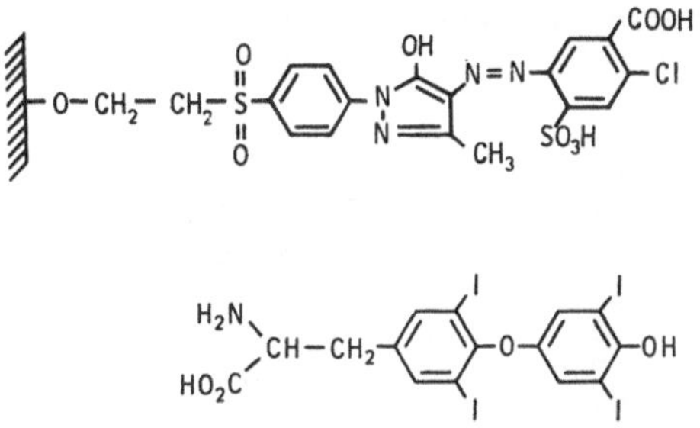

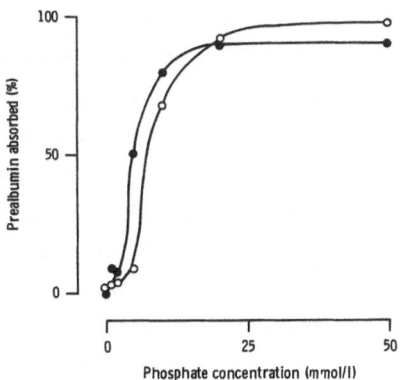

Figure 1. Structures of Remazol yellow GGL (upper) and thyroxine compared.

prealbumin content of the combined 3 ml of eluate was then measured to
determine the prealbumin adsorbed from the serum. Dialysis of serum
against water or 0.15 M NaCl did not remove the anticipated inhibitor of
binding, instead prealbumin had no interaction with the gel. Progressive
addition of phosphate, however, caused a progressive interaction with the
yellow gel and full adsorption was seen at concentrations of about 20 mM
and above (Fig. 2).

Figure 2. Adsorption of prealbumin from serum by Remazol yellow
GGL-Sepharose as a function of added phosphate concentration.
Serum dialysed against water (O) and against 0.15 M NaCl (●).

This phenomenon suggests an anion-binding site on the protein with an
allosteric communication with the dye-binding site to promote dye-protein
interactions. Whether or not phosphate is a natural ligand for that anion-
binding site is not clear as a number of other compounds likely to be found

TABLE 1

Adsorption of prealbumin from serum to the yellow gel in the presence of various anions at 20 mM concentration, pH 7.4. In all cases full recovery was achieved with 10% (w/v) ethanol in 20 mM phosphate buffer, pH 7.4.

Anion added	Adsorption (%)	Anion added	Adsorption (%)
none	2	oxalate	98
phosphate	92	malonate	98
bicarbonate	0	succinate	99
arsenate	60	glutarate	100
sulphate	95	adipate	97
ferricyanide	96	pimelate	70
tungstate	94	suberate	1
		formate	80
citrate	93	acetate	81
fumarate	100	propionate	70
maleate	35	hexanoate	17

in serum could also enhance binding (Table 1) as could some non-physiological anions. Short chain mono- and dicarboxylic acids were able to promote the interaction between Remazol yellow GGL and prealbumin but this ability decreased as the chain length increased. The structurally constrained isomers fumarate and maleate displayed an interesting stereospecific influence on binding which may reflect their interactions at the anion-binding site. Identification of this site with its allosteric communication with the central lumen where the dye and thyroxine bind, will be important for correlation with the effects of certain amino acid substitutions in the protein. Two of these, on the surface of prealbumin, also affect thyroxine binding in the central lumen [2], but further, cause a lethal deposition of prealbumin as amyloid in tissues. One other substitution on the surface also results in amyloid deposition but is without effect on thyroxine binding; yet another which enhances thyroxine binding does not associate with amyloid [3]. Understanding these intramolecular communications should shed light on factors responsible for these disastrous results from small changes in the protein.

It has been shown [4] that albumin mediates the binding of prealbumin to Remazol yellow GGL in aqueous two-phase systems of dextran and poly (ethylene glycol) using the dye attached to poly(ethylene glycol). This was re-examined using pure prealbumin and columns of yellow gel as described above. Pure prealbumin in 20 mM phosphate, pH 7.4, was less well adsorbed (49%) than when the experiment was done with serum (92%, Table 1).

TABLE 2

Adsorption of pure prealbumin to the yellow gel in the presence of
phosphate and added proteins.

Phosphate	Additive	Prealbumin adsorbed (%)
20 mM	None	49
20 mM	40 mg human albumin	82
50 mM	None	100
50 mM	40 mg human albumin	100
20 mM	10 mg human immunoglobulin G	65
20 mM	40 mg bovine albumin	81
20 mM	40 mg egg albumin	41

The addition of human albumin however increased adsorption to 82% (Table 2)
but full adsorption was achieved by increasing the phosphate concentration
to 50 mM with or without added albumin.   Thus, although albumin appears to
be able to enhance the binding between Remazol yellow GGL and protein,
phosphate (and perhaps other anions) clearly has a major role in
facilitating the interaction.   Bovine albumin and human immunoglobulin G
also enhanced binding but egg albumin did not.

IMMUNOGLOBULIN

Gradient elution with sodium chloride of the material adsorbed from serum
by the yellow gel revealed a large amount of IgG in addition to prealbumin
(Fig. 3).   Studies with pure IgG showed that the proportion of the total
that was adsorbed at pH 7.4 was independent of the load up to 17 mg per ml
yellow gel.   The proportion varied with the sample but was generally in the
range 50–60%.   Further work demonstrated that all IgG could be adsorbed if
applied to the yellow gel at pH 5.0 and this observation was used as the
basis for developing a separation method for serum IgG [5].   Following
precipitation of IgG by ammonium sulphate (2.5 M) the protein is dialysed
against 20 mM sodium phosphate, pH 5.0, and applied to the yellow gel.
Elution is then achieved with a gradient to 20 mM sodium phosphate, pH 7.4,
followed by a gradient to 300 mM NaCl in 20 mM phosphate, pH 7.4.   Two
broad bands of IgG result, one on each gradient, but assays for individual
antibody activities reveal peaks within the bands indicating that the bands
represent the overlap of the countless species of IgG in serum.

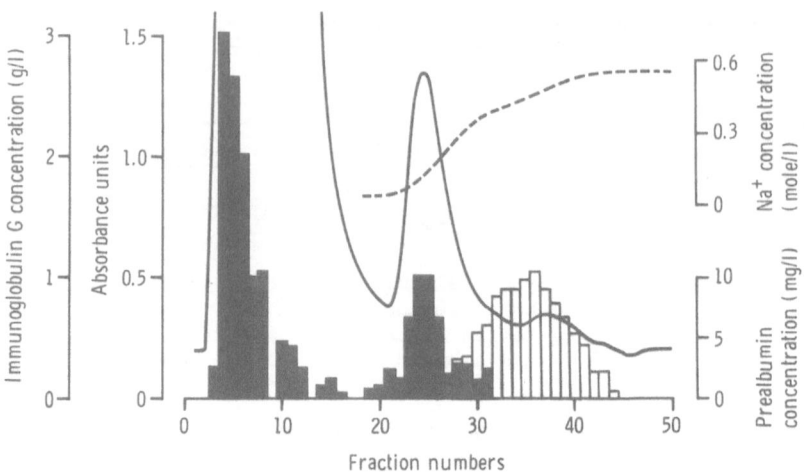

Figure 3. Chromatography of serum on Remazol yellow GGL-Sepharose at pH 7.4 eluting with an ionic strength gradient. Protein concentration, absorbance at 280 nm (——); IgG concentration (■); Prealbumin concentration (□); Sodium concentration (----).

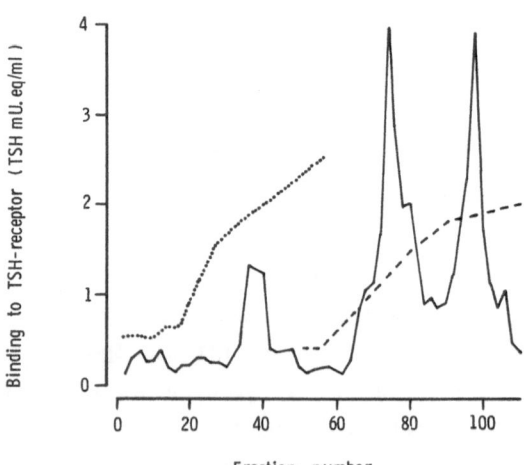

Figure 4. Separation of human thyrotrophin-receptor autoantibodies found in a patient with Graves' disease. IgG prepared by ammonium sulphate precipitation from 5 ml serum was chromatographed on a yellow gel column (1 x 20 cm) using a pH gradient (60 ml), followed by a NaCl gradient (120 ml), collecting 2 ml fractions. Three peaks of thyrotrophin-receptor binding activity were detected by their ability to inhibit ^{125}I-thyrotrophin binding.

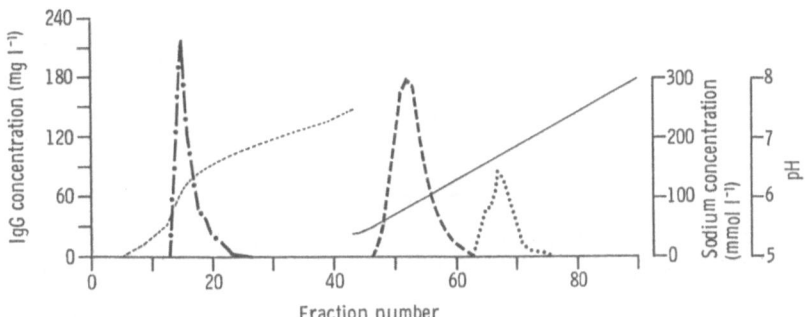

Figure 5. Composite drawing of the elution profiles of three myeloma
proteins chromatographed on the yellow gel. Serum samples
(100 μl) were chromatographed directly without precipitation
by ammonium sulphate.

Figure 4 shows the result of chromatographing a sample from a patient
with Graves' disease due to the presence of autoantibodies against the
thyrotrophin receptor; these bind at the hormone's site and mimic its
action in stimulating cellular activity. This patient was shown to have
three distinct species of autoantibody all reacting with the same epitope
by an assay involving the displacement of radioiodinated thyrotrophin from
the receptor. The separation probably could not be achieved by affinity
methods using the receptor as ligand. This cannot be tested as sufficient
quantities of receptor cannot be obtained but the yellow gel is clearly a
valuable alternative medium. Monoclonal antibodies secreted by mouse
hybridomas have not been examined in this system although polyclonal
antisera from chicken and rabbit in addition to human have been successfully fractionated. However in Figure 5 is a composite illustration of the
chromatography of three human monoclonal antibodies - those produced in
cases of myeloma. Here, the quantity of immunoglobulin produced by the
malignant cells was so far in excess that the normal IgG was reduced to an
undetectable background and each serum sample produced a single peak.

SUMMARY

The inexpensive adsorbent described here has a useful application in the extraction of prealbumin from serum but in addition the allosteric interactions may shed light on structure/function relationships and their influence in disease states. Of wider significance is the application of the yellow gel to the separation of immunoglobulins. Although work to date has concentrated on human autoantibodies, equivalent separations of polyclonal antisera from rabbits and chickens have also been made and there can be no doubt of the potential in the purification of monoclonal antibodies produced by hybridoma technology.

REFERENCES

1. Biellmann, J.F., Samama, J.P., Brändén, C.I. and Eklund, H. Eur. J. Biochem., 1979, **102**, 107-110.

2. Refetoff, S., Dwulet, F.E. and Benson, M.D. J. Clin. Endocrinol. Metab., 1986, **63**, 1432-1437.

3. Lalloz, M.R.A., Byfield, P.G.H. and Himsworth, R.L. Clin. Endocrinol., 1984, **21**, 331-338.

4. Birkenmeier, G., Tschechonien, B. and Kopperschläger, G., FEBS Lett, 1984, **174**, 162-166.

5. Worthington, J., Chan, C.T.J. and Byfield, P.G.H. FEBS Lett, 1987, **211**, 123-126.

Chapter 6

Blood Proteins Purification Using Dye–Ligand Affinity

DYE SERUM PROTEIN INTERACTION-ANALYSIS AND APPLICATION

G. BIRKENMEIER
Institute of Biochemistry, Karl-Marx-University, Leipzig,
Liebigstrasse 16, 7010-Leipzig
German Democratic Republic

ABSTRACT

The great demand for plasma proteins by the clinics and the industry presents a great challenge to the development of fast and effective purification procedures. Affinity techniques using synthetic dyes as biomimetic ligands gained special interest because most of the problems encountered with biological media can be circumvented. In this contribution the nature of the interaction of dyes with serum protein is described. Furthermore, the applicability of different affinity techniques such as dye-ligand affinity chromatography, dye-ligand affinity partition and dye-ligand affinity precipitation for separation of serum proteins is outlined.

INTRODUCTION

Human blood serum or plasma contains more than 100 different proteins only a small number of which have been fully characterized with respect to their molecular properties and biological functions. Purification of particular proteins from plasma often is a difficult task because of similarities in physicochemical properties and because many of them are present in low concentration. The demand by the clinics for highly purified individual human proteins and plasma fractions steadily increased over the last decades. Furthermore, there is a great need of purified and native plasma proteins to be used for affinity therapy, as substituents in culture media and more recently as tool for separation of enantiomers of pharmaceutical products and drugs. The commercial value of therapeutic proteins and the more rigorous quality criteria imposed on such products have placed greater demands on effective large scale purification technology. This has led to an intensive development of affinity techniques during the last years which has greatly improved protein purification (1). The literature abounds with examples of the use of synthetic dyes as biomimetic ligands for purification of enzymes and proteins (2,3). Since the discovery of the interaction of albumin (4) and blood

coagulation factors (5) with Cibacron Blue F3G-A the elaboration of a number of purification procedures for plasma proteins
using immobilized dyes has been launched. It is the aim of this
contribution to analyze the present knowledge on the dye-
protein interaction in molecular terms and to elaborate the
potential of dye mediated affinity techniques for separation of
proteins from human plasma or serum.

CHEMICAL BASIS OF DYE-PROTEIN INTERACTION

A number of textile dyes have emerged as "pseudospecific"
ligand used frequently in biochemical research and biotechnology in the last decade (2,3,6). This is because the dyes combine
three advantageous features: 1. the discriminating ability to
bind to selected proteins. 2. the ease of coupling to soluble
and insoluble supports, and 3. the low cost due to their large
scale production.
All the dyes share the common properties of possessing anionic
(sulphonated) groups and nonpolar (aromatic) ring systems in
the same molecule. This enables the dyes to complex with complementary structures at the surface of proteins by ionic,
hydrophobic or charge transfer forces. Roughly, the interaction
is classified according to the dominating forces as either
"hydrophobic" or "electrostatic". But in most cases dyes can
interact with protein in both modes whereas the contribution of
the single binding forces can be different. Irrespective of the
dominating interaction the binding of proteins to immobilized
dyes has been characterized as "specific" or "nonspecific". In
case of specific interaction the binding of dyes is said to
occur at sites where the natural ligand is bound in preference
to other regions. Either specific or nonspecific interaction
can be weakened by changing the pH and increasing the ionic
strength of the buffer and/or by adding substances counterac-
ting hydrophobic bonds. However, because of the higher binding
energies of bonds involved in specific interaction more drastic
changes of these parameters are necessary. Following this rule
the binding of serum proteins to dyes seems to be governed
mainly by nonspecific mixed ionic/hydrophobic interactions.
This explains why such a large number of different serum proteins are bound for instance to immobilized Cibacron Blue F3G-A
or Procion Red HE-3B at conditions of low ionic strength and
neutral pH.
    Only a few serum proteins were found to bind rather stron-
gly and/or specifically to certain dyes, e.g. albumin, prealbumin, thyroxine-binding globulin and Gc-protein (7-11). Noteworthy, just those proteins are involved in transport of small
molecular-weight compounds in the blood. For example, the
strong interaction of albumin with Cibacron Blue F3G-A is due
to binding of the dye to hydrophobic sites involved in binding
of long chain fatty acids (7). A more profound understanding of
the dye-protein interaction becomes apparent in case of prealbumin. This protein binds stoichiometrically Remazol Yellow GGL
at two different classes of binding sites one with high affinity ($K_H = 3.8 \mu M$) and one with low affinity ($K_L = 258 \mu M$) (9).
Furthermore, the dye strongly competes with the thyroid hor-

mones for common binding sites in prealbumin. Thus, the particular efficacy of that dye is due to mimicking the spatial
structural arrangement of the hormone. On the basis of this
findings an effective purification procedure for this proteins
has been elaborated as shown in the next chapter.

PURIFICATION AND SEPARATION OF PLASMA PROTEINS

Dye-ligand affinity chromatography

The available data on application of reactive dyes in chromatographic separation of plasma proteins are compiled in Table 1.
The outstanding properties of Cibacron Blue F3G-A in this
respect becomes apparent. The dye is known to bind with different affinity to plasma proteins which accordingly can be
grouped as follows:

1. Proteins with strong affinity (albumin and lipoprotein)
2. Proteins without or very low affinity (α_1-proteinase inhibitor, α_1-acid glycoprotein, α_2-macroglobulin, prealbumin,
   α-HS-glycoprotein)
3. Proteins with intermediate affinity (immunoglobulins,
   haptoglobins, Gc-proteins, α_1-antichymotrypsin, hemopexin)

Table 1
Application of dyes for isolation and fractionation of
plasma proteins

Dye	Protein
Cibacron Blue F3G-A	Albumin
	Lipoprotein
	Blood coagulation factors
	Complement factors
	α_1-Antichymotrypsin
	α_2-Macroglobulin
	α_1-Proteinase inhibitor
	α_1-Acid glycoprotein
	α-Fetoprotein
	Gc-Protein
	α-HS-Glycoprotein
	β-SP$_1$ -Glycoprotein
	Transferrin
	Immunoglobulins
Remazol Yellow GGL	Prealbumin
	Immunoglobulins
Procion Red HE-3B	Plasminogen

Albumin and lipoprotein were strongly bound to the immobilized
dyes under conditions where the remaining species were released. This has been exploited for isolation of human albumin
from different biological fluids and secondary sources to a

high degree of purity (12,13). This procedure is well established at laboratory level but developments to afford technology for large scale isolation of a high value product can be
anticipated. Selective removal of albumin from plasma is often
an essential step for the purification of those proteins having
similar physico-chemical properties such as proteins of the
α_1-fraction. Thus, α_1-acid glycoprotein and α_1-proteinase inhibitor passes through a highly substituted Cibacron Blue-Sepha-
dex column while other proteins are retained (14). Likewise,
α-fetoprotein, a clinically important oncofetal antigen could
be separated from albumin and the host of the proteins from
material of legal abortions by chromatography on Cibacron Blue-
Sephadex gels (15). The proteins of the third group show the
tendency to bind to the dye column with moderate affinity at
low ionic strength and at neutral pH. The bound species can be
fractionated by applying a gradient of salt or pH.

The effective binding of a protein to one type of dye does
not necessarily imply that other dyes behave similar. This
enables the application of so called tandem systems involving a
"negative" column and a "positive" column. The protein is first
passed through a negative column which retains contaminants and
then through a "positive" column which selectively adsorbs the
desired protein or vice versa. An example is given in Figure 1
and concerns the purification of prealbumin. This method utilizes the specific interaction of Remazol Yellow GGL and Cibacron Blue F3G-A with prealbumin and albumin, respectively (16).

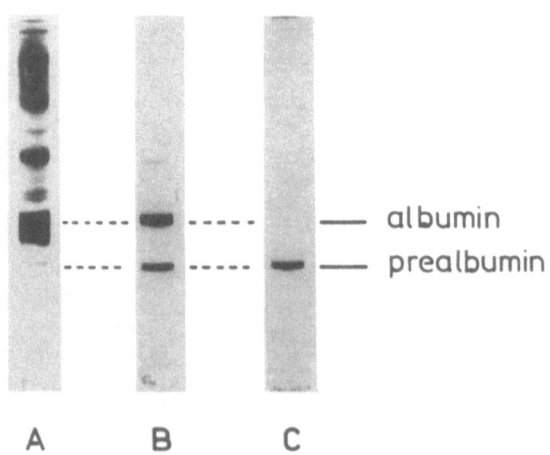

Figure 1: Disk-electrophoresis of prealbumin at different
stages of purification. A, human serum; B, eluate
after chromatography on Remazol Yellow GGL-Sephadex
G-100; C, break-through fraction after chromatography on Cibacron Blue F3G-A-Sephadex G-100. Starting
from 500 ml of serum 36 mg of prealbumin were
obtained.

The key to successful separation of various plasma proteins by mean of immobilized dyes appears to lie in the proper modulation of a number of variables listed in Table 2.

Table 2
Parameters determining the binding of proteins to immobilized dyes

Medium-dependent factors	Dye-matrix dependent factors	Protein-dependent factors
Ionic strength	Ligand concentration	Isoelectric point
pH value	Nature of matrix	Hydrophobicity
Buffer composition	Dye structure	Selectivity of binding sites
Temperature	Mode of attachement of the dye	Presence of competing proteins
Effector concentration		Protein concentration

In this respect chromatography on dye adsorbens closely parallels the approaches used in conventional biospecific affinity chromatography. However, the mixed electrostatic-hydrophobic characteristics of most reactive dyes suggested that manipulation of pH, ionic strength and especially of ligand concentration is a worthwhile exercise to optimize binding and elution of complementary proteins.

Dye-ligand affinity phase partitioning
The principle of this method is based on partition of protein between two aqueous phases composed of two different polymers usually polyethylene glycol (PEG) and dextran (17). The distribution of a definite protein between the PEG-rich upper phase and the dextran-rich lower phase is described by the partition coefficient, K, defined as the ratio of the protein concentration in the top and the bottom phase. The K-value is known to be influenced by a number of parameters including concentration and molecular weight of polymers, kind of salts and buffer added to the system, pH and temperature. Since partition of different proteins show individual response to these parameters separation can be achieved. In affinity partitioning ligands such as dyes are bound to one of the phase-forming polymer in order to steer the partition of proteins with binding sites for the respective ligand more selectively. The fundamentals of this method for application in serum protein separation has been reviewed recently (18). This technique turned out to be excellently suited for screening appropriate dye ligands. The efficacy of a polymer-bound dye ligand to extract a protein is

judged by the difference in the log K-values in systems with and without the ligand-PEG. In Figure 2 screening of ten dye-PEG derivatives with respect to their binding to albumin and prealbumin is presented. Ones more, the strong tendency of albumin to complex with various dyes but with preference to Cibacron Blue becomes obvious. In contrast, prealbumin displays binding only to Remazol Yellow GGL indicating a high selectivity of the interaction.

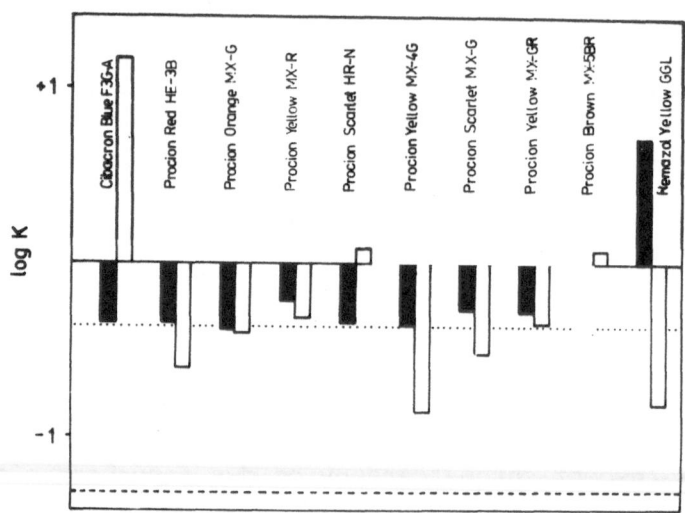

Figure 2: Effect of ten different dye-PEG derivatives on partition of prealbumin (black bars) and albumin (white bars). Phase systems were composed of 10% dextran T 500, 7.5% PEG 6000 including 1% of the respective dye-PEG derivative, 10 mM sodium phosphate, pH 7.0. Partition of albumin (‹– – –›) and prealbumin (‹····›) in the absence of dye-PEG is shown for comparison.

Moreover, this method offers the advantage of a precise study of the nature of the dye-protein interaction as well as of the optimum conditions for complex formation and dissociation. An example concerning the study of the effect of thyroid hormones on affinity partitioning of prealbumin is demonstrated in Figure 3. As seen, the naturally bound hormones thyroxine and triiodothyronine were able to compete with Remazol Yellow GGL for binding to prealbumin indicated by the reduced affinity partition effect (Δ log K).

To resolve complex mixtures of protein by partitioning in a single step is often inadequate. An improvement in separation is achieved by combining affinity partition with counter-current distribution (CCD)(19). In this method a mobile phase containing the PEG-bound ligand is moved against a series of tubes containing the stationary dextran-rich phase. The proteins of the sample system in the first tubes are thus continuously extracted according to their partition between the two phases. The diagrams in Figure 4 demonstrate separation of

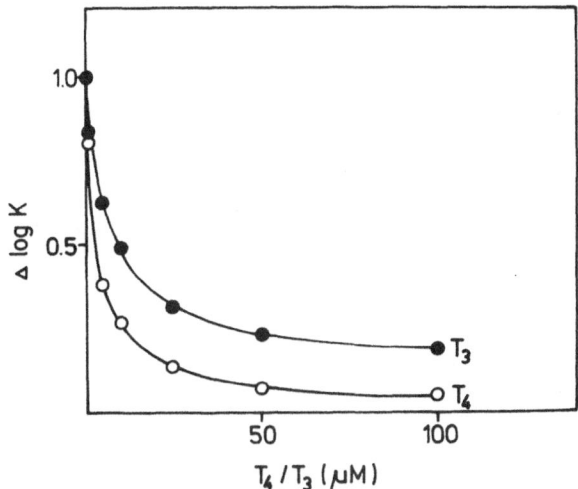

Figure 3. Influence of L-thyroxine (T 4) (○) and L-triiodothy-
ronine (T 3) (●) on the binding of prealbumin to
Remazol Yellow GGL-PEG. Phase systems were composed
of 10% dextran T 500, 7.5% PEG 6000 with and without
dye-PEG, 10 mM sodium phosphate buffer, pH 7.25, and
increasing concentration of the hormone.

whole serum by CCD in absence and presence of Cibacron Blue
F3G-A-PEG and Procion Yellow HE-3G-PEG. As seen effective separation of individual components could be achieved only in the
presence of dye ligands.

In terms of biotechnical application of CCD the time for
phase separation is still a critical factor for scaling up this
procedure. However, the use of centrifugal CCD-machines and the
application of magnetic polymers for speeding up phase separation makes it likely that phase partitioning may be included in
future development of fractionation procedures of plasma proteins.

Dye-ligand affinity precipitation

The purely technical difficulties of scale-up encountered with
chromatographic procedures are prompting alternatives to be
investigated. Affinity precipitation of proteins by dyes offers
an attractive possibility for future developments in this context. First description of using dyes in protein separation
traces back to the work of Horejsi and Smetana (20) who applied
rivanol for plasma fractionation. More recently, reactive dyes
were applied as precipitants in two different ways (Fig. 5).
Similar to the principle of antibody-antigen interaction bifunctional (bis)-dyes have been prepared by reacting of Cibacron Blue F3G-A on both amino groups of an aliphatic spacer. The
resulting compound was found to precipitate nucleotide-dependent enzymes as well as human serum albumin (21). The precondi-
tion for successful application of (bis)-dyes are the presence

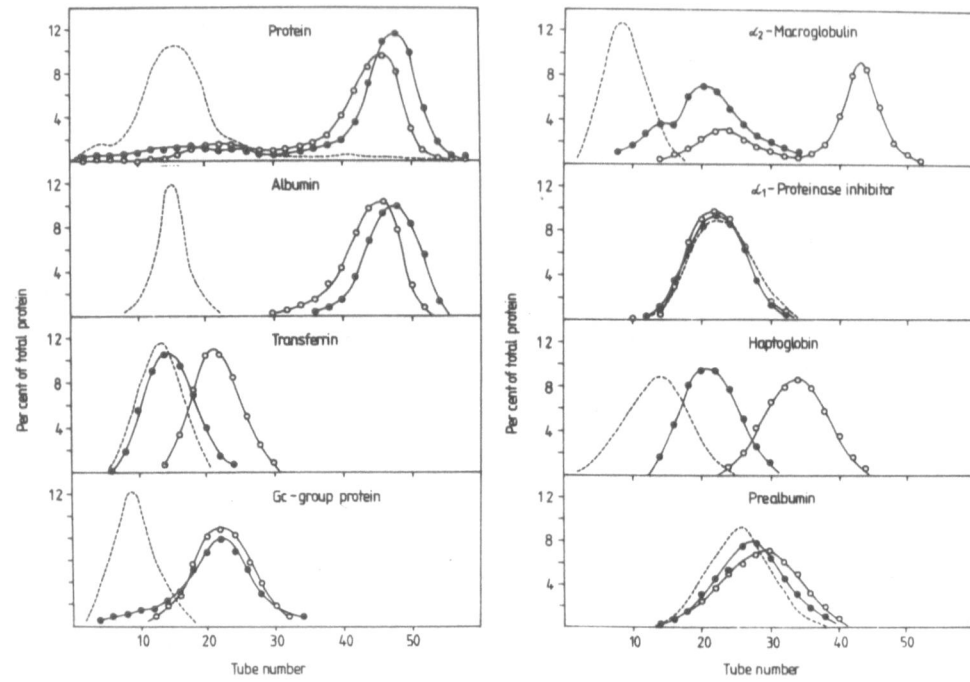

Figure 4: Thin-layer counter-current distribution of human
serum proteins in absence (←−−) and in presence of
Cibacron Blue F3G-A-PEG (●) and   Procion   Yellow
HE-3G-PEG (O). The two-phase system is composed
of 5% PEG 6000, 7.5% dextran T 70 and 20 mM sodium
phosphate, pH 7.0. Number of transfers, 58. 5% of
total PEG was replaced by dye-PEG.

of a number of specific dye binding sites in the protein.
Disadvantageously, a laborious chemical synthesis of the compounds including testing of the optimal chain length of the
spacer   must   precede any application.

Without such chemical efforts unmodified pure dyes were
found to have also the potential as precipitants for serum
proteins (22,23). The efficacy of precipitation was found to
be dependent on several important factors including the structure of the dye, the pH of the solution, the molar dye/protein
ratio and the intrinsic properties of the proteins. In general,
most dyes are endowed with the precipitating potentiality provided that they combine hydrophobic and anionic properties in
the same molecule. Usually precipitation of proteins is favoured by lowering the pH value and necessitates a certain
dye/protein ratio to become complete. This is demonstrated in
case of albumin, prealbumin and immunglobulin G (Fig. 6). As
seen these proteins can be precipitated by Cibacron Blue F3G-A
as well as by Remazol Yellow GGL. However, differences in the
precipitation curves of the two dyes become obvious. On the
other hand, ∝-1-acid glycoprotein which is known to   poorly
bind to different dyes escaped precipitation.

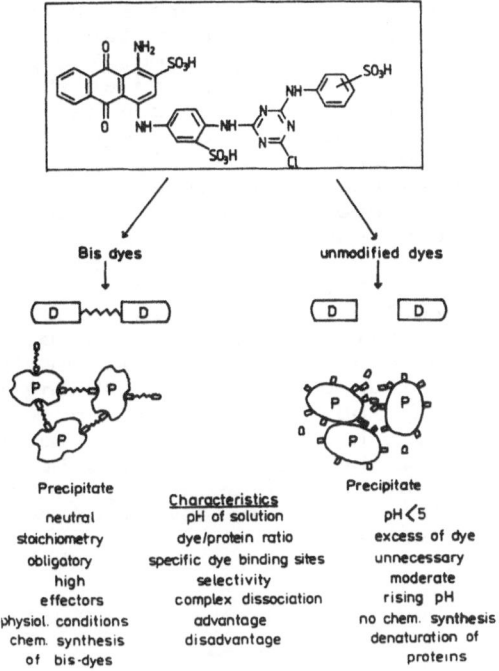

Figure 5:   Principle of precipitation of proteins by (bis)-dyes(left site) and unmodified dyes (right site). The given chemical structure represents Cibacron Blue F3G-A.

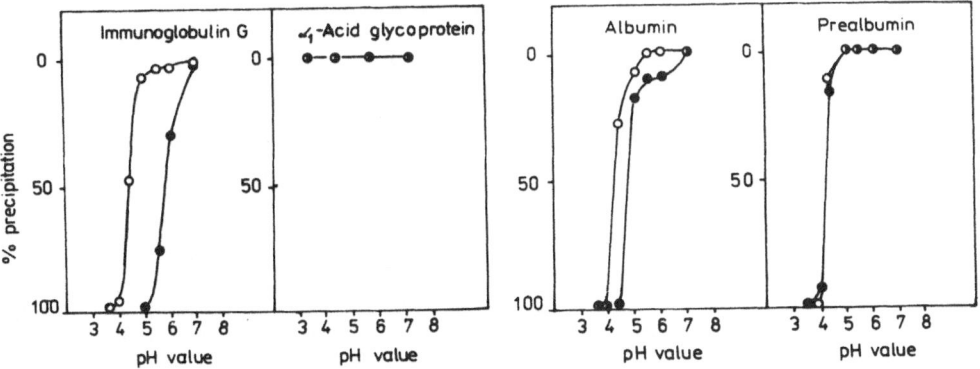

Figure 6:   Effect of pH on precipitation of different serum proteins by Cibacron Blue F3G-A (●) and Remazol Yellow GGL (○).

262

With certainty the mechanism of precipitation by unmodified dyes is far from being specific in terms of an imperative presence of specific dye binding sites. The presence and number of hydrophobic and cationic clusters at the protein surface mainly determine whether a protein can be precipitated or not by this mode. Nevertheless, it can be expected that selective protein precipitation renders possible because of the differences in charge distribution and degree of hydrophobicity among the different serum proteins. Indeed, albumin and prealbumin for example could be successfully separated by precipitating with Remazol Yellow GGL as shown in Figure 7. Albumin precipitates completely from the solution at a dye/protein ratio of about 45 whereas prealbumin remains in solution. The sensitivity of protein precipitation was found to be extremely high. Micro- to nanogram quantities of protein per milliliter could be precipitated and thus concentrated. Using this technique we have elaborated a method for separation of α_1-acid glycoprotein from human serum to a high purity and yield by one precipitation step (23). From the initial results it can be anticipated that precipitation by (bis)-dyes and by unmodified dyes looks set to play an role in the development of techniques for large scale isolation of proteins.

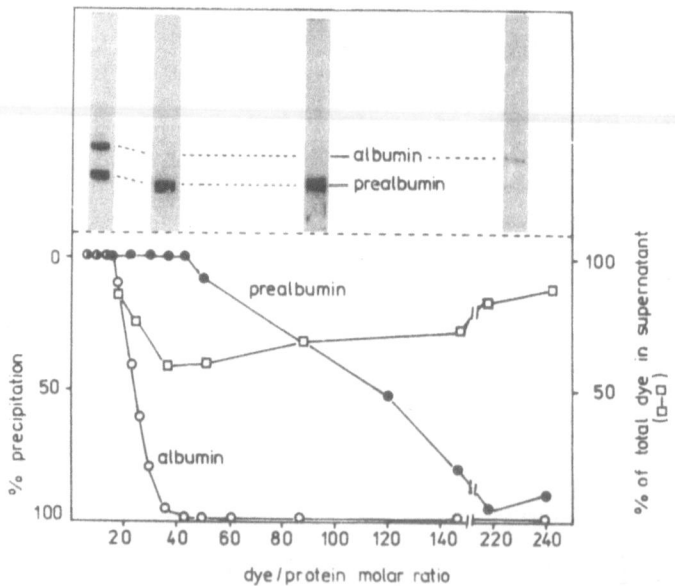

Figure 7:   Differential precipitation of albumin and prealbumin by Remazol Yellow GGL. The two proteins were dissolved in 25 mM sodium acetate puffer, pH 4.0 and a stock solution of the dye in water was added stepwise. After each addition aliquots were removed, centrifuged, and the supernatant was analyzed for the two proteins by electroimmunodiffusion and disk-electrophoresis. The dye/protein ratio refers to the total concentration of both in the reaction mixture.

REFERENCES

1. Lowe, C.R., and Dean, P.D.G., Affinity chromatography. Wiley, New York, 1974.

2. Kopperschlaeger, G., Boehme, H.-J. and Hofmann, E., Cibacron Blue F3G-A and related dyes as ligands in affinity chromatography. In Advances in Biochemical Engineering, ed., A. Fiechter, Springer, Berlin, 1982, pp. 101-138.

3. Subramanian, S., Dye-ligand affinity chromatography: The interaction of Cibacron Blue F3GA with proteins and enzymes. Crit. Rev. Biochem., 1984, 16, 169-205.

4. Travis, J. and Pannell, R., Selective removal of albumin from plasma by affinity chromatography. Clin. Chim. Acta, 1973, 49, 49-52.

5. Swart, A.C.W., Kop-Klaassen, B.H.M. and Hemker, H.C., Differential interaction of clotting factors II, VII, IX and X with Sephadex and dextran blue. Haemostasis, 1972/73, 1, 237-252.

6. Stead, C.V., The use of reactive dyes in protein separation processes. J. Chem. Tech. Biotechnol., 1987, 37, 55-71.

7. Metcalf, E.C., Crow, B. and Dean, P.D.G., The effect of ligand presaturation on the interaction of serum albumins with an immobilized Cibacron Blue 3G-A studied by affinity gel electrophoresis. J. Biochem., 1981, 199, 465-472.

8. Copping, S. and Byfield, P.G.H., Prealbumin: extraction from serum by Remazol Yellow GGL-Sepharose. Biochem. Soc. Trans., 1982, 10, 104-105.

9. Birkenmeier, G. and Kopperschlaeger, G., Interaction of the dye Remazol Yellow GGL to prealbumin and albumin studied by affinity phase partition, difference spectroscopy and equilibrium dialysis. Molec. Cell. Biochem., 1987, 73, 99-110.

10. Birkenmeier, G., Ehrlich, U. and Kopperschlaeger, G., Partition of purified human thyroxine-binding globulin in aqueous two-phase systems in response to reactive dyes. J. Chromatogr. 1986, 360, 193-201.

11. Chapius-Cellier, C., Gianazza, E. and Arnaud, P., Interaction of group-specific component (vitamin D-binding protein) with immobilized Cibacron Blue F3G-A. Biochem. Biophys. Acta, 1982, 709, 353-357.

12. Hanford, R., Maycock, W. and Vallet, L., Separation of human albumin by affinity chromatography. Chrom. Synth. Biol. Polym., 1977, 2, 288-293.

13. Harvey, M.J., Brown, R.A., Rott, J., Lloyd, D. and Lane, R.S., The purification of albumin by affinity chromatography. In Separation of Plasma Protein, ed., J.M. Curling, Pharmacia AB, Uppsala, 1983, 79-88.

14. Birkenmeier, G. and Kopperschlaeger, G., Application of dye-ligand chromatography to the isolation of α-1-proteinase inhibitor and α-1-acid glycoprotein. J. Chromatogr., 1982, 235, 237-248.

15. Huse, K., Himmel, M., Birkenmeier, G., Bohla, M. and Kopperschlaeger, G., A novel purification procedure for human α-fetoprotein by application of immobilized Cibacron Blue F3G-A as affinity ligand. Clin. Chim. Acta, 1983, 133, 335-340.

16. Birkenmeier, G., Tschechonien, B. and Kopperschlaeger, G., Affinity chromatography and affinity partition of human serum prealbumin using immobilized Remazol Yellow GGL. FEBS Lett., 1984, 174, 162-166.

17. Albertsson, P.-A., Partition of cell particles and macromolecules. 3. Edn. Wiley, New York, 1986.

18. Birkenmeier, G., Kopperschlaeger, G. and Johansson, G., Separation and studies of serum proteins with aid of aqueous two-phase systems containing dyes as affinity ligands. Biomed. Chromatogr., 1986, 1, 64-77.

19. Birkenmeier, G., Kopperschlaeger, G., Albertsson, P.-A., Johansson, G., Tjerneld, F., Akerlund, H.E., Berner, S. and Wickstroem, H., Fractionation of proteins from human serum by counter-current distribution. J. Biotechnol., 1987, 5, 115-129.

20. Horejsi, J. and Smetana, R., The isolation of γ-globulin by rivanol. Acta Med. Scand., 1956, 155, 65-70.

21. Hayet, M. and Vijayalakshmi, M.A.: Affinity precipitation of protein using bis-dyes. J. Chromatogr., 1986, 376, 157-161.

22. Bertrand, O., Cochet, S., Kroviarski, Y., Truskolaski, A. and Boivin, P., Protein precipitation induced by a textile dye. Precipitation of human plasminogen in the presence of Procion Red HE-3B. J. Chromatogr., 1985, 436, 111-124.

23. Birkenmeier, G., Dye-affinity precipitation of serum proteins. FEBS Advanced Course, Leipzig, 1986.

RECOVERY OF ALBUMIN FROM COHN FRACTION IV USING IMMOBILISED PROCION BLUE-HB : DEVELOPMENT OF A FULL SCALE PRODUCTION PROCESS.

J. E. More, A.G. Hitchcock, S. Price, J. Rott and M. J. Harvey, Blood Products Laboratory, Elstree, Herts, UK.

Abstract

The high capacity, high specificity and relatively low cost of the pseudo-affinity adsorbent, Procion blue HB-Sepharose CL 6B, has been exploited in the development of a production scale system for the recovery of human albumin from Cohn Fraction IV and other albumin containing source materials. Whilst the system has been established for the purification of albumin, other applications such as provision of albumin-depleted plasma fractions for use in the purification of α_1-antitrypsin and transferrin have followed.

The highly reproducible chromatographic characteristics of the adsorbent have allowed the process to be readily scaled-up through laboratory and pilot scale to the final production system. Optimisation of the coupled ligand concentration for the production scale adsorbent (40L bed volume) allowed a capacity of greater than 25mg albumin per ml adsorbent to be achieved without compromising product purity or process characteristics.

The system has been adapted to computer controlled automation which has permitted multiple cycles to be carried out without supervision. Processing conditions relevant to the industrial application of this full scale process eg. adsorbent regeneration, sterilization and pyrogen-free operation have been defined. The application of specific and non-specific albumin desorption conditions has further served to demonstrate the versatility of the system in that minor plasma proteins, eg. growth factors can be fractionated by a two-step elution procedure.

Introduction

In laboratory systems, the immobilised triazine dye Procion blue HB and its analogue Cibacron blue 3GA, have been widely used for both the purification and depletion of albumin from a number of sources including plasma, Cohn Fraction II and III supernatants and Cohn

Fraction IV precipitate (1). The latter, a by-product of cold ethanol fractionation, contains significant levels of albumin (~10% w/w) together with lesser quantities of a range of plasma proteins including transferrin and α_1-antitrypsin (2). The albumin in Fraction IV represents a net loss to the Cohn plasma fractionation process of between 10 and 15% (3).

Triazine dyes, offer considerable potential for large scale processing through their high capacity, high specificity, low cost and simple, readily automated, elution characteristics (4). A system for the efficient recovery of albumin from Fraction IV with a capacity sufficient to match the supply from the Cohn fractionation process has been developed. The provision of albumin depleted plasma source fractions has been a further application for triazine dye affinity chromatography (5).

Methods

1. Preparation and Filtration of Fraction IV Solution

Fraction IV paste, recovered during ethanol fractionation of human plasma (6), was suspended in 50mM Tris buffer, pH 8.3 at +2°C in the proportion 1kg paste to 4kg buffer. Following homogenisation, insoluble material was allowed to separate by flotation for 20 hrs at +2°C. Further clarification of Fraction IV solution was achieved either by depth filtration or by addition of ethanol (10% v/v).

2. Preparation and Elution of Procion blue-HB Sepharose 6BCL

Blue Sepharose in quantities up to 40L was prepared 'in-house' by the method of Atkinson and Harvey (7) and packed into a range of column sizes representing small (5, 13 and 150ml), pilot (14L) and full scale (40L bed volume 1000cm^2 XSA). These columns were equilibrated with 2 bed volumes (2BV) of 10mM sodium phosphate buffer, pH 7.5 containing 150 mM sodium chloride (PBS); a linear flow rate of 40 ml cm^{-1} hr^{-1} was maintained in this and subsequent stages. FrIV was loaded onto the adsorbent sufficient to saturate its capacity for albumin, non adsorbed protein was eluted with PBS (5BV); bound albumin was eluted using either 3M NaCl in 10mM sodium phosphate, pH 7.5 or 20mM sodium octanoate in PBS, pH 7.5. After 2 to 3 elution cycles the adsorbent was cleaned with 2 BV of 0.1M sodium hydroxide followed by extensive washing with pyrogen free (PF) water. With 14L and 40L blue-Sepharose columns, elution was accomplished using a micro-computer process control system as previously described (8).

Results & Discussion

1. Properties of the Blue-Sepharose Affinity Adsorbent

The pilot scale adsorbent (14L Bed Volume) was prepared at a coupled dye concentration of 2.4 mg/g adsorbent with a capacity for human albumin of ~18 mg/ml gel using 3M NaCl as eluent. A coupled dye concentration of 6 mg dye/g adsorbent was used for the full scale 40L column giving an improved capacity of ~27 mg albumin/ml adsorbent. Blue-Sepharose used for small scale experiments was the same as that for the large scale adsorbent.

Recovery of albumin could be optimised for yield and purity through the choice of eluent (Table 1); yields were a function of the selected eluent with only N-acetyl tryptophan being unacceptable. The non-specific eluent, 3M NaCl gave reasonable recovery of albumin but purity, routinely about 95%, was poor compared to the more specific eluents sodium salicylate, sodium octanoate (caprylate) and N-acetyl tryptophan. This was also reflected in the endogenous protease activity (using the broad specificity chromogenic substrate S2288) and, at least with salicylate and octanoate, in contaminant IgG levels (determined by radial immunodiffusion).

TABLE 1

Influence of Different Eluents on the Purity and Yield of Albumin Recovered from Procion blue Sepharose[1]

Eluent	3M NaCl	Salicylate (20mM)	Octanoate (20mM)	N-acetyl tryptophan (20mM)
Recovery (%)	81.5	100	88.0	47.8
Purity (FPLC)	95.4	97.0	98.6	99.4
Endogenous Protease activity (A_{405}/min/mg alb)	8×10^{-2}	1.3×10^{-2}	0.2×10^{-2}	1.2×10^{-2}
IgG content (%)	0.46	0.18	0.14	0.53

1. Column bed volume - 5 ml

Other factors influencing product yield and purity were also examined; among them, Fraction IV load volume, wash volume, eluent volume and effect of multiple elution cycles. The load volume was a function of albumin concentration in the Fraction IV solution and should be sufficient to saturate the column capacity. A close relationship between FrIV load volume and albumin recovery was evident although there was no influence on purity with $98 \pm 0.5\%$ (mean $\pm$ SD, n=5) for octanoate and $97 \pm 2.4\%$ (mean $\pm$ SD, n=5) for 3M NaCl eluted material. The latter, however, contained readily measurable protease activity (0.17-0.50 A_{405}/min/mg alb.) which was not detected in albumin prepared using the specific eluent. Increasing wash volume beyond the adopted five column volume wash did not affect the purity of eluted albumin although marginally lower recoveries observed with extended wash volumes suggested albumin 'leakage' from the column (data not shown). Recovery and purity of albumin were not substantially affected by eluent volume over the range 1-3.5 column bed volumes (e.g. for octanoate, rec. 82.8 $\pm$ 2.1%, purity 98.5 $\pm$ 1.0%, mean $\pm$ SD, n = 6); desorption of albumin in two column volumes was found to be suitable for process scale-up.

The relationship between column capacity and albumin purity and recovery was examined through a series of multiple elution cycles to determine the requirements for column regeneration. There was no change in capacity over 12 complete cycles, both recovery and purity of albumin were largely unaffected (91.6 $\pm$ 2.2% and 97.5 $\pm$ 0.8% respectively, mean $^+$ SD, n=12). No overall trend in IgG levels was evident however, protease activity (up to 6.6 x 10^{-2} A_{405}/min/mg protein) could be measured in the octanoate peak after 5 elution cycles. In subsequent large-scale process runs the column was regenerated/cleaned after not more than 5 complete cycles.

2. Process Scale-up and the Production System

Elution conditions determined in small-scale experiments were applied to pilot and large chromatographic systems. Scale-up utilised a 14L bed volume pilot column and ultimately a 40L column for the full-scale process. Given the highly reproducible characteristics of the adsorbent both stages of scale-up were readily adapted to microcomputer automated process control (8). The latter is presented schematically in Figure 1, which shows components of the control and monitoring systems for the chromatographic process. Conversion to automated process control has permitted continuous operation for up to 72 hours (5 complete elution cycles) with over 30 production runs carried out on the 40L column.

With scale-up of the chromatographic stage a number of operations associated with the overall production process were introduced. Briefly, these comprised of systems for the clarification of Fraction IV extract and finishing stages including desalting/concentration by ultrafiltration plus sterile filtration and filling operations.

Clarification of Fraction IV solution was probably the most significant factor limiting throughput in the full-scale process. Standard procedures based on asbestos depth filtration were both time consuming and manually intensive. Development of an alternative method based on the use of low concentrations of ethanol as a dispersant, has removed the need for a clarification stage by permitting the direct application of Fraction IV solution onto packed beds of Procion blue-HB Sepharose (9). A series of pilot and large scale chromatographic separations were carried out on the triazine dye adsorbent under micro computer control using asbestos filtered Fraction IV solution (no ethanol) or unfiltered Fraction IV solution in the presence of 10% ethanol (Table 2). Bound albumin was eluted using either 20mM sodium octanoate or 3M NaCl.

The presence of ethanol did not adversely affect adsorbent capacity, albumin recovery and purity with either eluent. With 3M NaCl as eluent the full scale column yielded 1,257 $\pm$ 78g (mean $\pm$ SD) in the absence of ethanol and 1152 $\pm$ 43g in the presence of 10%

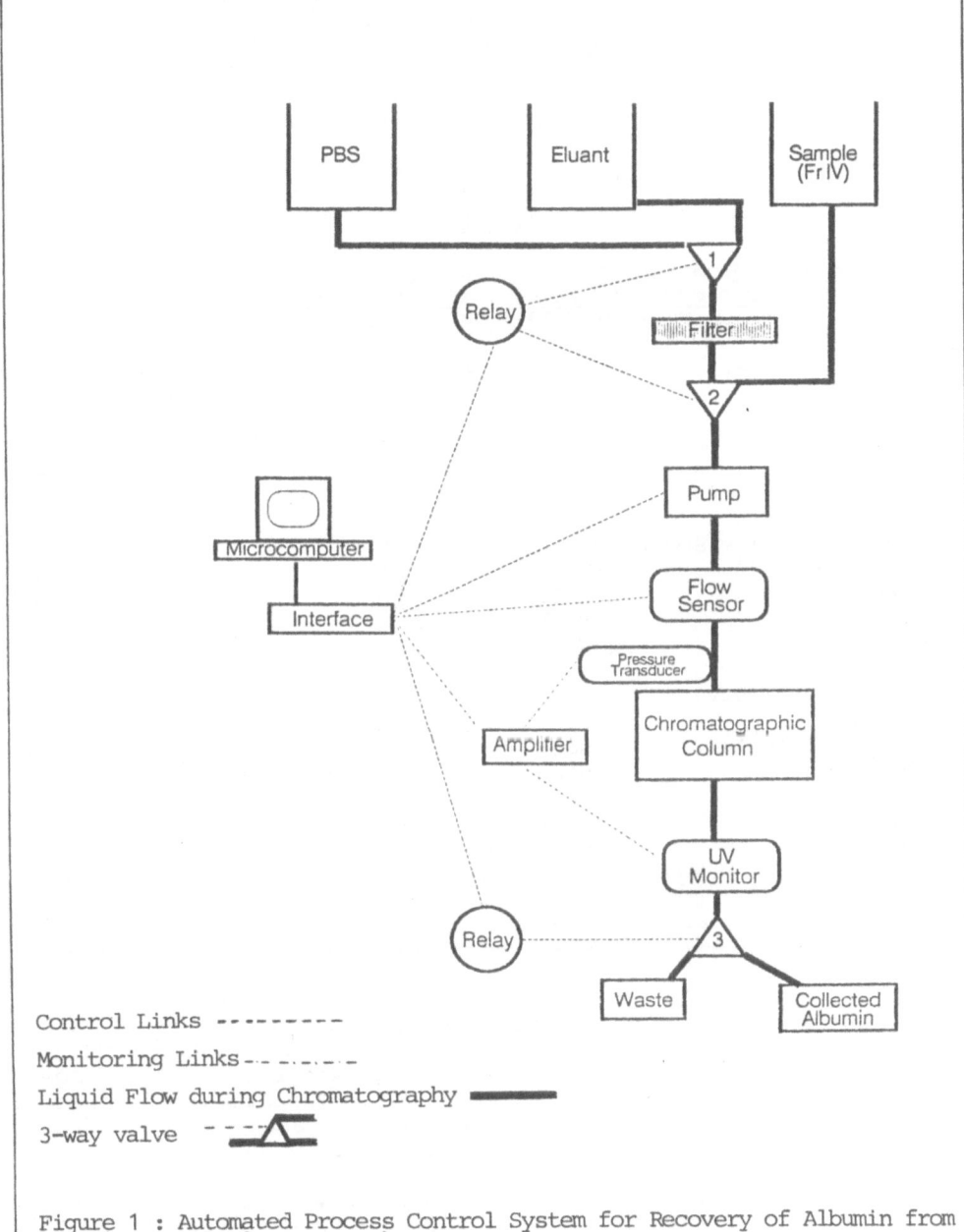

Control Links - - - - - - - - -
Monitoring Links - - - - - - - -
Liquid Flow during Chromatography ▬▬▬▬
3-way valve - - - - △

Figure 1 : Automated Process Control System for Recovery of Albumin from
Fraction IV using Procion blue HB Sepharose.

ethanol. Albumin purity was 94 $\pm$ 0.8% and 93.6 $\pm$ 1.2% respectively. While recovery using sodium octanoate was similar to that with 3M NaCl, purity was significantly improved at 98.7 $\pm$ 0.5%. The properties and structural integrity of the adsorbent were not compromised after multiple elution cycles demonstrating that the method can be readily used in conjunction with automated process control systems.

TABLE 2

Pilot and Large-Scale Purification of Albumin from Cohn Fraction IV in the Presence and Absence of Added Ethanol (10%) Using Procion Blue HB-Sepharose

Production Batch	Eluent	Albumin Recovery (g)	Purity (%)	IgG (%)
No Ethanol				
A/85/6A (pilot)	Sodium Octanoate	241	99.5	0.24
A/84/6B (pilot)	3M NaCl	286	99.0	N.D.
A/85/8B	3M NaCl	1356	94.0	0.43
A/85/9A	3M NaCl	1165	93.0	0.66
A/85/11B	3M NaCl	1250	95.0	0.76
+ 10% Ethanol (v/v)				
A/85/11A (pilot)	3M NaCl	311	94.4	0.76
A/85/12A	3M NaCl	1127	94.8	1.0
A/86/2B	3M NaCl	1120	92.5	0.65
A/86/3A	3M NaCl	1227	92.4	0.66
A/86/6A	3M NaCl	1136	94.8	0.43
A/86/4A	Sodium Octanoate	1082	98.5	0.24
A/86/5A	Sodium Octanoate	1173	99.3	0.07
A/86/7A	Sodium Octanoate	1180	98.2	0.07

Having removed the restraint of limiting quantities of source material through development of the ethanol dispersant method and with the introduction of product finishing stages, the full-scale production process could be established. Preparation of process documentation (batch records, SOPs etc.) completed the transfer of the system from development to the final manufacturing process.

Regeneration of Procion blue Sepharose was routinely carried out using 2 bed volumes of 0.1M NaOH (residence time 2hr), this treatment was able to both strip the adsorbent of non-specifically bound material and come some way towards sterilising the column, though it should not be assumed to be adequate to ensure complete inter-batch sanitation. Recognising the risk of HIV, Hepatitis B and NANBH transmission in plasma the stability of this adsorbent under conditions of steam sterilization was examined. With both the repacked sterilized adsorbent and the in situ NaOH treatment, extensive washing of the column with pyrogen free (PF) water was required to remove triazine dye released from the matrix. No loss in albumin capacity was evident after either treatment.

The estabished process has been used to prepare albumin solutions meeting B.P. specification. All removeable fittings and tubing were routinely steam sterilized prior to use, teflon valves and other fixed components were flushed with 0.1M NaOH, 0.4% w/v sodium hypochlorite and PF water before connection to the column. The adsorbent was washed with 2 bed volumes of 0.1M NaOH and cleared with 5 bed volumes of PF water prior to equilibration with freshly prepared buffer solutions. Under these conditions albumin solutions which met B.P. pyrogen test requirements (summed response in 3 rabbits <1.15°C) could be prepared.

3. Dye Leakage Studies

Perhaps the major limitation to the more widespread application of triazine dye affinity chromatography in the purification of 'biologicals' has been the dye leakage factor (10,11). The triazine dye adsorbent was prepared using [^{14}C] labelled Procion blue HB, coupled dye concentration was 4.5mg dye/g Sepharose 6B CL. Operating a 'continuous' mode of elution (4 days, 4 FrIV load cycles per day, 3 day PBS wash) over a period of 4 weeks, leakage of Procion blue HB decreased continuously over the series of runs. Dye leakage was not uniform within any given cycle, maximum loss being linked to Fraction IV load and albumin desorption stages (Fig.2).

Release of dye was highest for the run immediately after each 3 day PBS wash stage with dye levels 'tailing off' over the following 7 days. Overall leakage of dye was ~1.5 µg/day for Fraction IV elution cycles and ~1 µg/day for the PBS wash. Continuous flushing with PBS for 50 weeks caused the loss of 11.2 µg dye/week. Total leakage of dye over a 1½ year study was 618 µg or 1.4% of the initial coupled dye level; capacity of the adsorbent for albumin remained unchanged. Extended washing of the adsorbent with low ionic strength buffer or water increased dye leakage by at least a factor of 10 when compared to PBS.

4. Albumin depletion and development of differential elution conditions

In establishing the chromatographic process for the purification of albumin from Fraction IV, the converse situtation - albumin depletion - has merited consideration in view of interest in the other protein components. Efficient recovery of albumin from Fraction IV using Procion blue-Sepharose yields an unbound fraction substantially depleted of albumin which can be readily used as a source for other plasma proteins. Transferrin with a purity of at least 90% has been prepared from this fraction by ion exchange chromatography on Q-Sepharose. Purification of α_1 antitrypsin by triazine dye pseudo-affinity chromatography required the use of

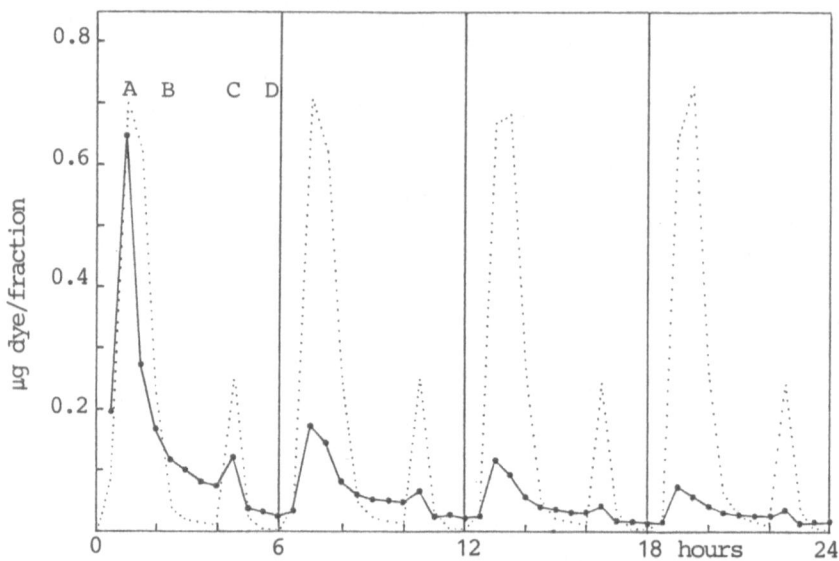

Figure 2 Leakage of dye from a [^{14}C] Procion blue HB-
Sepharose column over 4 consecutive elution
cycles.

NB. Continuous line - leakage of [^{14}C] dye;
dotted line - optical density of eluate.

A. Fr.IV load : 1.0hr   C. Albumin desorption : 0.6hr
B. Wash   : 2.7hr   D. Equilibration   : 1.7hr

'high dye' adsorbent with coupled dye levels of >25 mg/ml gel.
Elution under conditions of low ionic strength and pH yielded 80% of
the loaded α_1 antitrypsin with a purity of >60%.

Provison of enriched fractions of insulin like growth factor -
l/somatomedin-C (IGF-1/Sm-C) has been a further area of study.
Although an established source of this growth factor, Fraction IV
has proved difficult to utilize directly giving low yield after
extraction and with arduous and extensive purification schemes
generally being adopted (12).

Following the observation of very low or negligible IGF-1
levels in the sodium octanoate eluted albumin fraction as opposed to
high total levels in 3M NaCl eluted material, the notion of
differential elution of bound components was examined. Octanoate or
salicylate were used to elute albumin from Cohn Fraction IV loaded
Procion blue Sepharose (Table 3). Albumin recovered in each case
was of high purity and substantially depleted of IGF-I (<0.5 u/ml).

TABLE 3

IGF-1 Activity in Octanoate, Salicylate and 3M NaCl eluted fractions from Cohn Fraction IV loaded Procion blue Sepharose

Fraction	Protein (mg/ml)	IGF-1 (U/ml)	Potency	% IGF-1 loaded
Cohn Fraction IV solution	48.8	10.1	0.2	(100)
A. Octanoate peak (albumin)	14.5	0.3	0.02	1.5
3M NaCl peak	0.8	2.8	3.5	21.1
B. Salicylate peak (albumin)	8.8	0.3	0.03	1.5
3M NaCl peak	1.0	4.4	4.4	34.3

Subsequent elution with 3M NaCl displaced an IGF-1 enriched fraction (>2.8 u/ml) which was of significantly enhanced potency. This 15-20 fold purification stage provides a readily available enriched source of IGF-1 for purification by a multi-step chromatographic process (details not presented).

Conclusions

Triazine dyes, in the form of Procion blue Sepharose have potential in the purification of plasma proteins. This has been realised at production scale by the operation of a 40L column yielding 1.1kg albumin per 14 hour automatic cycle. The versatility of this system is demonstrated by the purification of transferrin, α_1 antitrypsin and IGF-1, the latter by differential elution, from albumin-depleted Fraction IV.

Acknowledgments

Our thanks to N. Thompson and C. Moon for technical assistance.

References

1.   Schroeder, D. D., 1979, UK Patent No. 1,540, 165.

2.   Heide, K., Haupt, H., and Schwick, H. G., In 'Plasma Proteins: Structure, Function and Genetic Control' ed. F. W. Putnam, Academic Press 1977, pp. 545 - 617, Vol III.

3.   Eriksson, S., Berglof, J. H., Hamalainen, E and Suomela, H., In 'Separation of Plasma Proteins' ed., J. M. Curling, Fine Chemicals, Uppsala, Sweden, Pharmacia, 1983, pp. 89 - 94.

4.   Harvey, M. J., Brown, R. A., Rott, J., Lloyd, D., and Lane, R. S., In 'Separation of Plasma Proteins', ed. J. M. Curling, Pharmacia, Fine Chemicals, Uppsala, Sweden, 1983, pp. 79 -88.

5.   Travis, J., and Parnell, R., Clin. Chem, Acta, 1973, **49**, 49-52.

6.   Kistler, P., and Nitschmann, H., Vox Sang., 1962, 7, 414 - 424.

7.   Atkinson, A., and Harvey,. M. J., 1979 UK Patent No. 2, 015, 53

8.   McFarland, C. D., Price, S., Brown, R. A., and Harvey, M. J.,
      Biochem Soc. Trans, 1984, 12 1098-1099.

9.   More, J. E., Hitchcock, A. G., Thompson, N. and Young, J. L.,
      Abstracts and Proceedings, 'Int. Symposium on Biotechnology of
      Plasma Proteins', Nancy, France, INSERM, 1988, p3.05.

10.   Harvey, M. J., In 'Methods in Plasma Protein Fractionation',
      ed. J. M. Curling, Academic Press, 1980, pp. 189-200.

11.   Clonis, Y. D., In 'Reactive Dyes in Protein and
      Enzyme Technology', ed. Y. D. Clonis, A. Atkinson, C. J.
      Bruton, and C. R. Lowe, MacMillan Press, London, 1987, pp 33-
      49.

12.   Morrell, D. J., Ray, K. P., Holder, A. T., Taylor, A. M.,
      Blows, J. A., Hill, D. J., Wallis, M., and Preece, M. A.,
      J. Endocrinol., 1986, **110**, 151-158.

THE INTERACTION OF HUMAN α-FETOPROTEIN WITH REACTIVE DYES AND THE ISOLATION OF THE PROTEIN BY AFFINITY CHROMATOGRAPHY

GERD BIRKENMEIER, KLAUS HUSE AND GERHARD KOPPERSCHLAEGER
Institute of Biochemistry, Karl-Marx-University,
7010 Leipzig, Liebigstrasse 16
German Democratic Republic

ABSTRACT

Human α-fetoprotein is the major serum protein in developing fetus. It is suggested to be involved in binding and transport of fatty acids and other small molecular weight compounds. The protein is of important clinical significance because of its diagnostic value as oncofetal protein. Isolation of AFP from biological fluids is encountered by the close relatedness to albumin in many physicochemical and functional aspects. However the two proteins were found to have different properties with respect to binding to reactive dyes. On this basis an effective purification procedure for AFP from fetal material employing dye-ligand chromatography was elaborated.

INTRODUCTION

Human α-fetoprotein (AFP) is the major serum protein of the fetal blood. In mammals AFP is produced by the yolk sac and the fetal liver. Its primary function is still unclear but suggestions that it may play a certain role in binding and transport of polyunsaturated long chain fatty acids, estrogens, bilirubin and retinoids became manifest (1,2). Furthermore, an immunosuppressive action on T-cell dependent immune reaction is discussed (3). Accumulated data suggest that AFP has properties similar to albumin especially with regards to its binding and transport function. The close relatedness between the two proteins can further be deduced from similar molecular weights, isoelectric points, amino acid sequence homology and from immunological cross reactivity (4).

The clinical significance of AFP emerges from its diagnostic value as oncofetal protein being synthesized by fetal organs and by different malignant tumours. Elevated levels of AFP above the normal average in fetal or maternal blood or in amniotic fluid are suggestive of pathological processes. The determination of AFP concentration in body fluids turned out to be a valuable diagnostic and prognostic tool in detection of abnormal gestation (e.g. anencephaly, spina bifida) and cancer diseases (hepatocellular cancer, germ cell cancer) (5,6). For studies of its biological function and for the elaboration of immunoassay procedures the availability of pure AFP in substantial quantities is essential. A special problem of purification arises from the close relatedness of AFP and albumin and from the large concentration difference between both proteins in biological fluids. In this report a simple procedure for the isolation of AFP involving dye affinity chromatography and ion exchange chromatography is described.

MATERIALS AND METHODS

Substitution of Sephadex G-100 with Cibacron Blue F3G-A was performed as described in (7). Quantitation of AFP and of albumin was accomplished by rocket immunoelectrophoresis (8). Detailed description of purification of AFP has been shown previously (9). Fetal material was obtained by legal surgical abortion before the 12th week of gestation.

Partition experiments were performed in aqueous two-phase systems composed of dextran T 500 or T 70, polyethylene glycol (PEG) 6000 including liganded PEG, protein and buffer of desired concentration, respectively. The strength of the ligand – protein interaction is characterized by the Δ log K-value referring to the difference in the logarithm of the partition coefficients, K, in the presence and the absence of liganded PEG (10).

TABLE 1

Effect of 0.5% dye-PEG on partition of AFP and albumin expressed in terms of Δ log K units. The two-phase systems were composed of 10% dextran T 500, 7.5% PEG 6000 (including 0.5% dye-PEG), 10 mM Na-phosphate, 20 mM NaCl, pH 6.0.

Dye-PEG derivatives	Δ log K at 0.5% dye-PEG	
	Albumin	α-Fetoprotein
Cibacron Blue F3G-A	2.91	0.73
Cibacron Brilliant Blue FBR-P	2.59	0.47
Procion Blue MX-3G	1.93	0.54
Procion Navy H-ER	1.70	1.12
Procion Blue MX-R	1.70	0.49
Procion Red HE-7B	1.36	0.65
Procion Red HE-3B	1.33	0.84
Procion Red H-3B	0.67	0

RESULTS AND DISCUSSION

AFP was found to display binding to different immobilized dyes
as judged from affinity partition experiments (Tabl. 1). The
overall affinity expressed in terms of Δ log K-values of AFP to
the respective dyes was less compared with albumin. The large
difference in binding to Cibacron Blue F3G-A favoured the
selection of this dye for affinity chromatography.

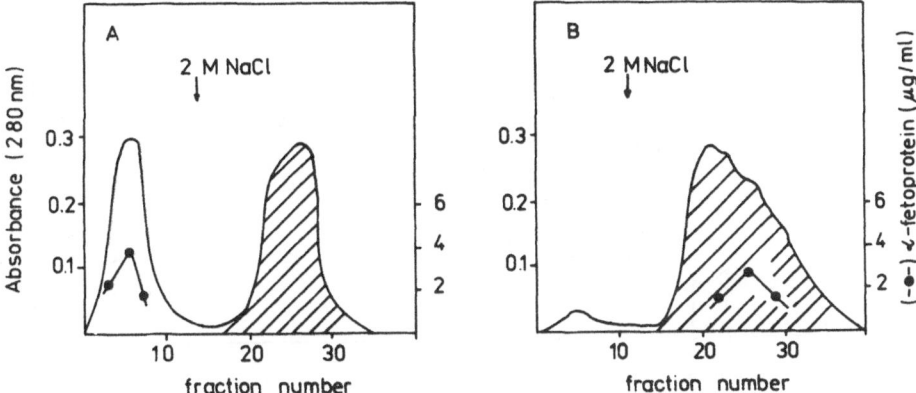

Figure 1.   Chromatography of human cord serum on differently
             substituted   Cibacron   Blue   F3G-A-Sephadex G-100
             gels.   The columns were equilibrated with 10 mM Na-
             phosphate,   pH   7.0.   After loading the sample   the
             material bound was eluted by 2 M NaCl (arrow).   The
             shadded area indicates the presence of albumin.   A,
             ligand concentration,   3 μmol/g; B, ligand concen-
                          tration 143 μmol/g.

Preliminary analytical studies revealed that the correct set of
dye ligand concentration of the gel was the key to successful
separation of AFP from albumin by affinity chromatography
(Fig. 1).   It turned out that only low substituted gels are
endowed with separating abilities.   At higher ligand concentration of the gel both AFP and albumin were strongly adsorbed to
the dye in addition to the host of serum proteins (11).

        As   source   of AFP we favoured fetal material obtained by
artificial abortion because of its availability and the high
level of AFP.   After dialysis against 10 mM Na-phosphate, pH
6.8, the starting material was subjected to DEAE-chromatography
to remove abounding hemoglobin and immunoglobulins (Fig.   2).
The pooled eluate obtained after salt gradient elution (0.04 –
0.2 M NaCl) was subsequently chromatographed twice on Cibacron
Blue F3G-A-Sephadex G-100 equilibrated with 20 mM Na-phosphate,
50 mM NaCl, pH 8.0. By setting the ligand concentration of the

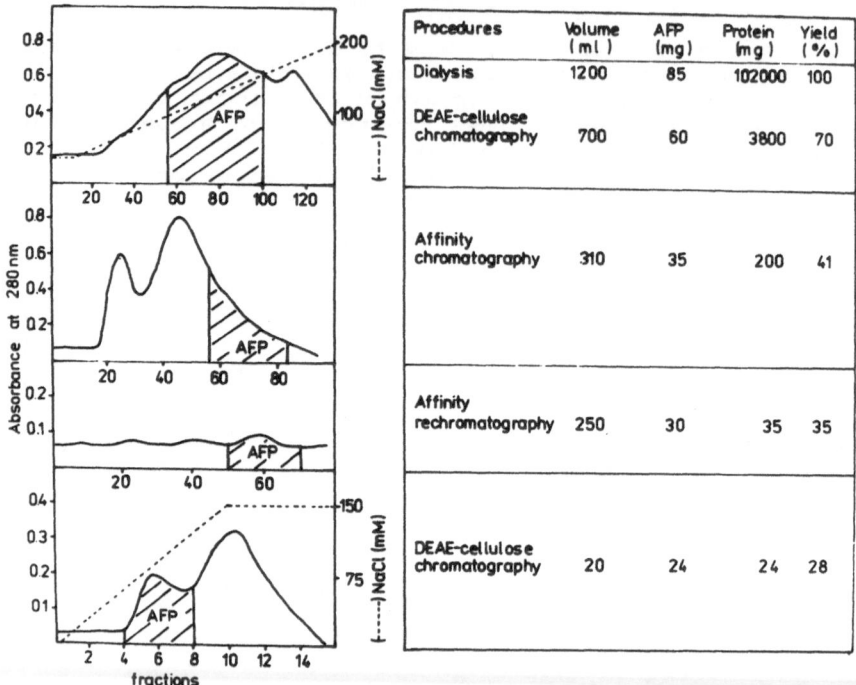

Procedures	Volume (ml)	AFP (mg)	Protein (mg)	Yield (%)
Dialysis	1200	85	102000	100
DEAE-cellulose chromatography	700	60	3800	70
Affinity chromatography	310	35	200	41
Affinity rechromatography	250	30	35	35
DEAE-cellulose chromatography	20	24	24	28

Figure 2. Elution pattern of human AFP on DEAE-cellulo
(a,d) and on Cibacron Blue F3G-A-Sephadex (b,c)

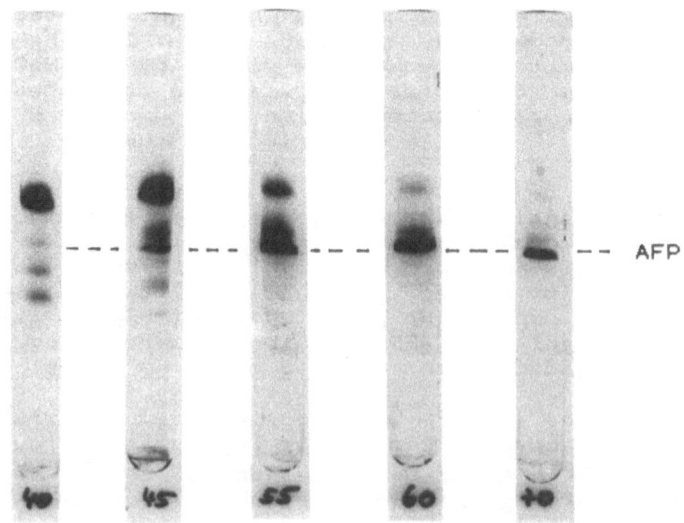

Figure 3. Polyacrylamide gel electrophoresis of fractic
resulting from affinity rechromatography on Cit
cron Blue F3G-A-Sephadex G-100 gel.

gel to about 100 μmol/g optimal conditions were created under which AFP was only loosely bound and thus separated from albumin and partially from the unbound fraction constituted by prealbumin, α-1-acid glycoprotein, α-1-proteinase inhibitor and transferrin (Fig. 3). The proper modulation of the ligand concentration and of the buffer composition was found to be a critical step with respect to the yield of the preparation and the purity of AFP. Minute contaminations of proteins and traces of liberated dye were removed by final ion exchange chromatography.

The final yield of the protein was about 25-30% of the starting material resulting in about 25-30 mg of pure AFP (Fig. 2). The procedure described permits isolation of AFP without using immunoadsorption or concanavalin A chromatography which was found to alterate the functional properties and natural microheterogeneity of AFP (12).

The purified AFP was found to be native with respect to its receptor-mediated endocytosis by hepatocytes and to the ability to bind long chain aliphatic compounds. The latter property was proved by affinity phase partitioning using PEG-bound fatty acids of different chain length (Fig. 4). The binding to the aliphatic hydrocarbons increased with increasing chain length reaching a maximum at PEG-myristate. In this respect the binding properties of AFP strongly resemble that of human albumin. These findings are in good agreement with the suggested function of AFP as carrier of fatty acids in the developing fetus (1).

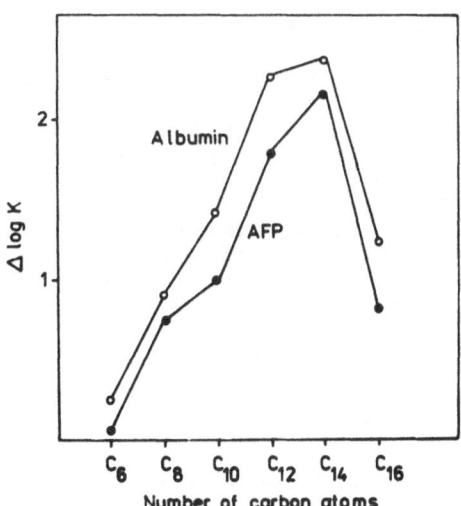

Figure 4. Effect of PEG-bound fatty acids of different chain length on partition of albumin and α-fetoprotein. The two-phase systems were composed of 8% dextran T 70, 8% PEG 6000 (including 2% of liganded PEG), 100 mM K-sulphate, 2 mM Na-phosphate, pH 7.1.

REFERENCES

1. Aussel, C. and Masseyeff, R., Human alpha-fetoprotein-fatty acid interaction. Biochem. Biophys. Res. Commun., 1983, 115, 38-45.

2. Aussel, C., Presence of three different binding sites for retinoid, bilirubin and estrogen or arachidonic acid on rat alpha-fetoprotein. Tumor Biol., 1985, 6, 179-193.

3. Murgita, R.A. and Tomasi, T.B., Suppression of the immune response by AFP. I. The effect of mouse AFP on the primary and secondary antibody response. J. Exp. Med., 1975, 141, 269-286.

4. Morinaga, T., Sakai, M., Wegmann, T. and Tamaoki, T., Primary structures of human α-fetoprotein and its mRNA. Proc. Natl. Acad. Sci. USA, 1983, 80, 4604-4608.

5. Seppala, M., Fetal pathophysiology of human α-fetoprotein. Ann. N.Y. Acad. Sci., 1975, 259, 59-73.

6. Tatarinov, Y.S., α-Fetoprotein in the laboratory testing for cancer. Gann, 1979, 18, 300-308.

7. Birkenmeier, G. and Kopperschlaeger, G., Application of dye-ligand chromatography to the isolation of α-1-proteinase inhibitor and α-1-acid glycoprotein. J. Chromatogr., 1982, 235, 237-248.

8. Laurell, C.-B., Quantitative estimation of proteins by electrophoresis in agarose. gel containing antibodies. Anal. Biochem., 1966, 15, 45-52.

9. Huse, K., Himmel, M., Birkenmeier, G., Bohla, M. and Kopperschlaeger, G., A novel purification procedure for human α-fetoprotein by application of immobilized Cibacron Blue F3G-A as affinity ligand. Clin. Chim. Acta, 1983, 133, 335-340.

10. Birkenmeier, G., Kopperschlaeger, G. and Johansson G., Separation and studies of serum proteins with aid of aqueous two-phase systems containing dyes as affinity ligands. Biomed. Chromatogr., 1986, 1, 64-77.

11. Birkenmeier, G., Usbeck, E., Saro, L. and Kopperschlaeger, G., Triazine dye binding of human α-fetoprotein and albumin. J. Chromatogr., 1983, 265, 27-35.

12. Calvo, M., Naval, J., Lampreave, F. and Pineiro, A., Pitfalls in the isolation of α-fetoprotein by solid-phase immunoadsorption. J. Chromatogr., 1985, 328, 392-395.

Chapter 7

Dye–Ligand Affinity Chromatography for the Purification of Plant Proteins

DOWNSTREAM PROCESSING OF POTATO PROTEINS

L. JERVIS
Department of Biology, Paisley College,
High Street, Paisley, PA1 2BE,
Renfrewshire, Scotland, U.K.

ABSTRACT

Plant proteins are used widely in the food industry, either for processing, conditioning or supplementing human and animal dietary materials. Many plants have considerable potential as sources of enzymes and other proteins for use in medicine or the production of fine chemicals and pharmaceuticals. A number of plants are suitable as host organisms for the production of recombinant healthcare proteins. Unlike plant proteins used in the food industry, the new range of enzymes and proteins will require considerable purification before use. Aspects of the problems of downstream processing of plant proteins are discussed, using potato tuber-specific proteins as examples.

INTRODUCTION

Plants are exploited widely by man. They provide an enormous variety of products ranging from food and fuel through to building materials and medicines. However, the use of plant proteins, although considerable in scale, is limited in range. The potential for expansion of this range of applications is substantial, thanks primarily to the biochemical versatility of plants.

Food Industry

The direct use of plant proteins on a large scale is limited to the food industry. Proteins, particularly from soya bean, are used widely to supplement both human and animal food (1). Soya protein is also used as an emulsifying agent (2). Plant enzymes such as papain, bromelain, ficin and α-amylase are used to modify and improve the texture and quality of many items of food and drink (3). They are also used to produce protein and starch hydrolysates (3). Downstream processing of proteins in this area is limited to those treatments needed to give an acceptable final product. Protein fractionation is not justified unless it gives a better, or an extra, product at little extra cost. The low cost of dye-ligand adsorbents and their potential for very specific protein binding makes their use in this respect very attractive.

Medicine

A few plant proteins, notably lectins, are used on a small scale in medicine and in associated high technology areas. Soya bean agglutinin for example has been used to remove mature T cells from human bone marrow and for the isolation and enrichment of haematopoietic stem cells (4). Plant toxins such as ricin and abrin have potential for use in cancer chemotherapy as immunotoxins (5). Plant proteases have been used for wound cleaning (6).

It is clear that plant proteins for use in medicine require extensive purification, particularly if they are intended for intra-venous or intra-muscular administration. Products for topical application may need less rigorous processing but the requirements of regulatory bodies must be met. The costs associated with extensive downstream processing can be acceptable for medical products - particularly where their efficacy is unquestionable. However, it is important to ensure that the processing of proteins for therapeutic use does not result in their inadvertent contamination with potentially hazardous chemicals. Concern has been expressed about the use of dye-ligand chromatography for therapeutic agents because of low level ligand leakage (7).

Biotransformations

Plants produce a large number of complex secondary metabolites, many of which are used as pharmaceuticals or colouring or flavouring agents (8,9, 10). The biosynthetic capabilities of plants make them potentially valuable sources of enzymes for use as industrial biocatalysts. Unfortunately, many of the most useful enzymes are present in vivo at very low levels. The cost of their isolation, purification (if necessary) and development to efficient biocatalyst stage is prohibitive, even for large volume or very high value products. In most cases it will not be possible to compete with traditional extraction methods, or with plant cell culture technology. However, the high level synthesis of plant enzymes as recombinant proteins could lower biocatalyst costs to accept-able levels, and open up much wider prospects for plant enzyme technology. Downstream processing requirements would be diverse, but complete purification should rarely be required. The elimination of enzymes that produce unwanted products should be sufficient.

Recombinant Proteins

The production of recombinant proteins, especially for therapeutic use, in bacterial hosts is subject to a number of problems that have been discussed widely (11,12,13). Some of these problems apply to downstream processing, especially where recombinant proteins are recovered as inclusion bodies (14-17). Alternative hosts such as yeast or cultured animals cells also have difficulties associated with their use (18,19). Plants have not yet been used for large-scale production of recombinant proteins, but rapid progress is being made. Most effort is directed towards the improvement of protein quality in staple food crops, or into increasing disease and pest resistance (20-23). However, the potential for producing therapeutic proteins or industrial biocatalysts is recognised (24,25). There are a number of plant-vector systems available that allow the expression of recombinant proteins, (26-28). If these can be used as the basis for the production of high levels of soluble recombinant proteins, downstream processing will be simplified and costs minimised.

Potato Proteins

Potatoes are cultivated widely on a large scale. Annual World production is in excess of 300 million metric tons, containing about 3 million metric tons of recoverable soluble protein. Processing technology is well developed, with the Dutch starch industry alone consuming about 1% of total World production. The soluble protein fraction produced during starch manufacture is often discharged as effluent but some is recovered for use in animal feed. This protein has the same nutritional value as whole egg protein and has excellent emulsifying properties (29,30). Potato tubers contain a set of abundant, tissue-specific proteins that together account for over 60% of the total soluble tuber protein (31). The most abundant fraction is patatin, a family of dimeric glycoproteins that have esterase activity (32). The other abundant proteins are protease inhibitors (33). Tubers also contain a lectin of potential value in medicine (34).

Potatoes are amenable to transformation by Agrobacterium vectors (26,27, 28). Both patatin genes and protease inhibitor genes have been isolated, cloned and sequenced (35,36). Chimaeric patatin genes have been used to transform both potato and tobacco (37,38), whole transgenic plants being regenerated from tissue slices in each case.

The abundance of certain tuber proteins simplifies their purification and both patatin and protease inhibitors can be purified by affinity chromatography (39,40), the latter on dye-ligand adsorbents. Both groups of proteins are attractive candidates for use as "affinity tails" to aid the purification of fused recombinant proteins. These features, combined with the ability of plants to glycosylate proteins and their lack of infective agents that are pathogenic to animals (41), make potatoes an attractive host for recombinant protein production.

Downstream Processing Problems

Potato tuber extracts present typical problems of plant protein recovery. They combine low concentrations of protein with high levels of phenolics and glycoalkaloids, (40,42). Removal of cell debris and other particulate material is a requirement if a fixed-bed adsorption step is to be undertaken at an early stage (43). Fluidized bed adsorption processes (44) can deal with large volumes of dilute protein solution containing suspended particulate material, but they require specialised adsorbents. Although they are inherently less efficient than fixed-bed adsorption (45), their potential advantages warrant further investigation. Other techniques that can deal with un-clarified extracts include precipitation with salts or other agents (46) and aqueous two-phase partitioning (47). These techniques are expensive for large volume processing and have associated waste disposal or solvent/polymer recovery problems. New, inexpensive, polyampholytic acrylic co-polymers have been used for two-phase protein partitioning (48). These should be economical for large-scale use. The technique is rapid and simple. It is suitable for scale-up and for continuous operation in countercurrent distribution systems. The latter are used widely in the pharmaceutical industry and are easily adapted. It is probable that two-phase aqueous partitioning will have a major impact on large-scale protein recovery within a short time. In the immediate future, however, adsorption processes offer the best solution to the problem of protein recovery from plant extracts.

These might be of fixed-bed or fluidised bed types but new methods of extract clarification such as borate affinity flocculation (48a) should ensure continued preference for the former.

FIGURE 1

POTATO TUBER PROTEIN PROCESS CYCLES

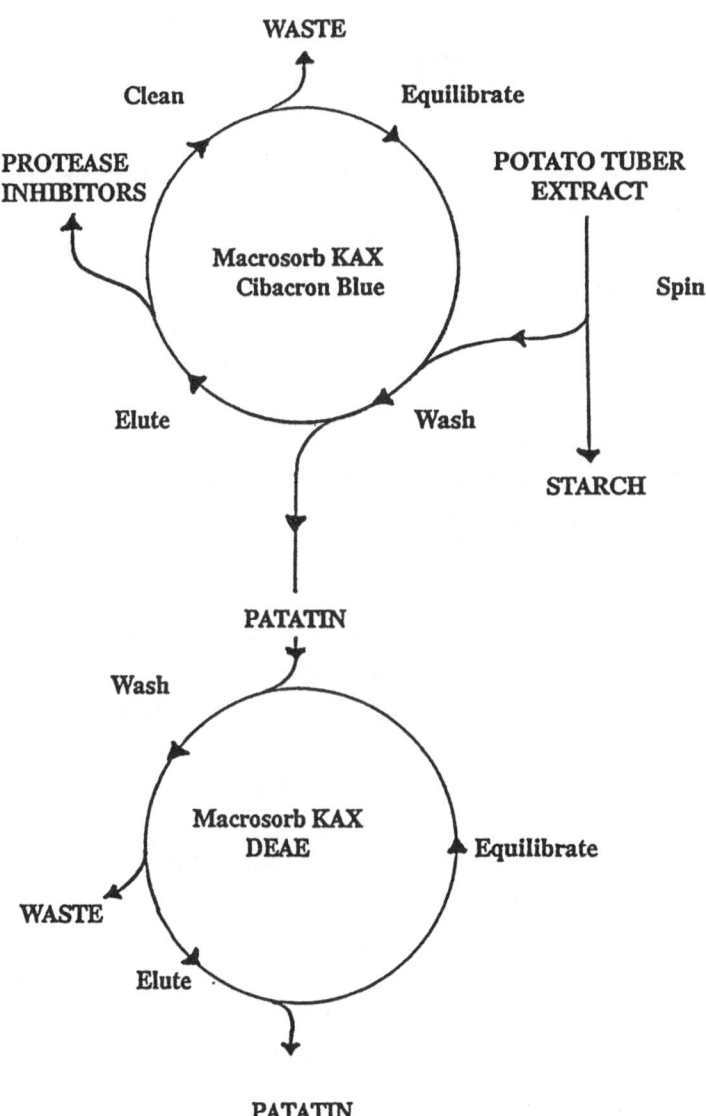

MATERIALS AND METHODS

Potato tubers were washed, peeled, diced and macerated in two volumes of 25mM sodium phosphate buffer, pH7.5 (buffer A). The brei was passed through a basket centrifuge lined with muslin to remove large cell debris. The resulting suspension was centrifuged at 10,000 x g for 10 minutes to remove particulate material. The clear supernatant was pumped through two fixed-bed chromatography columns (5cm x20cm each) connected in series. The first column was packed with Macrosorb KAX–Cibacron Blue (particle size 100–300 microns) and the second column was packed with Macrosorb KAX – DEAE (particle size 100–300 microns). Sample loading was carried out at a linear flow rate of 25cm. hr^{-1}. The effluent stream from each column was monitored at 254nm. An in-line three-way valve allowed sampling of the effluent stream from the first column. After loading, the columns were washed with buffer A at a LFR of 50cm. hr^{-1} until the absorbance at 254nm had fallen to its starting level. The columns were then disconnected and eluted separately, each at a LFR of 25cm. hr^{-1}. The Cibacron Blue column was eluted with buffer A containing 1.0M NaCl and the DEAE column was eluted with buffer A containing 0.3M NaCl. Samples of clarified crude extract, effluent streams and eluted materials were subjected to reducing SDS-PAGE.

The columns were cleaned in situ between each cycle with one column volume of 4M urea at a LFR of 100 cm. hr^{-1} and re-equilibrated with buffer A at the same flow rate. After every 10 cycles the columns were cleaned with 0.1M NaOH instead of 4M urea. When not in use they were stored in 20% ethanol. The complete process cycle is shown in Figure 1.

RESULTS AND DISCUSSION

Of the major tuber proteins, patatin did not bind to the dye-ligand adsorbent but the protease inhibitors did. Thus each group of proteins was substantially purified by this material. Salt elution of the column allowed recovery of the protease inhibitors, which could be desalted and purified from higher molecular mass proteins by gel filtration through Sephacryl S-200 HR. The patatin did bind to the DEAE column and was eluted with 0.3M NaCl. It was then resolved easily from minor contaminants by lectin affinity chromatography on concanavalin A-agarose (39).

The complete process cycle time with a 1 litre sample volume was
approximately 5 hours. The limiting factor was the maximum speed of the
available peristaltic pump. Smaller columns have been operated at twice
the linear flow rates described here without affecting the efficiency of
the process. The principal disadvantage of the Macrosorb materials was
their relatively large and polydisperse particles. Inefficient packing
of such particles gives rise to non-ideal chromatographic performance
(49). The overall effect was that chromatographic beds of Macrosorb
needed to be about twice the volume of corresponding Sepharose FF beds
to process a given amount of protein solution. However, we have found
that, with plant extracts, the Macrosorb materials are less affected by
polyphenolics than Sepharose FF. The latter retains dark brown material
very strongly, even after chemical scrubbing. This affects adsorbent
capacity to a significant extent and its advantage over Macrosorb in
this respect is soon lost. The Macrosorb materials can also be used in
fluidised beds, due to their high density, and this should allow their
use with unclarified extracts (50). If this mode of adsorption is
efficient with plant extracts, materials of the Macrosorb type may allow
the cost-effective recovery of many plant proteins.

Dye Ligand Leakage

There was no detectable loss of dye molecules from the Macrosorb KAX-
Cibacron Blue, even after NaOH scrubbing. Where plant proteins are
intended for food or drug use, it will be essential to demonstrate that
ligand leakage does not occur if a purification process is to be
licensed (43). If it is not possible to prevent ligand leakage
completely, (7), then efficient methods of removing contaminating dye
from protein preparations must be developed. We have found that
Cibacron Blue binds very tightly to highly cross-linked Sephadex such
as G10 or G25, and to dialysis membrane. These observations could form
the basis of simple and efficient methods of dye removal from
contaminated proteins.

CONCLUSIONS

Although plants contain many proteins of considerable commercial potential,
few of these are fully exploited. More efficient methods of plant protein
recovery and purification are required. These methods must be inexpensive
if the recovered proteins are to find a market.

Potato tubers contain several abundant proteins that are used to a limited extent in the food processing industry. These proteins have a wider potential market as protease inhibitors and biocatalysts. They are easy to purify by dye-ligand chromatography and by lectin affinity chromatography. Their ease of purification by these techniques gives them potential for use as a basis for the large-scale production of recombinant proteins. Potatoes can be transformed efficiently using Agrobacterium vectors containing heterologous genes that can be expressed at a high level in tubers.

Plants, such a potatoes, could form the basis of a new, biotechnology-based, agricultural industry in which recombinant proteins and other high value products are produced on agricultural land that is surplus to requirements for food production. Such an industry must be backed-up by well developed downstream processing technology.

REFERENCES

1. Whitaker, J.R., Covalent attachment of essential amino acids to proteins to improve their nutritional and functional properties. In, Protein Tailoring for Food and Medical Uses, R.E. Feeney and J.R. Whitaker, Eds., Marcel Dekker Inc., New York, 1986, pp 41–74.

2. Holm, F. and Erikssen, S., Emulsifying properties of undernatured potato protein concentrate. J.Fd. Technol., 1980, 15, 71–83.

3. Adler-Nissen, J., Enzymic Hydrolysis of Food Proteins, Elsevier Applied Science Publishers, London, 1985.

4. Sharon, N., Biomedical aspects of lectins. Biochem. Soc. Trans. 1988, in the press.

5. Vitetta, E.S., Fulton, R.J., May, R.D., Till, M. and Uhr, J.W. Redesigning Nature's poisons to create anti-tumor reagents. Science, 1987, 238, 1098–1104.

6. Bickerstaff, G.F., Enzymes in industry and medicine, Edward Arnold Ltd., London, 1987.

7. Jervis, L., Polymers in affinity chromatography. In Synthesis and Separations using Functional Polymers, Eds. P. Hodge and D.C. Sherrington, John Wiley and Sons Ltd., Chichester, 1988, pp. 265–304.

8. Fowler, M.W., Process possibilities for plant cell cultures. In Plant and Animal Cells. Process Possibilities. Eds. C.Webb and F. Mavituna, Ellis Horwood Ltd., Chichester, 1987, pp. 21–32.

9. Holden, M.A., Hall, R.D., Lindsey, K. and Yeoman, M.M.,
    Capsaicin biosynthesis in cell cultures of Capsicum frutescens. In
    Plant and Animal Cells : Process Possibilities. Eds. C. Webb and
    F. Mavituna, Ellis Horwood Ltd., Chichester, 1987, pp.45-63.

10. Anderson, L.A., Phillipson, J.D. and Roberts, M.F., Alkaloid
    production by plant cells. In Plant and Animal Cells : Process
    Possibilities. Eds. C.Webb and F. Mavituna, Ellis Horwood Ltd.,
    Chichester, 1987, pp. 172-192.

11. Berman, P.W. and Lasky, L.A., Engineering glycoproteins for use as
    pharmaceuticals. Trends in Biotechnol. 1985, 3, 51-53.

12. Bebbington, C. and Hentschel, C., The expression of recombinant
    DNA products in mammalian cells. Trends in Biotechnol. 1985, 3,
    314-317.

13. Kingsman, S.M., Kingsman, A.J., Dobson, M.J., Mellor, J. and
    Roberts, N.A., Heterologous gene expression in Saccharomyces
    cerevisiae. Biotechnol. and Gen. Eng. Revs. 1985, 3, 377-416.

14. Hartley, D.L. and Kane, J.F., Properties of inclusion bodies from
    recombinant Escherichia coli. BioChem. Soc. Trans., 1988, 16,
    101-102.

15. Fish, N.M. and Hoare, M., Recovery of protein inclusion bodies.
    Biochem. Soc. Trans., 1988, 16, 102-104.

16. Uhlen, M., Moks, T. and Abrahamsèn, L., Protein engineering to
    optimise recombinant protein purification. Biochem. Soc. Trans.,
    1988, 16, 111-112.

17. Marston, F.A.O., Angal, S., Lowe, P.A., Chan, M. and Hill, C.R.,
    Scale-up of the recovery and reactiviation of recombinant proteins.
    Biochem. Soc. Trans., 1988, 16, 112-115.

18. Kingsman, S.M., Kingsman, A.J. and Mellor, J., The production of
    mammalian proteins in Saccharomyces cerevisiae. Trends in
    Biotechnol., 1987, 5, 53-57.

19. Ramabhadran, T.V., Products from genetically engineered mammalian
    cells : benefits and risk factors., Trends in Biotechnol., 1987,
    5, 175-178.

20. Bryant, J.A., Transgenic plants with agriculturally important
    traits., Trends in Biotechnol., 1987, 5, 240-241.

21. Harrison, B.D., Mayo, M.A. and Baulcombe, D.C., Virus resistance
    in transgenic plants that express cucumber mosiac virus satellite
    RNA., Nature, 1987, 328, 799-802.

22.   Gerlach, W.L., Llewellyn, D. and Hareloff, J.,   Construction of
      a plant disease resistance gene from the satellite RNA of
      tobacco ringspot virus., Nature, 1987, 328, 802–805.

23.   Hilder, V.A., Gatehouse, A.M.R., Sheerman, S.E., Barker, R.F.
      and Boulter, D.,   A novel mechanism of insect resistance
      engineered into tobacco., Nature, 1987, 300, 160–163

24.   Long, E., Potatoes grow into medicine factories., The Sunday
      Times, November 1, 1987.

25.   Anon., Protein levels could get a boost. Farming News, 9
      October, 1987.

26.   Schell, J., Van Montagu, M., Hernalstreens, J.P., De Greve, H.,
      Leemans, J., Konez, C., Willmitzer, L., Otten, L. and Schroder,
      G., Gene vectors for higher plants. Beltsville Symp. Agric. Res.,
      1983, 7, 197–213.

27.   Ooms, G., Bossen, M.E., Burrell, M.M. and Karp, A., Genetic
      manipulation in potato with Agrobacterium rhizogenes., Potato
      Research, 1986, 29, 367–379.

28.   Shah, D.M., Tumer, N.E., Fischhoff, D.A., Horsch, R.B., Rogers,
      S.G., Fraley, R.T., and Jaworski, E.G.,   The introduction and
      expression of foreign genes in plants. Biotechnol and Gen. Eng.
      Revs., 1987, 5, 81–106. Lebensm.-Wiss.u.-Technol., 1978, 11, 109–115.

29.   Knorr, D., Protein quality of the potato and potato protein
      concentrates. Lebensm.-Wiss.u.-Technol., 1978, 11, 109–115.

30.   De Noord, K.G., Recovery of protein in potato starch manufacture.
      In Chemical Engineering in a Changing World, ed W.T. Koetsier,
      Elsevier Scientific Publishing Co., Amsterdam, 1976, pp 199–214.

31.   Park, W.D., Potato tuber proteins as molecular probes for
      tuberisation., Hort.Science, 1984, 19, 37–40.

32.   Racusen, D., Lipid acyl hydrolase of patatin. Can. J.Bot.,
      1984, 62, 1640–1644.

33.   Ryan, C.A. and Hass, G.M., Structural, evolutionary and
      nutritional properties of proteinase inhibitors from potatoes.
      In, Antinutrients and Natural Toxicants in Foods, R.L.Ory (Ed.),
      Food and Nutrition Press Inc., Westport CT, 1981, pp 169–185.

34.   Desai, N.N. and Allen, A.K., The purification of potato lectin
      by affinity chromatography on an N,N!., N" - triacetylchitotriose
      - Sepharose matrix. Anal. Biochem., 1979, 93, 88–90.

35.   Bevan, M., Barker, R., Goldsbrough, A., Jarvis, M., Kavanagh,
      T. and Iturriaga, G., The structure and transcription start site
      of a major potato tuber protein gene, Nucleic Acids Research,
      1986, 14 4625–4638

36. Sanchez-Serrano, J., Schmidt, R. Schell, J. and Willmitzer, L., Nucleotide sequence of proteinase inhibitor II encoding cDNA of potato (Solanum tuberosum) and its mode of expression. Mol. Gen. Genet., 1986, 203, 15-20.

37. Jefferson, R.A. and Bevan, M., Regulated expression of a chimeric patatin-glucuronidase fusion in tubers and induced internode cuttings of a transformed potato. J. Cell Biochem. Suppl. 0, 1987 11, 57.

38. Rosahl, S., Schell, J. and Willmitzer, L., Expression of a tuber-specific storage protein in transgenic tobacco plants : demonstration of an esterase activity. EMBO Journal, 1987, 6, 1155-1159.

39. Racusen, D. and Foote, M., A major soluble glycoprotein of potato tubers. J. Food Biochem. 1980, 4, 43-52.

40. Jervis, L. and Robertson, E.R. Multiple purification of plant enzymes. In Plant and Animal Cells : Process Possbilities. Eds. C. Webb and F. Mavituna, Ellis Horwood Ltd., Chichester, 1987, pp. 216-228.

41. Cartwright, T., Isolation and purification of products from animal cells. Trends in Biotechnol., 1987, 5, 25-30.

42. Woolfe, J.A., The Potato in the Human Diet, Cambridge University Press, 1987.

43. Chase, H.A. Purification of protein products from cell culture by fixed-bed adsorption. In, Plant and Animal Cells : Process Possibilities, C.Webb and F. Mavituna, Eds. Ellis Horwood Ltd., Chichester, 1987, pp. 205-215.

44. Burns, M.A. and Graves, D.J. Continuous affinity chromatography using a magnetically stabilized fluidized bed. Biotechnology Progress, 1985, 1, 95-103.

45. Buijs, A. and Wesselingh, J.A. Batch fluidized ion-exchange column for streams containing suspended particles. J. Crhomatogr. 1980, 201, 319-327.

46. Scopes, R. Protein Purification : Principles and Practice. Springer-Verlag, New York, 1982.

47. Hustedt, H., Kroner, K.H., Menge, U. and Kula, M.-R. Protein recovery using two-phase systems. Trends in Biotechnol. 1985, 3, 139-143.

48. Hughes, P. and Lowe, C.R. Purification of proteins by aqueous two-phase partition in novel acrylic co-polymer systems. Enzyme Microb. Technol. 1988, 10, 115-122.

48(a) Bonnerjea, J., Jackson, J., Hoare, M. and Dunnill, P., Affinity flocculation of yeast cell debris by carbohydrate-specific compounds. Enzyme Microb. Technol., 1988, 10, 357-360, 1988.

49. Janson, J.-C. and Hedman, P., Large-scale chromatography of
     proteins. Adv. Biochem. Eng. 1982, 25, 43-99.

50. Bite, M.G., The use of "Macrosorb" Kieselguhr-agarose composite
     media in biotechnological extraction processes. Eurochem '86 -
     Process Engineering Today, 1986, pp. 137-142.

ACKNOWLEDGEMENTS

The author wishes to thank Dr. M.A. Parrish of Sterling Organics Ltd.,
Newcastle-upon-Tyne, U.K. for generous gifts of Macrosorb chromatographic
materials.

A NEW PROCEDURE FOR PURIFICATION OF POP CORN INHIBITOR BASED ON IMMOBILIZED DYE CHROMATOGRAPHY

ETIENNE ALGIMAN, YOLANDE KROVIARSKI, SYLVIE COCHET, ALAIN TRUSKOLASKI°,
PIERRE BOIVIN and OLIVIER BERTRAND
INSERM U 160, and U 24 (°) Hopital Beaujon, 92118 Clichy Cedex, France

ABSTRACT

A new purification procedure for purification of Pop Corn Inhibitor, the trypsin and activated factor XII inhibitor present in corn, is presented. It is shown that Polenta is a more convenient starting material than the more commonly used sweet corn kernels. Complete purification of the inhibitor is obtained through use of one chromatography step on Procion red HE-3B immobilized on agarose followed by a reverse phase chromatography step. Advantages of this procedure compared to those using immobilized trypsin or immobilized anhydro trypsin are discussed.

INTRODUCTION

Pop Corn Inhibitor (PCI) is a trypsin inhibitor found in sweet corn kernels (1). It was proven to be also a very specific inhibitor for activated Factor XII devoid of inhibitory effect on kallikrein another enzyme of the contact activation system of the coagulation cascade (-2- this allows to use this inhibitor to assay independently the two enzymes present in a mixture eventhough it does not exist a chromogenic substrate specific for òne single enzyme -3-)

Published purification procedures to date relied upon either the association of several non specific chromatographic techniques (ion exchange gel filtration ...-2-) or upon affinity chromatography on immobilized trypsin (1, 4). The drawback of this latter procedure lies in the fact that a significant portion of the inhibitor is produced in a nicked state (-1-2,4-6,- nicked inhibitor is only a poor inhibitor for Factor XIIa -2-). Use of an inactive derivative of trypsin like anhydro trypsin has been proposed -2- but this derivative (which is commercially available, but at a very high cost) is rather difficult to prepare (7).

Immobilized Procion Red HE-3B has been shown to be able to retain trypsin inhibitors from extracts prepared from various plant sources (8). Based on this finding we have devised a new procedure for purification of PCI which associates :
- Extraction of PCI from Polenta, a non conventionnal starting material which was found more convenient to use than corn seeds.
- Chromatography of the extract on Procion Red HE3B agarose, and elution of the PCI by Urea containing buffer.
- Final chromatography step on a reverse phase column.

MATERIALS AND METHODS

Procion Red HE 3B was a much appreciated gift of I.C.I. France (Clamart). Dye was immobilized onto Sepharose CL 4B (Pharmacia Uppsala Sweden) according to established methods as detailed in (9). Level of dye incorporation was found by an acid hydrolysis method (10) to be 1.0 μmole of dye per milliliter of settled gel. Chemicals were bought from Merck (Darmstadt FRG) or Sigma (St Louis, MO).

Sweet corn kernels (Rogers Seeds, Jubilee Brand) were bought from Caillard (Les Ponts de Cé, France). Sample of Opaque 2 strain was obtained through the courtesy of Dr. Pollacsek (INRA Clermont Ferrand France). Polenta, an Italian traditionnal meal was bought from local groceries several brands were tested results presented below have been obtained with a preparation processed by Molino Boccardi (Morano Po, Italy).
A sweet corn flour of unknown origin was obtained from a local grocery.

Measurements of inhibitory activity towards trypsin. Inhibitory activity of PCI towards trypsin was measured by incubating 10 mn at 37° C serial dilutions of samples with a constant amount of trypsin in 50 mM sodium acetate buffer pH 5.4 containing 20 mM $CaCl_2$ and assaying thereafter residual trypsin activity using Nα-Benzoyl-L-Arginine-p- nitroanilide as a substrate (buffer conditions for inhibitory assays are stringent, hence sample buffer was exchanged before assay by gel filtration). One inhibitory Unit (Inh.U.) is defined below as the amount of PCI inhibiting one International Unit of trypsin.

Protein assays. Protein assays have been made according to Bradford (11) or for purified material by absorbance measurement at 280 nm using $E_{1\ cm}^{1\%}$ of 20 (1).

Preparation of the extracts. For the trials on the small scale, 10 g of the sweet corn kernels were finely ground in a household coffee grinder, suspended in 40 ml of buffer A (10 mM KOH adjusted to pH 6.5 with solid MES containing 20 mM NaCl and 2mM $MgCl_2$, -12-) and 8 ml of toluene and let shake overnight in the cold room, (10 g of Polenta and corn flour were treated likewise but without grinding). Aqueous phases were then adequately clarified (see below). Preparations of Polenta extracts on a larger scale were made similarly with the same ratios of extracting buffer and toluene to dry weight. 750 g were usually processed at a time.

RESULTS AND DISCUSSION

Choice of a convenient starting material for PCI purification. Results are shown in table 1. Results obtained with the two corn kernels samples showed that amount of extracted activity was much higher with the Opaque 2 mutant. This result is similar to those described by others (13) and would suggest to use opaque 2 mutant as a rich source for PCI purification. Unfortunately this mutant corn is of limited availability. Worth to notice is the fact that with Jubilee brand corn kernels, one centrifugation step followed by two filtration steps on paper were necessary for adequate clarification of the extract while the other extracts were clarified through one single filtration step. Polenta extraction gave the highest PCI specific activity and a reasonable amount of PCI per gram of dry product. Moreover Polenta did not need grinding before extraction, was easily available at a low cost, furthermore, extract was easily clarified. Hence Polenta was preferred as a starting material for PCI purification.

TABLE 1

Results obtained with various starting materials which can be used for PCI purification

Extracted material	Volume of clarified extract (ml)	Protein conc. (mg/ml)	Activity (Inh.U/ml)	Specific activity (Inh.U/mg)	Total extracted activity (Inh U)
Sweet corn kernels Jubilee brand	11	1.13	0.082	0.072	0.92
Sweet corn Kernels Opaque 2 strain	18	1.57	1.187	1.119	3.36
Sweet corn flour	18	0.30	0.034	0.113	0.61
Polenta	18	0.28	0.043	0.153	0.77

Chromatography of Polenta extract on immobilized Procion Red HE 3B.
This chromatography step is conducted in an automatic machine assembled in a cold room from easily available parts, similar to one described before (15). 750 ml of filtered extract prepared as was said before are loaded at 330 ml/h on the immobilized dye column (5 cm internal diameter, length 10 cm) equilibrated in buffer A. Thereafter the column is rinsed for 250 minutes with Buffer B (10 mM KOH adjusted to pH 6.5 by solid MES containing 2M NaCl). Then flow rate is automatically reduced to 240 ml/h and column developed for 300 minutes with Buffer C (of same composition as Buffer B but containing also 6M Urea —flow rate reduction is made necessary by higher viscosity of Buffer C—). Column effluent is at the same time diverted to a fraction collector. Thereafter column will be reequilibrated with Buffer A (at 330 ml/h) and a new purification cycle will begin. Machine works cyclically as long as volume in extract

reservoir does not fall under 750ml.
Photography of the recorder trace of the immobilized dye chromatography
step is shown in figure 1.

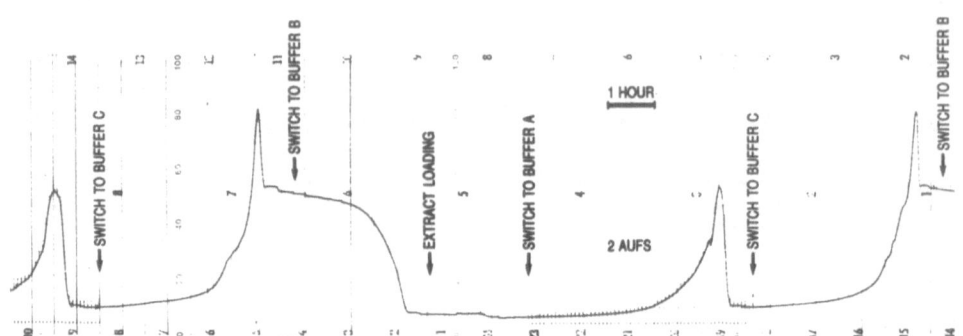

Figure 1. Recorder trace of the automated immobilized dye chromatography

Reverse phase chromatography :
Fractions corresponding to the optical density peak eluted from the
immobilized dye column with buffer C are pooled filtered through a 0.2
micrometer filter and loaded at a flow rate of 10 ml/mn into a reverse
phase column (Magnum 20 ODS 3 -Whatman Clifton NJ- 2.2 cm internal
diameter, lenth 50 cm) equilibrated in 0.1 % Trifluoroacetic acid in
water (solvent A). After completion of loading (500 ml of partially
purified PCI can be loaded at a time, this volume corresponding roughly
to the output of two consecutive cycles of Procion Red HE 3B
chromatography), column is rinsed for 20 minutes with solvent A and then
developped at 5 ml/mn by a gradient between solvent A and solvent B (0.1%
Trifluoroacetic acid in water in 50% water acetonitrile mixture -v:v- ; 0
to 60% B in 5 mn, then 5 to 100% B in 120 mn).
Photography of the recorder trace of the reverse phase chromatography
step is shown in figure 2, pooled fractions corresponding to pure
inhibitor are shown by a double arrow.

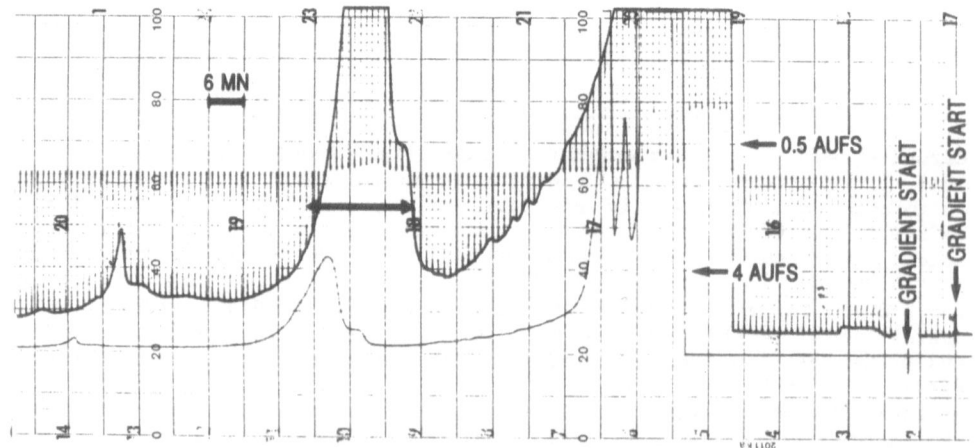

Figure 2. Reverse phase HPLC for purification of PCI.

Some comments have to be done on quantitative results of the purification procedure shown in table 2 :

TABLE 2
Quantitative results of one typical purification procedure
of PCI from Polenta.

	Vol.	Protein conc. (mg/ml)	Activity (Inh U/ml)	Specific activity (Inh U/mg)	Total activity (Inh U)	Yield (%)	Purif. factor (fold)
Polenta extract	1500	0.57	0.078	0.137	117	100	1
Red HE-3B agarose	470	0.23	0.209	0.92	98.3	84	6.7
R.P.L.C.	90	0.31	0.975	3.18	87.8	75	23.2

A greater amount of PCI is routinely extracted from polenta when extraction is done on a large scale, probably because of a more efficient shaking (compare with results shown in table 1). Purification factor obtained through the immobilized dye chromatography step is lower than the one obtained by an affinity chromatography step on immobilized trypsin (6). However this chromatography step on immobilized dye allows to get rid of some unwanted proteins as well as of UV absorbing non proteic components which preclude the use of reverse phase HPLC as a single step procedure for PCI purification. Anyway one single affinity chromatography step on immobilized trypsin does not allow to produce PCI of satisfactory purity and has to be associated to a further HPLC step (4-6). An obvious drawback of trypsin agarose chromatography is partial proteolysis of PCI during the chromatography.
Indeed elution conditions from the immobilized dye column as well as mobile phases used for the HPLC step are rather harsh, fortunately PCI does resist very satisfactorily to such treatments (as do also numerous other proteinase inhibitors -15-) : activity of PCI is fully restored after exposure to urea or acidic conditions after buffer exchange.
The optical density peak eluted from the reverse phase column, is not symetrical, nevertheless SDS gel electrophoresis does not reveal any differences in migration between aliquotes of first and last fractions of the peak. Micro heterogeneity of PCI has been demonstrated (5) and offers a likely explanation to the observed shape of the peak.
Final product of the described purification procedure shows only one band on a polyacrylamide SDS gel under reducing conditions (see figure 3). It is the fully active virgin form of PCI.

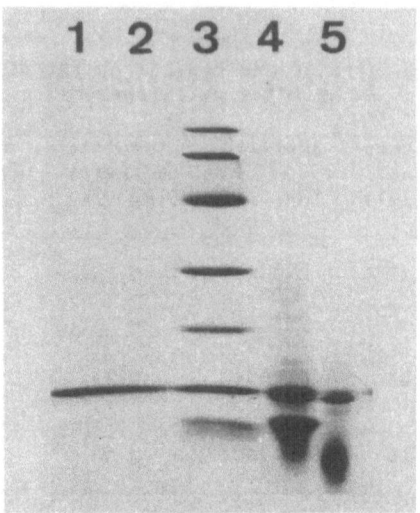

Figure 3 : SDS polyacrylamide 10 to 33 % gradient gel. Lanes 1 and 2 purified PCI, lane 3 Molexular weight markers (from top to bottom 94, 67, 43, 30, 20.1, 14.4, and 6.6 kD), lane 4 Polenta extract, lane 5 partially purified PCI obtained by immobilized dye chromatography.

CONCLUSIONS

This new purification procedure seems to us to give another exemple of the usefulness of immobilized dyes for protein purification. Immobilized Procion Red HE-3B was seen able to replace a not totally satisfactory true affinity chromatography media (immobilized trypsin) or a very expensive one (immobilized anhydro trypsin). Low cost and stability of the immobilized dye were also advantageous. Easy automation of the immobilized dye chromatography step did allow us to produce more than 100 milligrams of pure PCI each week using relatively small scale columns and with a minimal demand on worktime.

ACKNOWLEDGEMENTS

Sincere thanks are expressed by the authors to I.C.I. France for the generous gift of dyes and to Maurice Pollacsek for making available the Opaque 2 corn kernels.

REFERENCES

1. Swartz, M.J., Mitchell, H.L., Cox, D.J., Reeck, G.R., Isolation and characterization of trypsin inhibitor from opaque-2 corn seeds. J. Biol. Chem. 1977, 252, 8105-7.

2. Hojima, Y., Pierce, J.V. and Pisano, J.J., Hageman factor fragment inhibitor in corn seeds: purification and characterization. Thromb. Res. 1980, 20, 149–62.
3. G. Tans, T. Jansseen-Claessen, J. Rossing and Griffin J.H., Studies on the effect of serine protease inhibitors on activated contact factors. Application in amidolytic assays for factor XIIa, Plasma kallikrein and Factor XIa. Eur. J. Biochem. 1987, 164, 637–42.
4. Lei, M.G., and Reek, G.R., Combined use of trypsin agarose affinity chromatography and reverse phase chromatography and reversed phase high performance liquid chromatography for the purification of single chain protease inhibitor from corn seeds. J. Chromatogr. 1986, 363, 315–21.
5. Mahoney, W.C., Hermodson M.A., Jones, B., Powers, D.D., Corfman, R.S. and Reek, G.R., Amino acid sequence and secondary structural analysis of the corn trypsin inhibitoor of trypsin and and activated Hageman factor. J. Biol. Chem. 1984, 259, 8412–6.
6. Algiman, E. Bertrand, O., Unpublished results.
7. Yokosawa, H. and Ishii, S.I., Anhydrotrypsin: new features in ligand interactions revealed by affinity chromatography and thionine replacement. J. Biochem. 1977, 81, 647–56.
8. Kutty, A.V.M. and Pattabiraman, T.N., Behavior of plant amylase and trypsin inhibitors during dye-ligand chromatography on blue sepharose and red sepharose. Biochem. Arch. 1987, 3, 231–40.
9. Kroviarski, Y., Cochet, S., Vadon, C., Truskolaski, A., Boivin, P., Bertrand, O., New strategies for the screening of a large number of immobilized dyes for purification of enzymes. Application to purification from human hemolyzate. J. Chromatogr. (in press).
10. Qadri, F. and Dean, P.D.G, The use of various immobilized triazine affinity dyes for the purification of 6-phospho gluconate dehydrogenase from Bacillus Stearothermophilus. Biochem. J. 1980, 191 53–62.
11. Bradford, M.M, A rapid and sensitive method for the quantitation of microgram quantities of protein utilizing the principle of Protein dye binding. Anal. Biochem. 1976, 72, 248–54.
12. Scopes, R.K., Strategies for enzyme isolation using dye-ligand and related adsorbents. J. Chromatogr., 1986, 376, 131–40.
13. Halim, A.H., Wassom, C.E. and Mitchell, H.L., Trypsin Inhibitor in Corn (Zea mays L.) as influenced by genotype and moisture stress. Crop Science 1973, 13, 405–7.
14. Bertrand, O., Cochet, S., Kroviarski, Y., Truskolaski, A. and Boivin, P., A low cost modular apparatus of large applicability for automation of chromatographies. Proteins purification technologies ed. A. Faure GRBP Paris 1986, pp. 205–7.
15. Kassel, B., Naturally occuring Inhibitors of Proteolytic enzymes In Methods in Enzymology Proteolytic Enzymes part A ed. G. Perlmann and L. Lorand. Academic Press New York 1970 pp. 839–906.

Chapter 8

Dyes in Molecular Biology

INFLUENCE OF DYE BINDING ON THE INTEGRITY AND FUNCTION OF NUCLEIC ACIDS AND NUCLEOPROTEINS: COMPLEXATION OF POLYSOMES WITH COOMASSIE BLUE

M.R. VEN MURTHY[1], GEORGE LEVESQUE[1], ADI D. BHARUCHA[1], RENE CHARBONNEAU[1] JEAN-LOUIS VIALLARD[2] and BERNARD DASTUGUE[2]

[1]Department of Biochemistry, Faculty of Medicine, Laval University, Québec, Canada G1K 7P4; [2]Laboratoire de Biochimie Médicale, Faculté de Médecine, Université de Clermont Ferrand 1, Clermont Ferrand Cedex, France

ABSTRACT

Polysomes prestained with low concentrations of Coomassie blue gave rise to visible bands representing aggregates of different sizes when centrifuged in a sucrose density gradient and showed profiles similar to unstained polysomes. The visibility of the bands rendered them easy to recuperate and analyze. The stained polysomes were more active than the unstained specimens in protein synthesis, including the synthesis of tissue specific proteins *in vitro*. This enhanced activity was abolished when messenger RNAs were isolated from the prestained polysomes and translated *in vitro*. It is suggested that Coomassie blue may exert its influence by acting on one or more of the protein components involved in stabilizing polysomal structure and function.

INTRODUCTION

In the course of our work on the regulation of protein synthesis in brain, we have been interested in the effects of binding of substrates, analogues and other ligands, including dyes, on the structure and

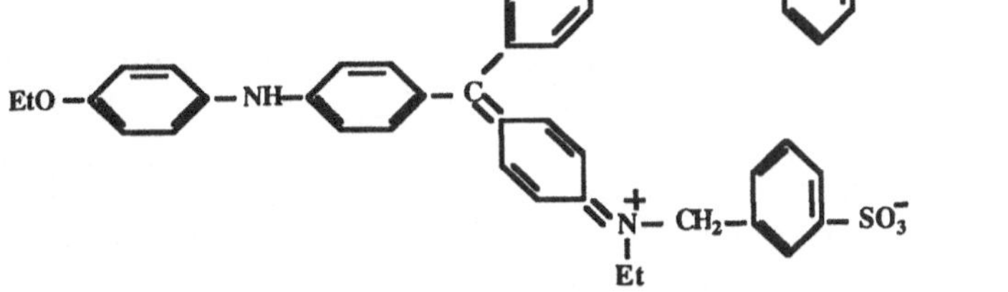

Figure 1. Structure of Coomassie blue. Et = Ethyl.

biological activity of nucleic acids and proteins (1-3). Coomassie blue is a sulfonic acid dye (Figure 1) and belongs to the magenta family of dyes. It binds to proteins producing an intense blue color (absorption maximum around 549 nm) and hence it is widely employed as a stain for revealing protein bands in a variety of analytical supports (4) and also for quantitative estimations of proteins in biological samples ((5). We report here on the binding of Coomassie blue to polysomes and the effects of this binding on the sedimentation behavior and biological activities of polysomes and their mRNA.

MATERIALS AND METHODS

Coomassie brilliant blue R-250 was from British Drug Houses, Montreal. L-(4,5-^{3}H) Leucine and (α-^{32}P) dATP were from Amersham, Ontario. Rabbit reticulocyte lysate and AMV reverse transcriptase were purchased from BRL, Madison and from Life Sciences, Florida respectively.

Preparation of polysomes and mRNA

Total polysomes (mixture of free and originally membrane-bound polysomes) were prepared, as described previously (6) from brains of adult sprague-Dawley rats, after treatment of the postmitochondrial cytoplasmic extract with detergents. mRNAs were extracted from polysomes by affinity chromatography of total polysomal RNA on columns of oligo(dT)-cellulose.

Coomassie blue staining of polysomes

Coomassie blue (4 mg/ml) was dissolved in 25 mM Hepes-K+, pH 7.2 and 4 mM Mg-acetate (Buffer A) containing 0.25 M sucrose. After centrifuging to remove any undissolved material, it was mixed with an equal volume of polysome suspension (50 OD units/ml) in the same buffer and left at 4° C for 1 hr. Polysomes stained with Coomassie blue in this manner will be referred to in the text as CB polysomes.

Analysis of polysomes by sucrose density gradient centrifugation

Polysomes were layered on 20-45% linear sucrose density gradients in Buffer A and centrifuged at 27000 rpm for 4.5 hr in a Beckman SW 27 rotor at 4° C. The visible bands of CB polysomes were collected by piercing the tube at the bottom of the desired band, with the needle (18 gauge) of a syringe and drawing out the liquid until the band was fully recovered. The fractions were diluted with two volumes of buffer A and the polysomes sedimented by centrifugation at 100,000 g for 1 hr. For unstained polysomes, the sedimentation profile was recorded by means of an ISCO recording spectrophotometer equipped with a flow cell and set at 254 nm. Fractions of the gradient corresponding to the desired peaks were pooled and the polysomes were recovered by centrifugation as above.

Immobilization of polysomes in the sucrose density gradient
In some experiments, the polysome bands resulting from sucrose density centrifugation were immobilized by a method described previously (7). The gradient buffer was modified to include, in addition to the components of Buffer A, the following: 4.75 % acrylamide, 0.25% bisacrylamide, 0.025% (v/v) TEMED and 5 µg of riboflavin. All manipulations were carried out in subdued light until the end of centrifugation. The tube was then placed in an upright support and the gradient illuminated uniformly by a fluorescent lamp. The gradient gelled completely in approximately 30 min. The bottom of the tube was excised with a sharp blade and the cylindrical gel was sectioned into longitudinal slices (8). The gel slices were fixed in 12.5% trichloracetic acid (TCA) and then fully stained by Coomassie blue according to Diezel et al (9).

Northern blotting of polysomal RNAs and hybridization
The gel slices containing immobilized polysome bands were denatured for 2 hr with 8 M urea in Tris 90 mM, boric acid 90 mM, Na_2EDTA 2.5 mM, pH 8.3. The polysomal RNAs were transferred from the gels to nitrocellulose filters by electroblotting in the same buffer. The filters were hybridized (10) with (^{32}P)-labelled cDNA prepared from rat brain mRNA by reverse transcription as described later. The radioactive spots on the filters were visualized by autoradiography.

Protein synthesis *in vitro*
The capacity of polysomes or mRNA isolated from these polysomes to synthesize proteins *in vitro* was determined by measuring the incorporation of (^3H)-L-leucine into total proteins or into specific proteins using the reticulocyte lysate translation system, in the presence of 3 mg/ml of polysomes or 40 µg / ml of mRNA. The reaction was carried out at 30° C for 30 min according to the BRL protocol.

For measuring the synthesis of neuron specific enolase (NSE), the reaction mixture was first immunoprecipitated by NSE antiserum. The precipitated NSE was quantitated by polyacrylamide gel electrophoresis and radioactivity assays (11). NSE and NSE antiserum were prepared as described elsewhere (12,13).

Synthesis of cDNA from rat brain mRNA
Rat brain mRNA was reverse transcribed in the presence of AMV reverse transcriptase according to a protocol detailed earlier (6). At the end of reaction (40 min at 42°C), 2 ul aliquots were taken out and deposited on to Whatman DE 81 filter discs. The discs were washed in 5% K_2HPO_4, dried and counted. The rest of the reaction mixture was passed through a column of Sephadex G-100 and the cDNA in the excluded volume was precipitated by ethanol for use as a hybridization probe for polysomal mRNA.

RESULTS

Sucrose density gradient patterns in liquid medium
When brain CB polysomes were subjected to sucrose density gradient centrifugation, a fraction of these particles were found to sediment to the bottom of the tube in the presence of the dye, probably due to aggregation. The rest were resolved into a number of faintly blue parallel bands against a background of less colored sucrose medium (Figure 2). The contrast was sufficient to be able to distinguish the

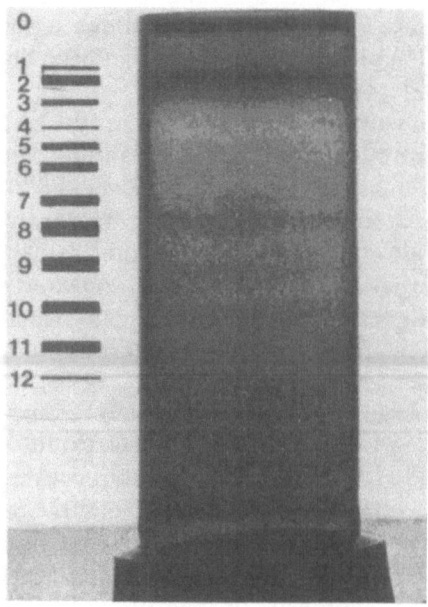

Figure 2.   Rat brain polysomes prestained with Coomassie blue and separated by sucrose density gradient centrifugation. The numbers at the left indicate the visible blue bands.

band demarcations and to collect the bands, individually, at least in the top portions of the tube, by means of a syringe. Alternatively, the distribution of stained polysomes in the gradient could be monitored by passing through a flow cell of a recording spectrophotometer set at 254 nm. Although Coomassie blue has a strong absorption at this wavelength, the polysomal peaks are still clearly revealed due to the much higher absorption of polysomes. In this manner, ultraviolet absorbance peaks corresponding to different sizes of polysomes, from monomers to dodecamers, could be easily identified (data not shown).

Sucrose density gradient patterns in gelled medium

Although ultraviolet scanning in a flow cell is a widely employed and sensitive method for localizing the positions of polysomal aggregates in a sucrose density gradient, the fluid dynamics in the flow cell and the considerable time lapse between successive analyses may lead to poor resolution of the peaks and uncertainty when comparing different preparations of polysomes. In addition, when it is desired to analyse the polysomes further, the gradient fractions corresponding to each peak have to be collected, pooled and recentrifuged at high speed in order to recover the polysomes. We have developped a simple modification of the sucrose density gradient procedure which permits entrapment of the polysomes in a rigid matrix after centrifugation. The principle involves incorporating, into the sucrose gradient, chemical components that do not affect either the characteristics of the gradient or of the polysomes during centrifugation, but bring about gelation of the gradient only when an appropriate signal is given. With this in view, the following additions were made to the sucrose solutions composing the gradient: acrylamide, bisacrylamide, riboflavin (polymerization catalyst) and TEMED (polymerization accelerator). In the usual TEMED-ammonium persulphate catalyzing system, the free radicals necessary for initiating polymerization are produced immediately after all the gel ingredients are brought together. In contrast, initiation of polymerization in the TEMED-riboflavin system requires exposure to light which, in the presence of traces of oxygen, brings about photodecomposition of riboflavin and production of free radicals (14). Accordingly, all manipulations of the tubes containing acrylamide and riboflavin are first carried out in diffuse light and when the centrifugation is terminated, the gradients are gelled by exposure to direct intense light. The gelation occurs rapidly (30 min) before the polysomes aligned into bands can disperse in the sucrose medium through diffusion. The gel cylinder is then sectioned into longitudinal slices for further treatments or analyses. The details of the procedure are given in Materials and Methods. This method not only serves to stabilize the polysomes for as long as needed, but permits analysis of polysomes and their constituents (rRNA, mRNA, ribosomal proteins) by techniques applicable to polyacrylamide gel systems, for example, a) identification of specific proteins by fluorescent antibodies; b) analysis of ribosomal proteins and RNAs by combining centrifugation in the first step and electrophoresis in the second; c) transfer of RNAs to appropriate filters by Northern blot and testing with gene or cDNA probes.

When the gel slices containing polysomes were fixed in acid and fully stained with Coomassie blue, there appeared a number of clear and distinct blue bands corresponding to different polysomeal aggregates (Figure 3). The unstained polysomes and CB polysomes gave

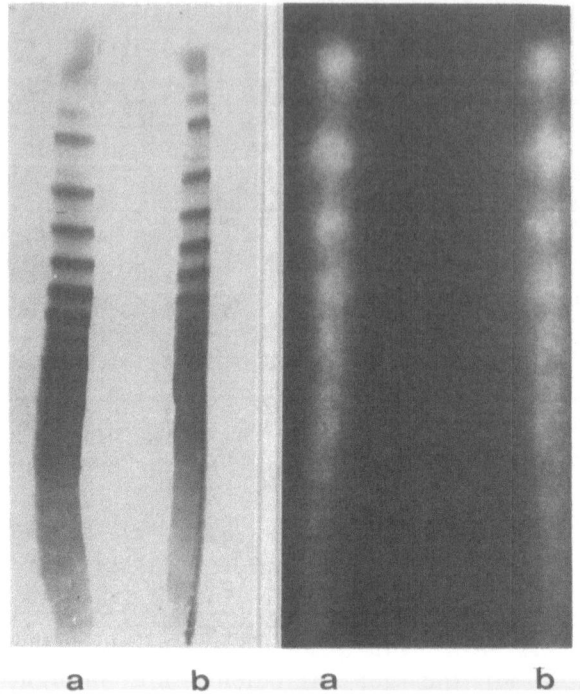

Figure 3 (Left). Slices of polyacrylamide gels containing bands of immobilized polysomes. (a) control (polysomes not subjected to prestaining with Coomassie blue); (b) polysomes prestained with Coomassie blue.

Figure 4. (Right) Autoradiography of Northern blots of brain polysomes following hybridization with brain cDNA. (a) and (b) as in Figure 3.

very similar patterns including the number, resolution and compactness of bands, indicating that the dye did not cause any significant deterioration in the structure of polysomes or their fragmentation to intermediate sizes. There also did not appear to be any selective loss of specific polysomal aggregates due to Coomassie blue binding.

The possibility of polysome degradation by Coomassie blue was further verified by making Nothern blots of gel slices containing immobilized brain polysomal bands. The Northern blot filters were hybridized with (^{32}P)- cDNAs prepared by reverse transcription of total brain polysomal mRNAs. Autoradiography of the filters showed several spots where mRNA-cDNA hybridization had taken place (Figure 4). The pattern of spots were very similar between the control and CB polysomes.

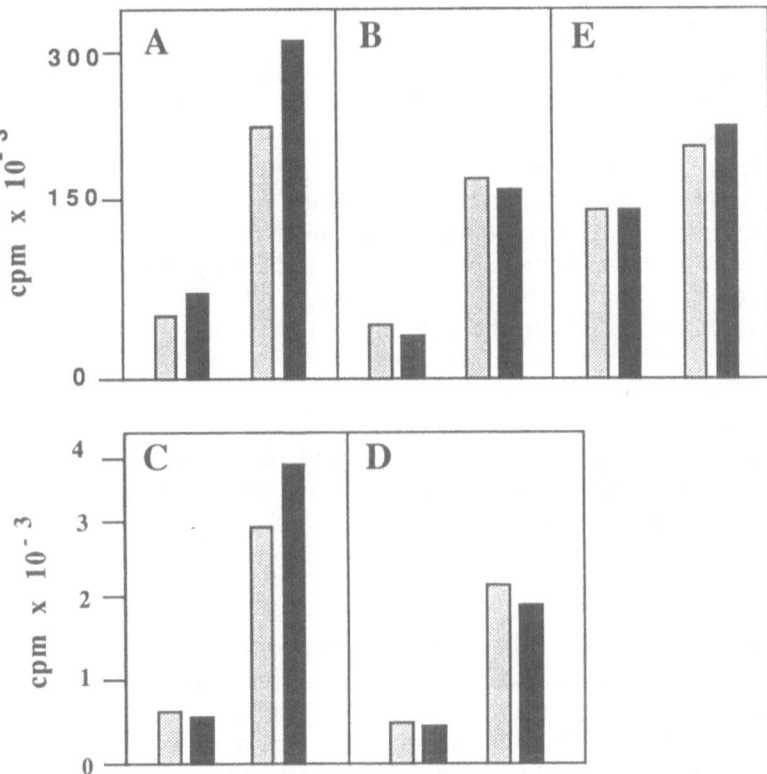

Figure 5. Effects of Coomassie blue prestaining of polysomes on the synthesis of total protein by polysomes (A) and by mRNA (B), on the synthesis of neuron specific enolase by polysomes (C) and by mRNA (D), and on the synthesis of cDNA by mRNA (E). In each figure, the two left columns and the two right columns show the activities of light polysomes (monomers to pentamers) and heavy polysomes (hexamers and above), respectively. The grey histograms represent the controls (unstained polysomes or mRNAs extracted from unstained polysomes) and the black histograms represent PB polysomes or mRNAs extracted from PB polysomes.

Protein synthesis *in vitro*

The sucrose gradients were divided into two fractions: the top portion containing pentamers and lighter aggregates (light polysomes), and the bottom portion containing hexamers and heavier aggregates (heavy polysomes). The polysomes were recovered by centrifugation and were tested for protein synthetic activity *in vitro* using the reticulocyte lysate system. The results showed that staining of polysomes with Coomassie blue enhanced their capacity for total polypeptide synthesis (Figure 5-A). This was true for both the light and heavy polysome fractions. In a similar manner, synthesis of neuron specific enolase (NSE) was also higher in the heavy CB polysomes (Figure 5-C) as compared to the heavy unstained polysomes. The light polysomes, whether pretreated with Coomassie blue or not, did not synthesize NSE. We have shown previously that the message for NSE is present only in those brain polysomes which are larger than pentamers (15).

mRNAs isolated from CB polysomes by phenol extraction exhibited an ultraviolet spectrum identical to mRNAs isolated from unstained control polysomes (data not shown), indicating that no residual dye remained with the mRNA. In contrast to the parent polysomes, these mRNAs, when tested in the reticulocyte system, showed no differences in regard to either total protein synthesis (Figure 5-B) or NSE synthesis (Figure 5-D).

cDNA synthesis

mRNAs isolated from both control and CB polysomes were equally competent in cDNA synthesis in the presence of AMV reverse transcriptase (Figure 5-E).

DISCUSSION

Generally, Coomassie blue staining of proteins is carried out under denaturing conditions, in acid pH in the presence of ethanol or methanol. However, when polysomes were treated with Coomassie blue under mildly alkaline conditions, the dye was found to bind to these particles strongly enough to stay with them during sucrose density gradient centrifugation and give rise to well separated blue bands (Figure 2). The visible bands could be recovered without resorting to a flow scanning ultraviolet spectrophotometer. Staining with Coomassie blue appeared to enhance the ability of polysomes to synthesize proteins, including NSE, *in vitro*. However, this superiority was abolished when the polysomes were deproteinized by phenol extraction. Thus, there were no differences between mRNAs of prestained and control polysomes in regard to synthesis of total protein or NSE or in the formation of cDNA by reverse transcription. This would indicate that the positive effects of Coomassie blue may be related to its action on the protein components of polysomes and not on the mRNA.

Coomassie blue binds to proteins by means of reversible electrostatic interactions between the sulfonic acid dye and the NH_3^+ groups of proteins, probably due to the formation of ion pairs between the dye anion and the positively charged amino groups of lysine and arginine (16). The dye-protein complex is further stabilized by van der Waals forces between the aromatic moieties of the two reactants (17). Such a complexation, particularly if it involves the internal amino acid residues, could lead to changes in the three dimensional structural characteristics of proteins and presumably also in their function. However, the similarities in the sucrose density gradient profiles of control and CB polysomes (for example, the number, width, compactness and resolution of bands in Figure 3) would suggest that Coomassie blue staining did not produce large enough conformational changes to affect the sedimentation behavior of polysomes significantly. It is possible, therefore, that most or all of the basic amino acids implicated in the binding of Coomassie blue to polysomes, under the conditions of prestaining, were those situated at the surface of the particles.

Coomassie blue may exert its effect by acting on other components of the reaction mixture, for example, by stimulating those that are essential for protein synthesis (initiation, elongation or translation factors etc) and/or by inhibiting those that may reduce protein synthesis (RNases, proteases etc). One important difference between brain polysomes and mRNA when translated in an *in vitro* reticulocyte lysate system is that, with polysomes, these particles themselves furnish at least some of the homologous initiation and other protein factors required for translation, whereas, with mRNA, these factors are exclusively derived from a heterologous source, i.e; the reticulocytes. We have shown previously that efficient translation of brain mRNA in vitro requires tissue specific initiation factors (18) coupled with an optimized cation concentration (19). It has been reported that the complexation of proteins with Coomassie blue promotes a site directed nucleation of cations, such as silver (16), a phenomenon responsible for the Coomassie blue- silver staining of proteins. It is possible that the enhanced protein synthesis by Coomassie blue stained polysomes as compared to purified mRNA is due the effects of the dye on the protein translation factors and/or on the ionic concentration in the translation medium.

REFERENCES

1. Malhotra, L.C., Murthy, M.R.V. and Chaudhary, A.D. Separation and purification of small quantities of specific RNA species by polyacrylamide gel electrophoresis using prestained RNAs as markers. Anal. Biochem;, 1978, **86**, 363-370.
2. Murthy, M.R.V. and Carrier-Malhotra, L. Progressive unfolding of the conformational states of transfer RNA and ribosomal 5S RNA by methylene blue binding. in Protein-dye interactions, ed. M.A. Vijayalakshmi, Elsevier Applied Science Publishers Ltd, 1989, pp
3. Murthy, M.R.V., Hanna, N., Bharucha, A.D., Charbonneau, R., Viallard, J.L. and Dastugue, B. Structure and biological activity of polysomes stained with Coomassie blue. FEBS Letts., 1985, **191**, 131-135.

4.   Hames, B.D. and Rickwood, D. (Eds). Gel electrophoresis of proteins: a practical approach. IRL Press, Washington, 1981.

5.   Bradford, M.M. A rapid and sensitive method for the quantitation of microgram quantities of protein using the principle of protein dye binding. Anal. Biochem., 1976, **72,** 248-254.

6.   Murthy, M.R.V., Lévesque, G., Pandian, S., Viallard, J.L., Ogier, R. Cavagna, A.M. and Dastugue, B. Isolation of free and membrane-bound polysomes and mRNA highly active in translation and reverse transcription from small discrete regions of rat brain. Neurochem. Int., 1986, **8**, 381-387.

7.   Murthy, M.R.V., Bharucha, A.D. and Charbonneau, R. A novel method for sucrose density gradient fractionation of polysomes and mRNA. Nucl. Acids Res., 1986, **14**, 6337.

8.   Murthy, M.R.V. and De Grandpré , P. A longitudinal slicer for obtaining multiple uniform slices from cylindrical polyacrylamide gels. Anal. Biochem., 1986, **152**, 35-38.

9.   Diezel, W,. Kopperschlager, G. and Hofmann, E. Improved procedure for protein staining in polyacrylamide gels with a new type of Coomassie Brilliant Blue. Anal. Biochem., 1972, **48**, 617-620.

10.   Maniatis, T., Fritsch, E.F. and Sambrook, J. in Molecular cloning: a laboratory manual. Cold Spring Habor Laboratory, 1982.

11.   Murthy, M.R.V. Effect of cap analogues or cap removal on the translation of rat brain mRNA *in vitro.* J. Neurochem., 1982, **38**, 41-51

12.   Viallard, J.L., Murthy, M.R.V., Bétail, G. and Dastugue, B. Determination of neuron specific enolase by differential immunocapture. Clin. Chim. Acta, 1986, **161**, 1-10.

13.   Viallard, J.L., Murthy, M.R.V. and Dastugue, B. Preparation and purification of γγ enolase (neuron-specific enolase) using high performance ion exchange chromatography. Neurochem. Res., 1988, **13**, 31-35.

14.   Gordon, A.H. Electrophoresis of proteins in polyacrylamide and starch gels. Laboratory techniques in biochemistry and molecular biology, eds., T.S. Work and E. Work, North Holland, 1975.

15.   Murthy, M.R.V., Bharucha, A.D., Charbonneau, R. and Chaudhary, K.D. Synthesis of brain specific proteins *in vitro* using cerebral cortex polysomal components from young and old rats. In Mechanisms, regulations and special functions of protein synthesis in brain, eds. S. Roberts et al. 1978, Elsevier, pp 21-28.

16.   De Moreno, M., Smith, J.F. and Smith, R.V. Mechanism studies of Coomassie blue and silver staining of proteins. J. Pharm. Sci., 1986, **75**, 907-911.

17.   De St. Groth, S.F., Webster, R.G. and Datyner, A. Two new staining procedures for quantitative estimation of proteins on electrophoretic strips. Biochim. Biophys. Acta, 1963, **71**, 377-391.

18. Murthy, M.R.V., Couderc, J.L., Viallard, J.L. and Dastugue, B. Role of tissue specific factors in the translation of brain messenger nucleic acids *in vitro*. Neurochem. Internat., 1983, **5**, 385-394.
19. Murthy, M.R.V. Translation of brain mRNA in homologous, heterologous and mixed cell free systems. Neurochem. Internat., 1983, **5**, 395-403.

PROGRESSIVE UNFOLDING OF THE CONFORMATIONAL STATES OF TRANSFER RNA AND RIBOSOMAL 5S RNA BY METHYLENE BLUE BINDING

M.R. VEN MURTHY and LISE CARRIER-MALHOTRA
Department of Biochemistry, Faculty of Medicine
Laval University, Québec, Canada G1K 7P4

ABSTRACT

Methylene blue is a basic vital dye extensively used for staining of nucleic acids and other cellular components in many areas of biology. When 4S and ribosomal 5S RNAs were stained with low relative concentrations of methylene blue (less than 10 mole% of RNA nucleotides), the two RNAs gave rise to blue colored complexes which migrated in a manner identical to unstained controls during polyacrylamide gel electrophoresis. The visible bands could be trapped on DEAE-cellulose discs and isolated individually at a high efficiency of recovery (80-90%). The binding was reversible and the dye was removable by washing with ethanol and dialysis. The stained tRNAs exhibited an enhanced aminoacylation activity *in vitro*, which reverted to the level of the untreated control when the dye was removed. At higher relative concentrations of methylene blue, both 4S and 5S RNAs migrated as discrete multiple bands, with a progressive intensification of the cathodic bands at the expense of the anodic bands, as the proportion of the dye in the prestaining reaction increased. Our results suggest that the effect of methylene blue may consist in progressively unravelling the multi-layered three dimensional structures of 4S and 5S RNAs, thus producing a number of different conformational forms. If this is the case, methylene blue binding, followed by electrophoresis, may serve as a useful approach for studying the structural architectures of RNAs and ribonucleoproteins.

INTRODUCTION

Methylene blue (tetramethylthionin chloride) (Figure 1) belongs to the quinoneimine group of dyes and is a member of the thiazine subgroup. Its molecular weight (319) is comparable to that of an average ribonucleotide (340). It is a basic dye and interacts with nucleic acids as well as a number of other cellular components, making it a valuable tool in several areas of biology including histology, cytology, parasitology, microbiology etc. Next to haematoxylin, it is probably the dye most often employed for biological staining. It is a redox agent, acting as a reducing agent at low concentrations and an oxidant at

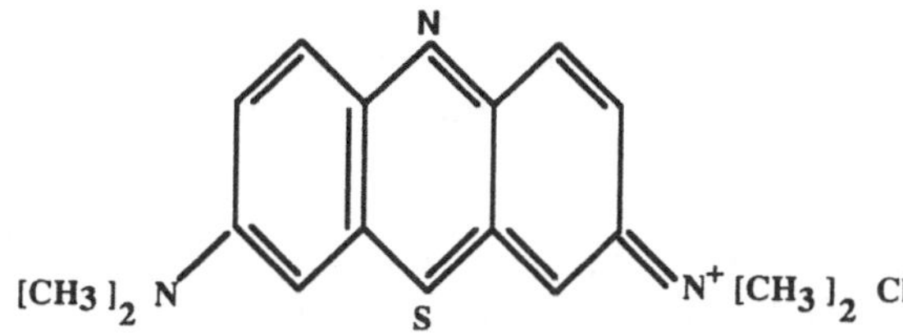

Figure 1. Structure of methylene blue.

higher concentrations. The dye is oxidized readily in alkali with an increase in metachromacy, while on reduction, a colorless leuco compound is produced. Although methylene blue is considered a nontoxic vital dye and has found application in clinical medicine, it has been shown to have a number of biological effects on a series of organisms and higher animals. All these and other aspects of methylene blue have been extensively reviewed by Barbosa and Peters (1).

Our initial interest in methylene blue was stimulated by the possibility that this dye could serve as a visible marker for following the migration pattern of different species of nucleic acids during polyacrylamide gel electrophoresis (PAGE). In this way, we hoped to be able to isolate and purify specific transfer RNAs from mixtures containing other nucleic acids for an eventual determination of their structure and biological activity. We report in this communication our results which show that the interaction of RNA with methylene blue may not only be used for the separation of one species of RNA from another, but may also offer an approach for studying nucleic acid structure.

MATERIALS AND METHODS

4S RNA (baker's yeast) and ribosomal 5S RNA (E. coli) were purchased from Plenum Scientific Research Inc., N.J. and Miles Laboratories, Ill., respectively. Methylene blue was from BDH Canada Ltd., Ont., DEAE-cellulose paper (DE-81) from Whatman and ^{14}C-L-amino acids from Amersham Canada.

Prestaining of RNA

Methylene blue and the RNA(s) to be prestained were each dissolved in borate buffer (Tris 90 mM, boric acid 90 mM, Na_2 EDTA 2.5 mM, pH 8.3) containing 10% sucrose. Forty ug of each RNA was mixed with varying amounts of the dye (0, 1.6, 2.4, 3.2, 6.4, 9.6, 12.8, 16 and 32 ug) and made up to 25 ul with buffered sucrose. The mixtures were left at 4°C for 1 hr in the dark prior to analysis. All manipulations of the RNA-dye complex, including electrophoresis and recovery procedures were carried out in subdued light and away from air currents, in order to avoid any photochemical degradation of RNAs (2).

The relative concentration of the dye with respect to RNA is expressed as mole %, i.e. the number of moles of methylene blue per 100 moles of RNA nucleotides present in the mixture. Assuming an average molecular weight of 340 per RNA nucleotide, the relative concentrations of methylene blue in the above mixtures, corresponded to 0, 4.2, 6.1, 8.4, 16.8, 25.2, 33.6, 42.0, 84.0 mole % (referred to in the text by the nominal values 0, 4, 6, 8, 16, 25, 33, 42 and 84 mole% for the sake of simplicity).

Electrophoresis

Electrophoresis was performed as described by Peacock and Dingman (3) in short cylindrical glass tubes (3 mm i.d. and 40 mm long) with gel concentrations of 9.5% acrylamide and 0.5% bisacrylamide. The borate buffer was used both in the preparation of the gels and for electrophoresis. The gels were equilibrated by pre-electrophoresis at 200v for 45 min. at 4°C. The RNA samples were then applied on top of the gels and electrophoresis was continued at 50v and 4°C. Electrophoresis was stopped when a bromphenol blue indicator, run simultaneously on a separate gel, reached the end of the gel.

Postelectrophoretic staining of RNAs on gels

The gels containing the prestained or unstained control RNAs were removed from the tubes and placed in 1M acetic acid for 15 min. The solution was then replaced by 0.2% methylene blue in 0.2M acetic acid and after 2 hr at room temperature, the excess dye was washed off with water.

Recovery of RNAs from gel

Following electrophoresis, if it was desired to recover a particular methylene blue stained RNA band from a gel containing both faster moving and slower moving components, the following protocol was employed (4).

Trapping of the RNA bands on DEAE-cellulose discs: Electrophoresis was allowed to continue until all the faster moving bands ran off from the gel. The current was stopped and a circular disc of Whatman DE-81 DEAE-cellulose paper previously soaked in borate buffer was applied flat to the anode end of the gel tube. Electrophoresis was restarted and was continued until the top boundary of the desired band passed through the gel and was adsorbed on the DEAE disc. When unstained control RNAs were used, a prestained RNA sample was run at the same time, on a separate gel, to serve as a reference for carrying out the appropriate manipulations.

Destaining and recovery of RNAs: The DEAE discs containing the RNA were rinsed once with water and then with ethanol until the blue color disappeared. RNA was eluted with 2M triethylammonium bicarbonate (TEAB) (pH 9.5), dialyzed against water to remove any residual dye and then lyophilized. The residue was dissoved in water and lyophilized two more times to eliminate all TEAB.

Aminoacylation of t-RNAs

The term 4S RNA will be used in this text to refer to cytoplasmic soluble RNA originally identified by sedimentation value. When amino acid acceptor activity is demostrated for a particular component of 4S RNA, that component will be referred to as transfer RNA (tRNA). The aminoacylation activities of specific tRNAs were determined *in vitro* by measuring the incorporation of corresponding (^{14}C)-L-amino acids into 4S RNA in the presence of aminoacyl-tRNA synthetases and ATP, according to the procedure described previously (5).

Electrophoretic separation of radioactive aminoacyl t-RNAs

4S RNA was aminoacylated as above and was then reisolated from the reaction mixture by phenol extraction (5). After electrophoresis of this RNA, the gel was sliced into fragments of approximately 1.5 mm, and radioactivity determined in each slice.

RESULTS

Effects of low concentrations of methylene blue on the electrophoretic behavior of 4S and ribosomal 5S RNAs

When 4S RNA prestained with 6 mole% of methylene blue was subjected to electrophoresis, two major blue colored bands were observed (Figure 2a, bands 1 and 2). There were also present three minor, faintly colored, slower moving bands (bands 3,4 and 5). Occasionally, band 3 was found to be split into two closely situated bands, as in Figure 2a. Probably, this was due to small inadvertent variations in the electrophoretic conditions employed. Complete

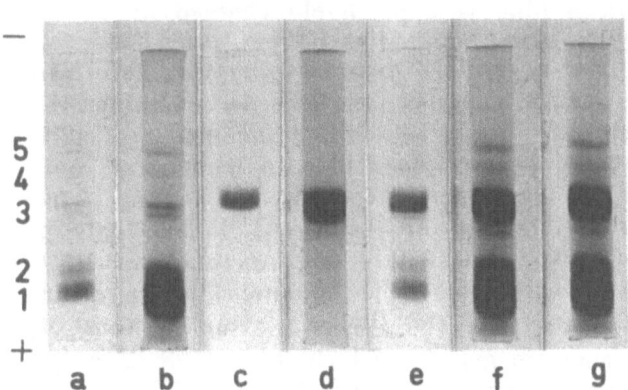

Figure 2. Electrophoresis of 4S and ribosomal 5S RNAs prestained with methylene blue (6 mole%). Columns a, c and e show prestained 4S, 5S and (4S+5S) RNAs; columns b, d, and f show the same gels after complete postelectrophoretic staining; column g shows untreated (4S+5S) RNAs after postelectrophoretic staining.

postelectrophoretic staining of the gel revealed all the above bands (Figure 2b), but with more intense coloration and greater width for each band. Based on their concentrations (95-98% of total RNA), electrophoretic mobilities and amino acid acceptor capacities, the major bands 1 and 2 were both identified as constituting various transfer RNAs. The minor bands are assumed to represent degradation products of ribosomal RNAs contaminating the 4S RNA preparation.

When ribosomal 5S RNA was prestained with 6 mole% methylene blue, the resultant colored complex migrated as a single compact blue band on electrophoresis (Figure 2c). Complete postelectrophoretic staining of the gel containing the 5S RNA also showed the same band, but wider and with a deeper color (Figure 2d), a phenomenon similar to that observed with 4S RNA. Postelectrophoretic staining also revealed the presence of one or two minor, fast migrating impurities in the 5S RNA, again probably corresponding to some degradation products of ribosomal RNAs.

Electrophoresis of prestained mixtures of 4S and 5S RNAS showed a series of bands that could be predicted by a combination of their individual patterns (Figures 2e and 2f), indicating that there was no interference between the two species of RNAs even in the presence of methylene blue. Further, mixtures of 4S and 5S RNAs prestained with 6 mole % methylene blue behaved in a manner identical to unstained RNA mixtures, both in the number of bands obtained and their mobilities (compare Figures 2f and 2g), indicating that prestaining, at least at this concentration of the dye, had no effect on those molecular parameters that determine the electrophoretic migration of RNA molecules.

Since methylene blue is a positively charged molecule, its migration toward the anode along with the RNAs would suggest that the dye interacts with the negatively charged polyelectrolyte giving rise to a complex with a net negative charge. An examination of Figure 2 reveals the following points concerning the nature of this interaction at a low concentration of methylene blue (6 mole%): a) only a portion of the total RNA complexes with the dye, since complete postelectrophoretic staining of the RNAs in the gels produce darker and wider bands; b) even within a given class of RNA (4S or 5S), certain subclasses may have a preferential kinetics of interaction with methylene blue since only the slower moving portions of bands 1 and 2 of 4S RNA (compare gels 2a and 2b) and of 5S RNA (compare gels 2c and 2d) are prestained while the faster moving portions of these bands become visible only after complete postelectrophoretic staining with an excess of the dye; c) the formation of the RNA-dye complex leads to only minimal modifications in net negative charge and conformation, since the complex migrates in a manner identical to the corresponding portion of the untreated RNA.

Isolation and recovery of prestained 4S and ribosomal 5S RNAs

The possibility of prestaining RNAs without affecting their independent normal electrophoretic mobilities suggested that this approach might be useful for the separation, purification and quantitative recovery of single RNA species from a mixture of cellular RNAs. A procedure was therefore developed, as described in Materials and Methods, for the recovery of individual RNA bands from the gel. This involved selective trapping of the desired band on to a DEAE-cellulose disc by controlled electrophoresis followed by extraction of the bound RNA from the disc. When bands 1 and 2 of 4S RNA and the single band of 5S RNA were isolated from the gel in this manner, efficiency of recovery was 80-90% of the RNA originally placed on the gel. When the recovered RNAs were subjected to a second electrophoresis, they showed the same migration as the parent bands. Furthermore, no new fast moving bands were detectable, indicating that no RNA degradation had occured during the purification procedure.

Biological activity of prestained 4S RNA

Although there was no visible degradation of RNA by methylene blue, it is conceivable that the dye may affect the biological activity of the RNA, either by the action of minute quantities of the dye which may possibly remain strongly attached to the purified RNA or by causing small, but critical modifications in the primary or higher structures of RNA. In order to verify this possibility, the following experiment was carried out. The prestained 4S RNAs (6 mole% methylene blue) were electrophoresed and the bands 1 and 2, (Figure 2a) were collectively trapped on DEAE-cellulose discs. RNA was then extracted from the DEAE-cellulose discs either directly without destaining (prestained tRNA) or after the application of destaining procedures (destained tRNA). The capacities of these tRNAs to ligate amino acids *in vitro* were compared to those of unstained control tRNAs (Table 1). The results show that the prestained tRNAs had significantly higher aminoacylation activities for all of the three amino acids tested (serine, leucine and alanine) as compared to the control. On the other hand, removal of the dye from the 5S RNA by washing and dialysis reduced all the three activities to the level of controls. It can be concluded from this a) that the continued presence of methylene blue is necessary for eliciting the favourable effect on aminoacylation, and b) that its action is reversible, i.e. the RNA molecule is left unaltered following a cycle of staining and destaining.

Effects of high concentrations of methylene blue on the electrophoretic behavior of 4S and ribosomal 5S RNAs

In an attempt to increase the proportion and types of RNAs interacting with methylene blue, we incubated 4S and 5S RNAs with progressively higher relative concentrations of the dye and verified the

TABLE 1
Effect of methylene blue staining on the aminoacylation
activity of tRNA

	tRNA		
Amino acid	unstained (control)	prestained	prestained + destained
	Amino acid incorporated: CPM/OD unit of tRNA		
(^{14}C) Serine	1375	1850	1445
(^{14}C) Leucine	1752	2145	1566
(^{14}C) Alanine	1642	2057	1710

4S RNA was prestained with 6 mole % methylene blue. Other
details were as described in Materials and Methods and in Results.

electrophoretic behaviours of the resulting complexes. It was found
that 4S RNA prestained with 4 mole% and 8 mole% methylene blue
gave the same blue bands 1 and 2, as in Figure 1a, but when the dye
concentration was increased to 16 mole%, a number of additional,
faintly colored, slower moving, bands appeared cathodic to band 2
(Figure 3, top). At the same time, the color intensities and widths of
bands 1 and 2 diminished. This indicated that the new cathodic bands
did not represent new species of 4S RNA capable of being stained at
the higher dye concentration, but were probably formed as a result of
changes in the mobilities of some of the RNA molecules already
present in bands 1 and 2. This was confirmed by the observation that,
with further increases in the relative concentration of the dye (25
mole% and 33 mole%), bands 1 and 2 continued to diminish with a
concurrent intensification of the cathodic bands. The mobilities of
these cathodic bands themselves were progressively reduced as the
dye concentration increased until finally, at 42 mole%, there were no
colored bands migrating in the gel.

Complete postelectrophoretic staining of the gels (Figure 3, bottom)
rendered more evident the progressive disappearance of the anodic
bands and the simultaneous appearance and intensification of cathodic
bands. This behavior applied not only to 4S RNA, but also to the RNA
impurities present in the 4S RNA preparation (bands 3, 4 and 5). At
the highest relative concentration of methylene blue tested (84
mole%), RNA failed to enter the gel. No RNA bands were visible even
when the polarity of the electric current was reversed (data not
shown) indicating that the absence of migration was not due to a
reversal of net charge.

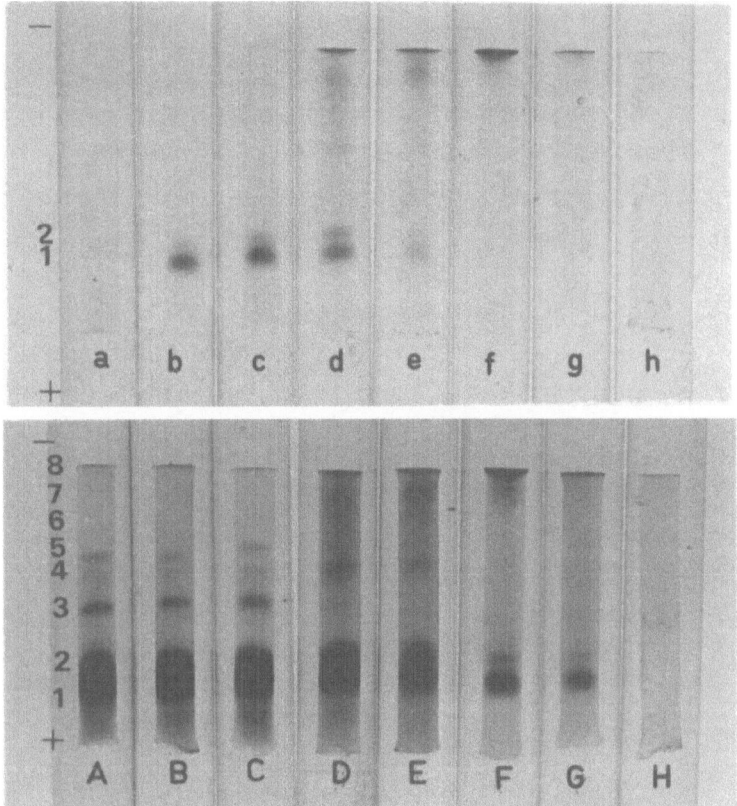

Figure 3. Effects of prestaining with different concentrations of methylene blue on the electrophoretic behavior of 4S RNA. Columns a to h (top) show 4S RNA prestained with 0, 4, 8, 16, 25, 33, 42 and 84 mole% methylene blue respectively; columns A to H (bottom) show the same gels after complete postelectrophoretic staining.

Another interesting observation concerned the different effects of methylene blue on the slow and fast moving components of each of the two 4S RNA bands. The slower moving components of the two bands were prestained by methylene blue giving rise to colored complexes which started exhibiting a multiplicity of bands at 16 mole% and stopped migrating altogether at 33 mole% of the dye (Figure 3, top). The faster moving components of the two bands did not become colored by prestaining at any of the concentrations of methylene blue tested; they were revealed only after complete postelectrophoretic staining (Figures 2 and 3). This suggested that either these faster moving molecules did not complex with methylene blue under the conditions of prestaining or if they did, the dye was in its reduced leuco form in the complex. This latter possibility appears less likely

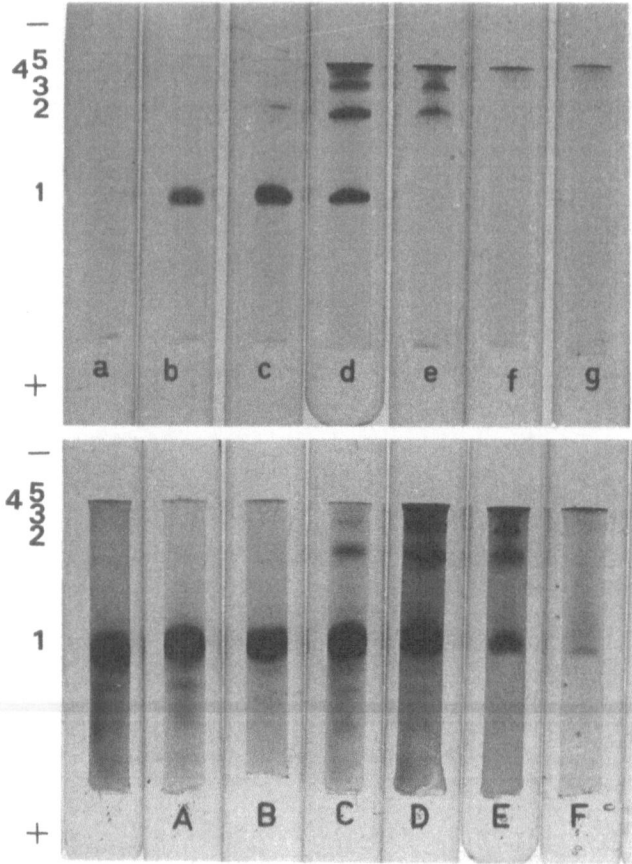

Figure 4. Effects of prestaining with different concentrations of methylene blue on the electrophoretic behavior of ribosomal 5S RNA. Columns a to g (top) show 5S RNA prestained with 0, 4, 8, 16, 25, 33 and 42 mole% methylene blue respectively; columns A to F (bottom) show the same gels after complete postelectrophoretic staining.

since the oxidized metachromasic form is expected to predominate in the alkaline pH used for prestaining and electrophoresis (1). In any case, methylene blue did not seem to affect their electrophoretic mobility until a concentration of 42 mole% (Figure 3G).

The interaction of 5S RNA with methylene blue was similar to that of 4S RNA, except that the number of cathodic bands formed were fewer and much more discrete (Figure 4). This could be the result of differences in the homogeneity and structures of the two RNAs. However, as in the case of 4S RNA, the 5S RNA also exhibited the presence of two subclasses, a) a slow moving component which gave colored complexes with methylene blue after prestaining (Figures 2c and 2d) and whose mobility was affected by relatively lower concentrations of the dye (16 mole%) (Figure 4d); and b) a fast moving component which was · revealed only after complete post electrophoretic staining and whose migration was affected only at higher concentrations the dye (33 mole%) (Figure 4F). The significance of these subclasses will be discussed later.

Separation of different transfer RNAs by PAGE after methylene blue binding

The differential reactivities of the slow and fast moving components of 4S RNA toward methylene blue suggested that this reaction could serve as a possible means of separating transfer RNAs from each other and isolating them in pure form. In order to verify the feasibility of this approach, 4S RNA was aminoacylated in the presence of radioactive alanine or serine, and was then subjected to electrophoresis, before or after prestaining with methylene blue. In order to avoid any possible degradation of tRNA, the relative concentration of the dye used for prestaining was kept at a level which was just enough to cause band multiplication (16 mole%). The locations of the aminoacyl-tRNAs in the gel were determined by scanning the gel for radioactivity. It is seen from Figure 5(A), that alanyl- and

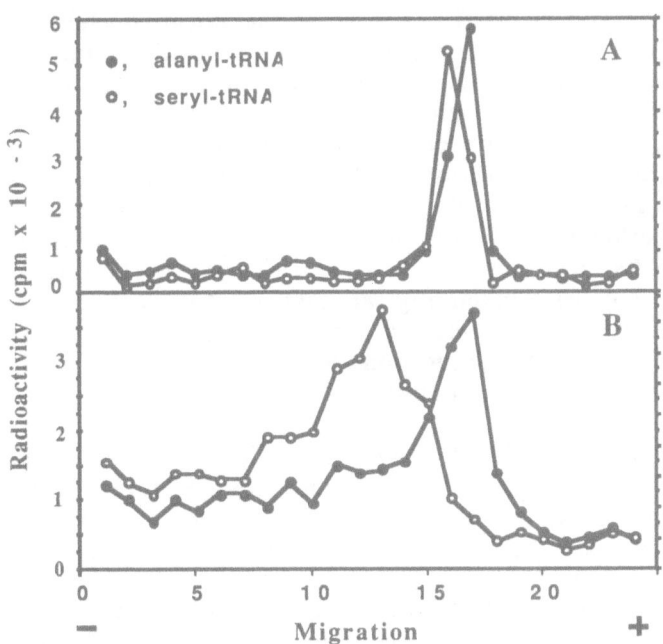

Figure 5. Effects of methylene blue prestaining on the migration of alanyl- and seryl-tRNAs. A, untreated control; B, after prestaining with 16 mole% methylene blue. Aminoacylation of 4S RNA with radioactive alanine and serine was carried out as described in Materials and Methods.

seryl-tRNAs migrated very similarly, with the former moving slightly ahead of the latter. However, when these two tRNAs were prestained with 16 mole% methylene blue, their radioactivity peaks separated considerably. This was brought about mainly by the retardation in the mobility of seryl-tRNA while ananyl-tRNA remained relatively unaffected. Some minor peaks of radioactivity cathodic to the main peaks were also observed which presumably corresponded to the cathodic bands noted earlier for the total 4S RNA.

DISCUSSION

The nature of RNA-methylene blue interaction

The electrophoretic migration of a polyelectrolyte, such as RNA, in a given gel system, would depend on a number of factors some of the most important of which are attributable to the structure of the RNA itself: for example, chain length, net charge and conformation. There are a number of possible ways in which methylene blue could affect these parameters in 4S and 5S RNAs and produce the effects observed in this work. Among them may be considered:

a) Charge effects: Methylene blue is a cationic dye (Figure 1) and it may be expected to neutralize a part of the negative charge in the RNA molecule. Since at the pH employed for prestaining and electrophoresis (pH 8.3), the phosphodiester linkages as well as all the four common ribonucleosides of RNA are negatively charged (6), the degree of charge neutralization may be expected to be directly proportional to the concentration of the dye. However, the following observations suggest that the reduction in net charge is not the sole reason for the effects of methylene blue on the electrophoretic behavior of RNAs: a) not all RNAs are equally susceptible to prestaining under the conditions used; b) prestaining at low relative concentrations of the dye (4-8 mole%) does not affect the migration of RNAs while considerably less than saturation levels of the dye (16-42 mole%) drastically impedes their mobility and even prevents their entry into the gel.

b) Aggregation of RNA: RNA molecules aggregate easily depending on their concentration, the ambient temperature, the levels and proportions of monovalent and divalent cations in solution and the possibility of degradation (7). Hydrogen bonds as well as base-base interactions could mediate RNA aggregation. However, when the apparent relative sizes of RNAs in the different bands were calculated (Table 2), it was found that the largest RNA-dye complexes differed from the smallest by only a small factor (3 for 4S RNA and 2 for 5S RNA) with a number of intermediate fractional sizes between these limits. It is possible that an RNA aggregate with a relative size between 1 and 2 could be produced by combining an intact RNA molecule with a degraded fragment, but this appears unlikely for the

TABLE 2

Electrophoretic mobilities and apparent molecular sizes of 4S and 5S RNAs prestained with methylene blue.

Band no	4S RNA		5S RNA	
	Mobility[a]	Apparent relative size[b]	Mobility[a]	Apparent relative size[b]
1	0.75	1.00	0.47	1.00
2	0.69	1.08	0.15	1.63
3	0.60	1.22	0.07	1.85
4	0.38	1.73	0.016	2.00
5	0.31	1.89	0	=>2.06
6	0.17	2.38	-	-
7	0.08	2.74	-	-
8	0	=>3.09	-	-
unstained control	0.74	1.00	0.47	1.00

[a] Ratio of the distance of migration of a given band (Figures 3 and 4) to the distancce of migration of bromphenol blue (the total length of the gel). [b]Molecular size is calculated assuming a linear relation between log (molecular weight) and electrophoretic mobility.

following reasons: a) electrophoresis of stained 4S or 5S RNAs in the presence of 8M urea did not reveal any bands corresponding to low molecular weight fragments (data not shown); b) when radioactive aminoacyl-tRNAs were prestained with methylene blue, there were no minor peaks anodic to the main peak of radioactivity (Figure 5). This would indicate that RNA aggregation is not an important factor in the appearance of mutiple cathodic bands, although it may play a role at very high dye concentrations.

c) Aggregation of methylene blue: Methylene blue also can form aggregates, particularly at high concentrations and in the presence of neutral salts (1). Interaction of RNA with methylene blue aggregates could produce complexes which would be expected to vary continuously in size and be revealed as an extended smear on the gel. However, the observed formation of discrete new bands instead of a Gaussian smear suggests that the molecular transformation of the RNAs by methylene blue may occur in a step-wise manner and may involve sequential interactions at a limited number of hierarchical structural organizations.

d) <u>Conformational changes of RNA</u>: It has often been noted that 4S and 5S RNAs can undergo reversible, conformational rearrangements and give rise to multiple forms when they are subjected to destabilization by treatments such as urea, heat or changes in ionic composition. The existence of these forms are distinguishable by their chromatographic behavior, their tendency to aggregate and their ability to interact with other biopolymers (9,10). Unlike heat and urea treatments which denature RNA by rupturing hydrogen bonds without actually combining with the RNA, the structural transformations induced by methylene blue apparently require the formation of an RNA-dye complex. There are several mechanisms by which methylene blue may affect the three dimensional conformation of RNAs, for example, by altering the net charge, by intercalation, by its property as a redox agent or by mediating a direct chemical change in the nucleic acid. Methylene blue sensitizes a highly specific oxidation of guanine residues in both RNA and DNA in the presence of light and oxygen (2). The rate of reaction depends on the degree of ionization of the substrate, the anionic form of guanosine being 30 times faster than the neutral form. The guanosine of RNA is present mainly in the anionic form under the buffer conditions used in this study for electrophoresis (pH 8.3). Such a photo-oxidation may therefore become a significant factor at high relative concentrations of the dye, although care was taken to avoid exposure to direct light and aeration in the present study.

Both 4S and 5S RNAs exhibit complex multi-layered three dimensional structures (11). It is conceivable that the effect of methylene blue consists in unravelling this structure in a sequential fashion, thus giving rise to a number of conformational forms which are increasingly less compact and consequently migrate slower than the native RNA. If this is the case, the interaction of methylene blue may serve as a useful approach for studying the structural architecture of RNAs and ribonucleoproteins.

Differentiation between RNA subclasses based on methylene blue interaction

The observation that, within the same main class of RNA, some molecules are stained while the others are resistant (Figures 3 and 4) would indicate the existence of structural heterogeneity in these RNAs. Indeed, 4S RNA is composed of a large number of transfer RNAs possessing unique as well as common structural features and may be expected to act differently toward methylene blue. The changes in the electrophoretic mobilities of alanyl- and seryl-tRNAs after prestaining with methylene blue (Figure 5) suggests that this interaction could, indeed, form a basis for separating different tRNAs in a mixture of these nucleic acids. On the other hand, ribosomal 5S RNA which constitutes a single species of molecules also showed the presence of two subclasses: a slow moving component which was readily affected by methylene blue and a fast moving component which required higher concentrations of the dye for interaction (Figures 2, 3 and 4). One possible explanation of this phenomenon may be that, although 5S RNA may be homogeneous in regard to its primary structure, some of the molecules in a given preparation may have suffered certain subtle conformational changes in the course of their isolation from ribosomes, thus leading to partial unfolding and increases in effective volume. These altered molecules may migrate more slowly and may permit the entry and fixation of methylene blue while the more compact native molecules migrate faster and are less accessible to the dye.

Enhancement of tRNA aminoacylation activity

The stimulation of *in vitro* aminoacylation activity by methylene blue is surprising in view of the observed inhibitory effect of this dye on a certain number of enzymes *in vivo* and *in vitro* (1) as well as on some antibodies (12). The dye could influence aminoacylation either by modifying the structures of the biopolymers (the tRNAs and the aminoacyl-tRNA synthetases) or by interacting with the small molecular substrates participating in this reaction (ATP, magnesium, amino acid). Although the involvement of the biopolymers is possible, it has also been reported that methylene blue can bind to ATP in the presence of neutral salts (1). Since the concentration of ATP and magnesium are critical for the aminoacylation of tRNA, it is possible that the action of methylene blue may be related to its property of binding to these molecules.

ACKNOWLEDGMENTS

We are grateful to Professor Kapil Chaudhary for his interest in this work and for helpful discussions and to Dr. Jean-Louis Couderc for his help in preparing the graphs. We thank the Medical Research Council of Canada for financial assistance.

REFERENCES

1. Barbosa, P. and Peters, T.M., The effects of vital dyes on living organisms with special reference to Methylene Blue and Neutral Red. Histochem. J., 1971, **3**, 71-93.
2. Simon, M., Photosensitized oxidation of purines and pyrimidines. Methods in Enzymol., Vol **12**, Part A, pp. 45-47.
3. Peacock, A.C. and Dingman, C.W., Resolution of multiple ribonucleic acid species by polyacrylamide gel electrophoresis. Biochemistry, 1967, **6**, 1818-1827.
4. Malhotra, L.C., Murthy, M.R.V. and Chaudhary, A.D., Separation and purification of small quantities of specific RNA species by polyacrylamide gel electrophoresis using prestained RNAs as markers. Anal. Biochem., 1978, **86**, 363-370.

5.   Murthy, M.R.V., Roux, H. and Thénot, J.P., Isoacceptor tRNAs for glutamate, glutamine, aspartate and asparagine in calf brain. J. Neurochem., 1974, **22**, 19-22.

6.   De Wachter, R. and Fiers, W., Two dimensional electrophoresis of nucleic acids. In Electrophoresis of nucleic acids: a practical approach, ed., D. Rickwood and B.D. Hames, IRL Press, Oxford, 1982, pp77-116

7.   Boedtker, H., Molecular weight and conformation of RNA. Methods in Enzymol., Vol. **12**, Part B, pp 429-433

8.   Loening, U.E., The determination of the molecular weight of ribonucleic acid by polyacrylamide gel electrophoresis. Biochem. J, 1969, **113**, 131-138

9.   Aubert, M., Scott, J.F., Reynier, M. and Monier, R., Rearrangement of the conformation of Escherichia coli 5S RNA, Proc. Nat. Acad. Sci., U.S.A., 1968, **61**, 292-299

10. Gartland, W.J. and Sueoka, N., Two interconvertible forms of tryptophanyl-sRNA in E. coli. Proc. Nat. Acad. Sci., U.S.A., 1966, **55**, 948-955

11. Reid, B.R., NMR studies on RNA structure and dynamics. Ann. Rev. Biochem., 1981, **50,** 969-996

12. Shojania, A.M. and Orr, K., The effect of toluidine blue and methylene blue in immunochemical reactions *in vitro*. Clin. Immunol. Immunopathol., 1987, **43**, 223-228

USE OF TRIAZINE DYES IN PURIFICATION OF T4 POLY-NUCLEOTIDE KINASE

R.P.MARCISAUSKAS, D.J.KARALYTE, O.F.SUDZIUVIENE,
I.-G.I.PESLIAKAS
ESP "Fermentas", All-Union Research Institute of
Applied Enzymology, Vilnius, Lithuanian SSR,USSR

ABSTRACT

The sorption capacity of Sepharose CL-6B and other matrices with linked triazine dyes: Cibacron blue F3G-A, Bright-red 6C, Procion red 2K, Bright yellow 53, Yellow light resistant 2KT and Orange 5K in respect to polynucleotide kinase T4 was determined. It was shown that Sepharose CL-6B with linked Bright-red 6C dye and Sepharose 4B with immobilized Blue dextran are effective substrates suitable for purification of polynucleotide kinase from nuclease contamination. The activity of DNAses after a procedure was 2×10^{-3} and 3.5×10^{-3} units of activity per unit of enzyme respectively. A method of purification of polynucleotide kinase from contamination of DNAses, RNAses and 5'-phosphatases and its isolation by means of Sepharose-6B with immobilized Bright-red 6C and phosphocellulose is suggested.

INTRODUCTION

Phage T4 polynucleotide kinase is widely used in contemporary studies of molecular biology and molecular genetics (1,2). For application in these fields polynucleotide kinase preparations should be free from nuclease and 5'-phosphatase contamination, though the production of enzyme with mentioned qualities is very tricky. Recently the structural gene of polynucleotide kinase was cloned (3).

Adsorbents with immobilized triazine dyes are now widely

used to purify various enzymes, to make them available and
easier to synthesize, to increase their chemical stability
and high sorption capacity which enables better preparative
application.

The reported work was directed at developing the application of these adsorbents for the purification of polynucleotide kinase from T4 am N82-infected E.coli cells.

RESULTS

T4-induced enzymes such as DNA- and RNA-ligase, DNA polymerase and polynucleotide kinase are produced simultaneously using the precipitation of nucleic acids with polymine P at
0.7 - 0.8 % concentration (4,5). Our investigations indicated
the equal partition of 3'-5'-exonucleases and endonucleases
between the polynucleotide kinase-containing liquid phase and
the pellet with ligases and DNA polymerase. This poses a problem in the course of purification. After polymine P treatment it is impossible to free the fraction from the nuclease
activity chromatographically using Blue dextran-Sepharose (6)
(Table 1). Therefore the possible use of other triazine dyes
to purify the polynucleotide kinase was under question and
sorption capacity of dye bound adsorbents based on Sepharose
and other matrices were established (Table 2).

Table 2 shows that the highest sorption capacity is displayed by Bright-red 6C-boundmacroporous silica gel and by
Sepharose 4B bound Blue dextran. Sepharose CL-6B with bound
Procion red HE-3B, Red-brown 2K, Cibacron blue F3G-A and
Bright-red 6C showed acceptable capability. The effectiveness
of these adsorbents in removing deoxynuclease and ribonuclease contamination from the polynucleotide kinase preparations
have been tested using the polymine fraction of the enzyme,
precipitated with ammonium sulphate and dialyzed against 10
mM Tris-HCl buffer pH 7.5 containing 10 mM $MgCl_2$ and 10 mM
2-mercaptoethanol.

Table 1 shows that the kinase/3'-5'-exonuclease ratio
was $2 \cdot 10^{-3}$ or $3.5 \cdot 10^{-3}$ using Bright-red 6C-Sepharose CL-6B
and Sepharose 4B-Blue dextran respectively. The preparation

TABLE 1
Activity of 3'-deoxyexonucleases, deoxyendonucleases
and ribonucleases in polynucleotide kinase preparations
purified using various adsorbents

Adsorbents with various dyes	3'-deoxyexonuclease activity, units per enzyme unit	Deoxyendonuclease activity	Ribonuclease activity, units per enzyme unit
Cibacron blue F3G-A-Sepharose CL-6B	$6 \cdot 10^{-3}$	+++	-
Blue dextran-Sepharose 4B	$3.5 \cdot 10^{-3}$	+	-
Bright-red 6C-Sepharose CL-6B	$2 \cdot 10^{-3}$	+	$1 \cdot 10^{-4}$
Bright-red 6C-macroporous silica gel coated with dextran	$1.7 \cdot 10^{-1}$	+++	$2.7 \cdot 10^{-3}$
Procion red HE-3B-Sepharose CL-6B	$1 \cdot 8 \cdot 10^{-2}$	++	-
Red-brown 2K-Sepharose CL-6B	$9 \cdot 10^{-2}$	+	$2.5 \cdot 10^{-4}$

+ - 25% superhelical form of pBR322 DNA changes into the nicked form during the incubation with 10 U of enzyme
++ - 50% superhelical form changes into nicked and linear form
+++ - 100% superhelical form changes into nicked and linear form

treated with Bright-red 6C-Sepharose CL-6B displayed only the traces of deoxyendo- and ribonucleases. Table 1 shows the crucial role of dye immobilization procedure and of nature of matrix in purification efficiency. Thus Cibacron blue F3-G-A bound immediately to Sepharose CL-6B demonstrates poorer nuclease selectivity than that immobilized via dextran bridge. Bright-red 6C bound to Sepharose CL-6B compared to macroporous silica gel is almost 100 times more effective in 3'-5'-exonuclease, 3 times - in deoxyendonuclease and almost 30 times - in ribonuclease removal.

TABLE 2
Sorption capacity of adsorbents with immobilized
triazine dyes[*]

Adsorbent	Dye concentration in adsorbent uM/ml	Sorption capacity, units enzyme/uM dyes	Sorption capacity, mg protein/uM dyes	Activity yield %
Cibacron blue F3G-A-Sepharose CL-6B	2.5	2520	2.6	90
Blue dextran-Sepharose 4B	0.076	10000	5.8	98
Bright-red 6C-Sepharose CL-6B	1.8	2000	2.6	91
Bright-red 6C-macroporous silica gel coated with dextran	1.7	21411	7.7	93
Procion red HE-3B-Sepharose CL-6B	2.0	4050	5.5	71
Red-brown 2K-Sepharose CL-6B	1.4	3430	3.9	75
Bright yellow 53-Sepharose CL-6B	5.0	720	1.92	62
Yellow light resistant 2KT-Sepharose CL-6B	5.0	1500	0.42	62
Orange 5K-Sepharose CL-6B	2.4	2900	1.4	12

[*]Enzyme preparation precipitated with polymine P

Thence in nuclease- and phosphatase- free polynucleotide kinase production the purification procedure should involve four operations (see Table 3).

Polynucleotide kinase has been isolated simultaneously with DNA- and RNA- ligases and DNA polymerase from E.coli cells infected with phage T4. Isolation procedure involved sonication of cells, precipitation with polymine P and ammonium sulphate and chromatography on Bright red 6C-Sepharose and phosphocellulose PII. When passed through the Bright red Sepharose column the target enzyme eluted at 0.6 - 0.9 M KCl gradient, while the 3'-5'-exonuclease activity stretched from 0.2 to 0.7 M. The final separation of 3'-5'-deoxynucleases

TABLE 3
Polynucleotide kinase purification procedure
(from 100g of cells)

Stage	Fraction vo-lume ml	Protein mg	Activity Total units	Activity Specific units/mg protein	Yield%
Crude extract	478	22325	230300	10.3	100
Polymine P and ammonium sulphate precipitation	98	2940	198000	67.3	85.9
Bright-red Sepharose chromatography	82	43	180400	4190	78.3
Phosphocellulose PII chromatography	14	1.05	95000	90666	41.3

was performed in the phosphocellulose PII column.

The preparation produced according to this scheme was free from deoxyendonucleases, 3'- and 5'-deoxyexonucleases, 5'-phosphatases and ribonucleases. Therefore the application of immobilized triazine dyes is very effective in the isolation and purification of polynucleotide kinase.

REFERENCES
1. Richardson, C.C., Bacteriophage T4 polynucleotide kinase. The Enzymes, vol. 14A, P.Boyer (Ed.), 3rd edition, Academic Press, New York, 1981, pp. 299-314.
2. Maxam, A.M., Gilbert, W., A new method for sequencing DNA Proc. Natl. Acad. Sci. USA, 1977, 74, 560-4.
3. Midgley, C.A., Murray, N.E., T4 polynucleotide kinase, cloning of the gene (pse T) and amplification of its product. Eur. Mol. Biol. J., 1985, 4, 2695-703.
4. Dolganov, G.M., Chestukhin, A.V., Shemyakin, M.F., A new procedure for the simultaneous large-scale purification of bacteriophage T4-induced polynucleotide kinase, DNA ligase, RNA ligase and DNA polymerase. Eur. J. Biochem., 1981, 114, 247-54.

5. Kanopka, A.E., Baronaite, Z.A., Marcisauskas, R.P., Laka-
   ciauskiene, R.V., Vaitkeviciene, R.S., The simultaneous
   purification of bacteriophage T4-induced DNA- and RNA-li-
   gases, polynucleotide kinase and DNA polymerase. In <u>Thesis
   of All-Union Conference on Enzyme Application in Biochemical Analysis</u>, Palanga, 1984, 25.
6. Nichols, B.P., Lindell, T.D., Stellwagen E., Donelson J.E.
   A rapid purification of T4 polynucleotide kinase using
   blue dextran-sepharose chromatography. <u>Biochim. Biophys.
   Acta</u>, 1978, 526, 410-7.

EPILOGUE

It is of course extremely stimulating and promising to see that the Dye-ligand affinity systems have developped in different spheres and this domain is becoming one of the milestones in the progress of Downstream Processing in Biotechnology.

However, only very little information is available on the toxicological aspects of the eventual traces of dye molecules or their fragments present in the final product. This situation is undoubtedly an obstacle to a real industrial exploitation of this fast developping field.

So, it was strongly felt that, such a meeting should highlight the importance of this aspect and bring forth the different angles at which this aspect should be dealt with in the near future. Hence, a discussion session was dedicated to this theme and in the following few lines, I try to summarize this interesting discussion.

First problem to be studied in detail is the dye leakage from the chromatographic adsorbents. The detection methods to evaluate the dye leakage are to be standardized. This leakage will depend of course on the pH of the chromatographic steps and that of the stocking solution.

An enzyme immuno-assay method was shown by Dr J.C. Pearson of Biotechnology Centre, Cambridge, U.K., to be more useful than the conventional spectrophotometric methods for the detection of very small amounts of dye leaking.

But a more important issue was raised concerning the dye polysaccharide conjugates released from the affinity columns. It was demonstrated in fact that at least some dye coming off the column is the result of the polysaccharide (agarose) hydrolysis, particularly at acidic pHs, where the dye is chemically coupled. The dye sugar conjugates may have certain properties to bind to cell surfaces, and might be toxic as opposed to the free dyes. A study of the toxicological aspects of these dyes should therefore be approached from two distinct angles.

- toxicity of the free native dyes
- toxicity of the dye-sugar conjugates

According to Dr S. Subramanian of MILES, Indiana, USA, the LD50 doses for rats and for Cibacron Blue F3GA is in the order of 5 mg dye kg body weight. However, this lethal dose limit should be considered with caution, as no data is available in terms of physiological elimination or accumulation of the dye molecules or its metabolites. At this stage it is also important to study the influence of dyes or dyes + metals on the cytochrome P. 450 system, for the scavenging of the dyes and their metabolites.

J.E. More of Blood Products Laboratory, Elstree, U.K., presented his data on the dye release studies done in the course of a preparative chromatography of albumin from the Cohn fraction IV. From a column running for about 18 months in a discontinuous manner, for $>$100 cycles, the albumin obtained will have a final contamination of 0.4 μg dye/g of albumin. The dye concentration was calculated from typical experiments, with radioactivity measurements by using C^{13} marked dyes. This works out to be about 10 μg of dye/a normal patient dose which is 25 g of albumin. While this dose did not show any significant toxicity, subacute doses over a long period of administration could have some renal or liver damage.

In order to evaluate the toxicity, due to dyes released into the purified therapeutic products, the importance of the studies in terms of genotoxic effects, cell toxicity effects on chromosomal damage were stressed.

Dr E. Boschetti of I.B.F. Biotechnics, France, presented his preliminary data on the effects of both free and immobilized dyes on the human cells. According to his data upto a level of 100 μg/ml of free dye no appreciable toxicity was observed. In terms of mutagenesis, the data were about three orders of magnitude less than the standard SOS test. In terms of chromosomal damage, a specific study realised on MRC-5 human cells for six passages (from 29th to 35th passage), showed that the polyploid level determined according to the recommendations of WHO was below 6/500, which is significantly below the acceptability level (max 17/500) recommended by WHO. However, much caution should be taken before concluding on this point. Dye molecule by itself is polyanionic and hence could penetrate easily the cells On the contrary, dye sugar conjugates would be able to recognize certain cell surfaces acceptors and help in coming in contact with the DNA molecule. Also, in terms of neurotoxicity estimation, the same will hold good.

So, still for free dyes it was strongly suggested that future studies should be oriented towards the dye-sugar conjugates. Studies of the effect of immobilized dyes on cell cultures in vitro did not evidence any perturbation on cell growth and morphology as stated by Dr E. Boschetti.

To sum up the salient features of this discussion session, Dr E. Stellwagen of Iowa State University, USA, evoked the following points.

1. Is toxicity related to the penetrability of the dyes into the cells ? Apparently, it is not, as per data obtained by Dr E. Boschetti.

2. Diffusion through plasma membranes is possible. This transport should be possible either through the biomimetic nature of the dyes or because of the metal coupling to dyes, thereby reducing the anionic charges of the dye.

3. The DNA intercalation and mutagenesis should be studied in more detail.

4. The essential point is that these evaluations should be done both with free dyes as well as with the dye-sugar conjugates (leached products).

To conclude, these toxicological evaluation will decide the future evolution of the dye affinity field, and its wider exploitation in the industrial scale in view to produce biologicals of therapeutic interest.

M.A. VIJAYALAKSHMI

INDEX OF CONTRIBUTORS